Hawkey's Atlas of Wild and Exotic Animal Haematology

DEDICATIONS

This book is dedicated to the memory of Dr Christine M Hawkey and to honour the legacy of her outstanding and pioneering contribution to the comparative haematology of wild and exotic animals. She was a firm believer in the importance of haematology as part of the clinical diagnostic laboratory techniques when haematology of wild and exotic animals was still in its infancy. The laboratory techniques she developed and the observations she made over the years with her gifted mind have been used as the foundation for the analyses widely used today in comparative wild and exotic animal haematology. Christine was our mentor, our teacher and a dear friend. She was always there when advice and guidance was needed. Her kind smile, looking at you over her reading glasses, was always so reassuring. Christine's legacy will live forever in the hearts of all of those privileged enough to have met and worked with her.

The new publication of this Atlas could not be produced without mentioning the significant contribution to medical photography of the co-author of the previous publication of this Atlas, Mr Terry B Dennett. Terry was an avid and devoted scientist photographer and Head of the Photographic Unit at the Institute of Zoology, Zoological Society of London, for many years. He passed away recently, but his knowledge and dedication to medical photography still live in the memory of all who knew him.

The previous publication of this Atlas was dedicated to the memory of the late Gordon Henderson MRCVS, pathologist at the Veterinary Science Research Group, the Institute of Zoology, Zoological Society of London, whose life was cut short in an unfortunate accident in Glencoe, Scotland, on 22 June 1986. Gordon was a soft-spoken person, with a kind smile, gentle personality and at the beginning of a promising career. The wild and exotic animal medicine world lost a valuable member on that fateful day on that remote mountain in Scotland.

Jaime Samour

To my wife, Merle, for inspiring me to be a better man, and to my sons, Omar and Adam, and daughters, Miriam and Yasmeen, for inspiring me to be a better father.

Mike Hart

To my wife, Jean, my daughter, Melanie, and my son-in-law, Simon, for their encouragement and support in this project, and for my two granddaughters, who have shown me how to enjoy life.

CONTENTS

Foreword ix

Preface xiii

Acknowledgements xv

About the Editors xvii

Contributors xix

In Memoriam xxv

1 General Introduction 1
   1.1 Introduction to Haematology 1
   1.2 History of Haematology 2
        1.2.1 Introduction 2
        1.2.2 Invention of the Microscope 2
   1.3 Blood Sample Collection 4
        1.3.1 Small Mammals 4
        1.3.2 Large Mammals 4
        1.3.3 Birds 4
        1.3.4 Reptiles and Amphibians 5
        1.3.5 Fish 5
        1.3.6 Invertebrates 6
   1.4 Blood Sampling 6

SECTION A   Basic Wild and Exotic Animal Haematology

2 Haematopoiesis 23

3 Normal and Abnormal Red Blood Cells 25
   3.1 Reversible Changes in Red Blood Cell Shape 26
   3.2 Inclusion Bodies 26
   3.3 Rouleaux Formation 28
   3.4 Artefacts 28
   3.5 The Blood Film 28
        3.5.1 Red Blood Cells 28

4   Normal and Abnormal White Blood Cells 31
    4.1   Granulocytes 31
    4.2   The Blood Film 32
    4.3   Normal and Abnormal White Blood Cells – Lymphocytes, Monocytes and Azurophils 33
            4.3.1   Lymphocytes 33
            4.3.2   Monocytes 33
            4.3.3   Azurophils 33

5   Normal and Abnormal Platelets and Thrombocytes 35
    5.1   Platelets and Thrombocytes 35

SECTION B   Atlas of Wild and Exotic Animal Haematology

6   Normal and Abnormal Blood Cells 39
    6.1   Haematology 39
            6.1.1   Normal Species Variation in Red Blood Cell Size 39
            6.1.2   Normal Variation in Red Blood Cell Shape 44
            6.1.3   Examples of Active Erythropoiesis in Small Mammals, Birds and Reptiles 51
            6.1.4   Common Artefacts Affecting Red Blood Cells 53
            6.1.5   Abnormal Variations in Red Blood Cell Morphology 56
            6.1.6   Species Variation in Normal Granulocytes 79
            6.1.7   Common Artefacts Affecting Granulocytes 102
            6.1.8   Pathological Responses Involving the Granulocytes 106
            6.1.9   Normal Variation in Lymphocytes Morphology 132
            6.1.10   Abnormalities in the Lymphocytes Associated with Disease 145
            6.1.11   Normal Monocyte Morphology 156
            6.1.12   Normal Azurophil Morphology 165
            6.1.13   Normal Variation in Platelet and Thrombocyte Morphology 171
            6.1.14   Variations in Platelets and Thrombocytes Associated with Disease 175
            6.1.15   Cellular Appearance in the Haemolymph of Invertebrates 179

SECTION C   Atlas of Wild and Exotic Animal Blood Parasites

7   Blood Parasites 183
    7.1   Introduction 183
    7.2   Blood Parasites 183

SECTION D   Appendix: Normal Haematology Reference Values in Selected Wild and Exotic Animals

Appendix 215

Bibliography 247

Index 253

FOREWORD

It is a pleasure and an honour to write this foreword to this Atlas produced by Jaime Samour and Mike Hart. Both authors are long-standing friends and colleagues and the subject of their text is close to my heart.

I recall that it was a great boost to my confidence as a young veterinary diagnostician to meet Dr Christine Hawkey, in the late 1970s, on one of my frequent visits to the London Zoo. As Jaime Samour explains in his preface, Christine was at the forefront of a scientific approach to avian and reptilian haematology – and I had taken a few faltering steps in the discipline in previous years, as I recount later. Christine encouraged me and she frequently reviewed and commented on blood smears that I had taken from a diverse selection of animals, ranging from goldfish to golden lion tamarins. As a result of our friendship and collaboration, Christine kindly invited me to write the Foreword to her book, edited jointly with TB Dennett and MA Peirce, *A Colour Atlas of Comparative Veterinary Haematology: Normal and Abnormal Blood Cells in Mammals, Birds and Reptiles*, which was published by Wolfe in 1989 (referred to in the Preface as the previous publication of this new Atlas).

The production of this volume, appropriately entitled *Hawkey's Atlas of Wild and Exotic Animal Haematology*, is a grand achievement. It is the culmination of years of discussion, planning and collation of material. It was clearly motivated by a wish both to keep alive the reputation and achievements of Christine Hawkey and to produce a reference text that would adequately meet the needs of those studying disparate species of non-domesticated species in the 21st century.

Much of the success of this present undertaking can be attributed to the different but complementary backgrounds and nature of the two editors. Jaime Samour is Central American by birth, now a truly international citizen, who is fired with energy, enthusiasm and Latin passion, determined to make the world a better place. Mike Hart is a quintessentially modest, often self-effacing Anglo-Saxon, but one with a wealth of knowledge and experience and a master of both the science and the art of laboratory investigation. Put together, these two have forged a synergism, the fruits of which are reflected in the succeeding pages of this magnificent tome.

Jaime Samour has recounted in his introduction how he first met Dr Christine Hawkey and the influence she had on his life. Those who knew and derived so much from her guidance and friendship identify with Jaime's tributes and are grateful that this book will continue to remind us of Christine and the debt that we owe to her. Christine's skills in the 'art' of haematology were legendary. She would look at a blood smear and comment on something that no one else had noticed or felt worthy of note. I recently had reason to recall Christine's skills, wisdom and 'sixth sense' for all things haematological when, putting together a book about the pathology and health of gorillas, I rediscovered an original note from her, commenting on some findings in a lowland gorilla, and was able to incorporate this in my text.

Although primarily a haematologist, Christine was well versed in laboratory techniques, and I am mindful of how often this was put to good use when she supervised students and young colleagues. An example of such guidance – and an example of Christine's eminently practical orientation – was the production of guidelines for the interpretation of laboratory findings in birds and mammals with unknown reference ranges, which was published in the *Veterinary Record* in 1984.

Terry Dennett was less well known than Christine outside the scientific confines of the Zoological Society of London, but he was co-author of the original 1989 book, and the importance of his input cannot be overemphasised. He brought his skills as a photographer and artist to ensure that the finished work was attractive to the eye – and accurate – in respect of the all-important images of cells and parasites.

I was touched that the original publication by Christine and Terry was dedicated to Gordon Henderson, a dear friend, student and colleague, whose death in a climbing accident was reported in the *Veterinary Record* of 28 June 1986. Those who were with him on Glencoe on that fateful day recollected afterwards how, after Gordon's fall, a magnificent red deer stag passed in front of them, as if to indicate that his spirit lives on. It certainly does in the pages of Jaime's and Mike's book, perpetuating the memory not only of Gordon but also of others who were profoundly influenced by Christine.

It is a delight to find that this Atlas is a truly comparative text, covering mammals, birds, reptiles, amphibians, fish and invertebrates. The inclusion of invertebrate animals may come as a surprise to some readers, but it must be remembered that these creatures make up 90% of the animal kingdom and are very important components of biodiversity. In respect of haematology, it was of course the pioneering research of Metchnikoff (1892) on crustaceans that paved the way for our understanding of cellular defence mechanisms in animals and the introduction into scientific parlance of the term 'phagocyte'. The examination of the blood of insects dates back several decades, and two seminal volumes appeared in the 1970s – *The Cellular Defence Reactions of Insects* by Salt in 1970 and *Insect Hemocytes*, edited by Gupta in 1979.

I mentioned earlier my long-held interest in blood as a diagnostic tool. When I graduated as a veterinary surgeon (veterinarian) in 1966, I went to Tanzania for a year. In East Africa the taking of blood and the making of blood smears was – and still is – routine whenever an animal is examined, largely because of the importance in that region of blood-borne protozoa in livestock. It therefore came naturally to me to do the same whenever I was presented with a wild mammal, bird or reptile. In those days all techniques had to be performed manually, and, in the case of avian samples, there was much extrapolation from earlier research on poultry.

By the early 1970s, following a further four years in Kenya, I had sufficient data to be able to publish some papers on basic haematological values of East African birds (mainly raptors) and various species of reptile. In 1972, for example, I published a short paper in the journal *Raptor Research*, giving haemoglobin and pack cell values for 'normal' East African birds of prey. The number of birds covered in the study was small (17, of 12 species); but haematological data were obtained for species for which, apparently, there were then no comparable figures.

I continued the practice of taking blood from all my patients when I returned to Britain. I was then working for the British Medical Research Council at Northwick Park Hospital in Harrow, from 1974–1977, and my medical haematological colleagues there went to great pains to assist in my emerging studies on the blood of animals. Blood samples from birds, as well as diverse other species of animals, were submitted to the hospital haematologists. They reported back on what they could, often commenting wryly on how different the samples were from those that they usually examined – from *Homo sapiens*. I, often in collaboration with my friend, the indefatigable Michael Peirce, and others, carried out detailed examination of the blood smears, looking for cellular changes and haemoparasites. This pattern continued when I joined the staff of the Royal College of Surgeons of England in 1979, and my interest and involvement in the subject persist to this day.

Hawkey's Atlas of Wild and Exotic Animal Haematology is a scholarly treatise with contributors from diverse countries and disciplines. It will be of great assistance not only to veterinarians and others who deal with 'wild and exotic' species in captivity but also to those who work with free-living animals and to those undergoing rehabilitation prior to release. Many are at the coalface of conservation biology and species protection.

Recent authoritative reports confirm just how many taxa are threatened by factors such as habitat destruction, malicious persecution, poisoning and infectious diseases. Plummeting numbers of mammals, reptiles, amphibians, birds and fish around the world are an urgent sign that nature needs life support. For example, the 'WWF Living Planet Report 2018' showed that population sizes of wildlife decreased by 60% globally between 1970 and 2014.

As populations decline, the need to monitor and manage them becomes crucial. It requires a multidisciplinary and interdisciplinary approach, with the input of skilled personnel from many disciplines, both professional and 'amateur', including broadly based ecologists and field naturalists. To these, however, need to be added others, especially specialists in such fields as laboratory medicine – including haematology – who can contribute to advantage their expertise and increasingly sophisticated techniques.

The advances in veterinary and biological knowledge crystallised in the text of this Atlas, and often depicted graphically in its fine illustration, will, without doubt, contribute to the welfare and conservation of many species. The importance of understanding and practising 'ecosystem health' has never been more relevant. We need to look in a holistic way at the many current threats to biodiversity, to species survival and to human well-being. The answer to diagnosing and treating most of these afflictions lies in identifying and correcting the multifarious insults that we are wreaking on our planet. Scholarly texts have an important part to play, especially where they combine scientific knowledge with practical advice, and Jaime and Mike's opus is an exemplary contribution.

The Bible (Leviticus 17:11) reminds us 'For the life of a creature is in the blood...' No greater commendation is needed for this Atlas.

John E Cooper DTVM, FRCPath, FRSB, Hon FFFLM, FRCVS
Wildlife Health, Forensic and Comparative Pathology Services (UK) and
The Durrell Institute of Conservation and Ecology (DICE),
The University of Kent, UK

PREFACE

It was a dark and cold November morning, when, shivering in front of the main entrance to the Animal Hospital at the world-renowned London Zoo, I rang the bell... and my life changed forever. I was there to spend six months in an internship programme for foreign nationals and to fulfil a professional dream. An Animal Nurse opened the door with an inquisitive look on her face. This was Jude Howlett. I explained who I was and she instructed me to come in and to sit in the mess room and wait... and then she left. I was just getting comfortable when a few minutes later, I jumped out of my chair when all of a sudden, a tall man with a beard burst into the room looking for something. He said hi, shook my hand and then he disappeared as fast as he came. This was Rob Hutton, another of the Animal Nurses working at the Animal Hospital at the time. Then other people came into my life, among them, John Ffinch, Jane Lawrie, the late Tony Fitzgerald, the Head Animal Nurse, and more significantly Janet Markham, another Animal Nurse, and Dave Spratt, the Histopathologist at the laboratory, both dear friends who played a very important part in my work... and life. Then I met others, including Mr John Knight, and my dear friend Frances Gulland, both Veterinary Officers at the Animal Hospital, and Mr David Jones, the Director of the Animal Hospital, one of my mentors and the man who was instrumental in carving my future career. Senior members of staff of the Zoo who made a significant impact in my work at the Zoo included Mr Peter Olney, Curator at the Bird Department, and the late Mr Dave Ball, Curator at the Reptile Department, and Dr Harry Moore, Dr Bill Holt, Dr Caroline Smith, all based at the Gamete Biology Unit at the Nuffield Institute, who helped me throughout my higher degree studies and so many others who in one way or another contributed to my formation at the Zoo.

I must say, I found it difficult to keep up with their English as I was not familiar with words such as 'wellies' (for Wellington rubber boots) or 'brolly' (for *umbrella*) or 'lift' (for *elevator*) or 'let's hit the road' (for *let's go*) or 'knackered' (for *tired*). Even now and after all these years, I still do not understand why in the English language you can say 'Mop the floor with a mop' but you cannot say 'Broom the floor with a broom'!

Then Mike Hart came into my life. My first thought when I saw him was of a 'geriatric hippie', as he appeared middle-aged with long hair and beard, wore an earring, sleeveless t-shirt, blue jeans and leather sandals. He was the assistant of Dr Christine Hawkey at the Haematology Unit. Mike and I became good friends, probably over our love for music, and worked together as we embarked in one of the most ambitious haematology research programmes ever undertaken, suggested by Dr Hawkey. I have already mentioned somewhere else that Mr David Jones wanted to carry out a programme of sex determination in monomorphic avian species in the collections of the Zoological Society of London at Regent's Park and at Whipsnade, using the newly applied endoscopy techniques. This programme later extended to almost every major zoological collection in the United Kingdom, from Paignton to Bristol, from Chester to Edinburgh. It was always with Mike, one of the Animal Nurses and I travelling and working together in this ground-breaking project all over the UK. Parallel to the sex determination programme, blood samples were collected from every clinically normal bird and reptile that was handled. Some of the species we examined had never been handled before as they were extremely rare in the zoo community in the UK at the time, such as the brolga crane (*Antigone rubicunda*) or the Major Mitchell's cockatoo, also known as the Leadbeater's cockatoo (*Lophochroa leadbeateri*) or the Aldabra giant tortoise (*Aldabrachelys gigantea*). Please remember I am referring to the early 1980s when we did not even know how to collect a blood sample from penguins, tortoises, crocodiles or snakes!

Mike used part of his laboratory to prepare and fix blood films, estimate haemoglobin and haematocrit and carry out the red blood cell and white blood cell counts. The staining of the films and differential counts were all carried out at the Haematology Unit later upon our return. This was the humble beginning of avian and reptilian haematology during the early 1980s that established the cornerstone of normal haematology values and our understanding of haemoresponses to pathological insults in birds and reptiles.

How many times I visited the Haematology Unit to sit in front of the microscope (well, I have to say, also to play 'Space Invaders' on the computer at the office during the lunch time break) to learn haematology, with Mike teaching me and very often under the watchful eyes of Dr Christine Hawkey, always looking at me above her reading glasses. They both taught me everything I know on haematology. I still have the drawings Mike made for me to understand the main structural differences of white blood cells as observed on the blood films.

I will always be grateful, in particular, to Jude Howlett, Rob Hutton, Janet Markham, John Ffinch, Dave Spratt and Mr David Jones for their friendship and understanding; to Dr Christine Hawkey for guiding me through my career; and to Mike 'Minky' Hart for all those trips together to buy second-hand tapes and for those seafood specials on brown baps we had in Camden Town, or for sharing dinners at the 'Witches Kitchen' or the burgers at the 'Great American Disaster' during our trips around the country working in zoos, but above all for being my friend.

I said this before and I am saying it again, I have what I have, I am who I am and I am where I am because of the friendship and love of all those beautiful people.

Jaime Samour
Abu Dhabi, United Arab Emirates

ACKNOWLEDGEMENTS

The editors would like to acknowledge Alice Oven, Senior Editor, Veterinary Medicine, CRC Press, Taylor & Francis Group, for her support to the inception of this Atlas and for her leadership throughout its production; Damanpreet Kaur, Editorial Assistant, CRC Press, Taylor & Francis Group; Kyle Meyer, Medical Specialist Project Editor, CRC Press, Taylor & Francis Group; Lillian Woodall, Project Manager, Deanta Global, for guidance during the typesetting and editing of this Atlas; and Shashi Shekharkumar, from Cenveo Publisher Services, Mumbai, India, for scanning the original slides used in the previous publication of this Atlas to produce the excellent digital images used in this new publication.

To all the contributors, from the four cardinal points of the world, for entrusting us with your work and for believing that Mike and I could produce a publication worthy enough to perpetuate the memory of Dr Christine Hawkey. Thank you for believing in us and for helping in the production of such a magnificent work, without which this could not have been possible.

ABOUT THE EDITORS

Jaime Samour is a veterinary surgeon who has devoted the best part of his professional life to falcon medicine. He graduated with honours at the Faculty of Veterinary Medicine in Veracruz, Mexico, in 1978. After a short period of time in his native El Salvador working in private practice, he returned to Veracruz and became an Associate Professor at the Faculty of Veterinary Medicine. In 1981, Jaime travelled to London to study and to carry out research projects in reptilian and avian medicine at the Institute of Zoology, Zoological Society of London. He learned and practised haematology under the supervision of Dr Christine Hawkey and Mike Hart. He was awarded a PhD degree on reproductive biology in birds by the Royal Veterinary College, University of London, in 1987. Immediately after, Jaime moved to the Middle East, where he has worked ever since, firstly in the Kingdoms of Bahrain and Saudi Arabia and finally in the United Arab Emirates. Jaime has edited five textbooks and produced 17 chapters in different textbooks on avian and exotic animal medicine, and more than 120 peer-reviewed papers in international journals. Jaime was recipient of the British Veterinary Zoological Society/Parke-Davies/Upjohn Award in 1998 and the Ted Lafeber Avian Practitioner of the Year Award in 2006. Jaime was awarded Diplomate status of the European College of Zoological Medicine in the Avian Specialty in 2004 and became Professor of Raptor Medicine at the Veterinary College, University of Parma, Italy, in 2018.

Michael George Hart worked at the Haematology Unit of the Animal Hospital, Institute of Zoology, Zoological Society of London, headed by Dr Christine Hawkey, from 1969 to 1991. After retiring from ZSL, he worked from 1991 to 2013 at various commercial veterinary diagnostic laboratories specialising in wild and exotic animal haematology in the UK. Mike is a founder member and former committee member of the Association of Comparative Haematology, former member of the editorial board of Comparative Haematology International, former member of the editorial board of the Association of Reptilian and Amphibian Veterinarians and former reviewer of exotic animal haematology for the *Veterinary Record*. He is an Honorary Member of the British Veterinary Zoological Society. He has published 37 peer-reviewed papers jointly or as a sole author and 5 non-peer-reviewed articles. Mike has made 22 oral presentations and has undertaken teaching commitments at colleges, conferences and institutions in the UK and overseas and several poster presentations, including one that won Best Poster. He has participated in field studies, including haematology of rare birds, in the UK, Zoos and biomedical surveys, including haematology of the Aldabra giant tortoise in the Seychelles Islands and of the Galápagos giant tortoise in the Galápagos Islands.

CONTRIBUTORS

Xavier Valls Badia completed his veterinary degree in 1992 at the Universitat Autònoma de Barcelona, Spain. He is co-founder of the Clínica Veterinaria Exotics de Barcelona. Xavier is a well-known speaker in national and international conferences and symposia on medicine and applied clinical diagnostic techniques in exotic species. He is author and co-author of several chapters of books specialised in exotic animal medicine. He currently runs the Departments of General Medicine, Surgery, Clinical Diagnostic Laboratory Analyses and Clinical Imaging in Reptiles and Amphibians at the Clínica Veterinaria Exotics de Barcelona.

Linda Bruins-van Sonsbeek graduated *cum laude* from Utrecht University, the Netherlands, in 2011. For the next three years she worked in a mixed animal practice, during which time she volunteered as a Veterinary Intern at the Ouwehand Zoo in the Netherlands. She became the Head Veterinarian at the Rotterdam Zoo in 2014. Linda has conducted research into the validation of assays to measure calcium metabolism and the effect of calcium and cholecalciferol supplementation on calcium status in plasma and urine of captive Asian and African elephants.

Scott B Citino graduated *summa cum laude* from the Ohio State University, College of Veterinary Medicine, in 1983. Has worked at Miami Metrozoo and the Smithsonian's National Zoo and is currently Senior Veterinarian at White Oak Conservation. Scott is Past President of the American Association of Zoo Veterinarians and of the American College of Zoological Medicine and Past Chair of the Wildlife Scientific Advisory Board for the Morris Animal Foundation. He is also a member of the International Rhino Foundation Advisory Board, the Sumatran Rhino Husbandry and Propagation Expert Advisory Board, the Scientific Advisory Board for the Cheetah Conservation Fund, the Scientific Advisory Board for the International Wildlife Institute, the Board of Directors of Wildlife Conservation Global, and is Technical Advisor for the Okapi Conservation Project. Scott received the American Association of Zoo Veterinarian's Emil P Dolensek Award for exceptional contributions to the conservation, care, and understanding of zoo and free-ranging wildlife and the Ohio State University, College of Veterinary Medicine's Distinguished Alumni Award in 2007.

Ana Margarita Woc Colburn graduated from Cornell University, Ithaca, NY, in 2003 and completed a Small Animal Medicine and Surgery Internship followed by an additional Internship in Zoological Medicine and Surgery at the Smithsonian National Zoological Park. After completing her residency, she became the Associate Veterinarian at the Nashville Zoo. She has a special interest in critical care and anaesthesia as well as in xenarthran medicine and is one of the Veterinary Advisors for the Pangolin, Aardvark and Xenarthra Taxon Advisory Group, the Giant Anteater SSP (Species Survival Plan Programme) and the Southern Tamandua SSP as well as being the Secretary for the Veterinary Scientific Advisory Group.

Guillermo Sanchez Contreras is a veterinarian mostly focused on marine mammals. After graduating from Alfonso X el Sabio University, Madrid, Spain, during which he carried out several internships in zoos and aquariums and earned an Erasmus Lifelong Learning Grant to carry out domestic animals rotating internship at the Veterinary Teaching Hospital of Helsinki University, Helsinki, Finland. He moved to the Netherlands to work at the Seal Rehabilitation and Research Centre in Pieterburen. There he got to deal with more than 2,500 seals of different species and participated in many different research projects. After five years in the field of rehabilitation, he moved to Malta to take care of the Veterinary Services at Mediterraneo Marine Park. He earned his Master's degree in Medicine and Surgery of Exotic Animals in 2018, and has been a very active member of the European Association for Aquatic Mammals (EAAM) since 2014. He believes education is the key to raise awareness for conservation.

Daniela Denk graduated from Munich Veterinary School, Germany, in 2006 and completed her pathology residency at Liverpool University, UK. In 2011 she became a member of the renowned Pathology Division of International Zoo Veterinary Group, focusing on diseases of avian, mammalian, reptile and fish species. She is a Diplomate of the European College of Veterinary Pathologists and an RCVS Recognised Specialist in Zoo and Wildlife Veterinary Pathology. Daniela is an expert in penguin pathology and has a special interest in *Yersinia pseudotuberculosis*. She continuously publishes and presents in the zoo and wildlife pathology fields.

Robert James Tyson Doneley graduated from the University of Queensland, Australia, in 1982. After working as an Associate Veterinarian/Locum in practices in Queensland and the UK, he opened his own practice in 1988. In 2010 he moved on to take up the role of Head of the Small Animal Hospital at the newly relocated UQ School of Veterinary Science at Gatton, where he is now an Associate Professor, Head of the Avian and Exotic Pet Service. Along the way he obtained his Fellowship of the ANZCVS, becoming Queensland's first specialist in avian medicine. He has published 3 textbooks on birds and one on reptiles, and more than 50 peer-reviewed papers in Australian and international journals. In 2018 he was awarded the TJ Lafeber Avian Practitioner of the Year award.

Brett Gardner is a zoo and wildlife veterinarian who graduated from the University of Pretoria, South Africa, in 2006. He worked for Johannesburg Zoo for seven years soon after graduating. In 2014 he started consulting for various zoos, aquaria and wildlife reserves. In 2017 he moved to Australia. He works for Werribee Open Range Zoo and continues consulting for various conservation organisations. Brett is currently in the process of doing a postgraduate degree looking into infectious abortion of Australian fur seals. He has a keen interest in remote fieldwork, field anaesthesia and running field laboratories.

Frances MD Gulland is a veterinarian focusing on determining the impact of human activities on wildlife health, especially marine mammals, and how marine mammals can serve as indicators of ocean health. She received a veterinary degree in 1984 and a PhD in Zoology in 1991 from the University of Cambridge, UK. She worked as a veterinarian and Research Fellow at the Zoological Society of London before moving to the US in 1994. She then worked at the Marine Mammal Center in California for 25 years on the veterinary care of stranded marine mammals, and conducted research into marine mammal diseases. She currently serves as Commissioner on the U.S. Marine Mammal Commission appointed by President Barack Obama.

Jean-Michel Hatt studied veterinary medicine at the University of Geneva and the University of Zürich in Switzerland. Following graduation, he worked at the Clinic of Zoo Animals, Exotic Pets and Wildlife at the University of Zürich and at the Royal Veterinary College at the University of London, UK, where he obtained a Master of Science in Wild Animal Health. Since 2001, Jean-Michel is Head of the Clinic of Zoo Animals, Exotic Pets and Wildlife at the University of Zürich and Chief Veterinarian of the Zürich Zoo. Jean-Michel is a Diplomate of the American College of Zoological Medicine and a Diplomate of the European College of Zoological Medicine (Avian) and has set up residency programmes for both these colleges. The Residency Programme for the American College of Zoological Medicine is the first outside North America. His main research focus is avian surgery and diseases related to nutrition in zoo animals.

Shady Heredia Santos is a veterinarian who graduated in 2017 from the University San Francisco of Quito, Ecuador. She has a Master's degree in Climate Change and the Environment from the University Andina Simón Bolívar, Ecuador. She has worked with wildlife in the past five years. Shady was recipient of a scholarship from the Peregrine Fund, USA, to carry out her veterinary thesis. She currently works at the Zoológico de Quito in Guayllabamba, which is part of the Grupo Nacional de Trabajo del Condor de los Andes in Ecuador.

Karen Jackson graduated from the University of Queensland, Australia, in 2002. Then she worked as a Veterinarian in a mixed animal practice and completed a small animal rotating internship before moving to Pennsylvania for five years to complete a Haematology and Transfusion fellowship, Veterinary Clinical Pathology Residency, and then lectureship. In 2011, Karen returned to Australia and worked in diagnostic pathology before moving back to academia first at the University of Adelaide and now at the University of Queensland. Karen is passionate about veterinary student education and diagnostic clinical pathology with specific interests in haematology of all species, transfusion medicine, immune haematology and oncology.

Benjamin Michael Kennedy is a veterinary surgeon who graduated from the Royal Veterinary College, London, UK, in 2016. Prior to this, he had completed a Bachelor's degree in Biochemistry and Genetics at the University of Nottingham, UK, and a Master's degree in Virology at Imperial College, London, UK. Benjamin is a member of the steering committee of the Veterinary Invertebrate Society, where he is the Editor of the *Veterinary Invertebrate Society* journal. Alongside small animal and exotic clinical work, he dedicates his time to pursuing research into invertebrate medicine. He regularly gives lectures at conferences and publishes articles on invertebrate medicine. In his spare time, he enjoys drawing, looking after his personal invertebrate collection and going on countryside walks with his whippet.

Helen McCracken is a zoo veterinarian with a special interest in clinical haematology and cytology. She graduated from Sydney University, Australia, in 1986, and then completed a Master's degree and Residency at the University of Melbourne and Melbourne Zoo in 1988, undertaking research in reptile disease. Following the Residency, Helen continued at Melbourne Zoo, becoming Senior Veterinarian in 1993. She developed expertise in avian and reptile haematology through clinical practice and study tours to the US, and over the past 25 years has trained many zoo residents and veterinary students in these skills, so that they may have the capacity for such diagnostics in remote locations and circumstances where pathology laboratory services are not available. In 1989, Helen established Melbourne Zoo's blood smear library that now has a collection representing over 400 Australian and exotic wildlife species.

Jaume Miguel Martorell Monserrat studied at the Autonomous University of Barcelona, Spain, from which he received a degree in Veterinary Medicine in 1993 and a PhD in 2006. He has worked in private exotic pet practice since 1995. Jaume has been a Professor at the Facultat de Veterinària de la Universitat Autònoma de Barcelona since 2000 where he is responsible for the Exotic Animal and Zoo Medicine subject and is the head of the exotic pet service in the 'Hospital Clinic Veterinari'. In 2011, Jaume was recognised as a *de facto* Diplomate of the European College of Zoological Medicine (ECZM) in the speciality of Small Mammals Medicine. Currently, he is the supervisor of the ECZM residency programme (small mammal), the Director of the ECZM alternative residency programme (reptile) and the exotic pet internship. He is the author of many papers in scientific journals and speaker at numerous national courses and conferences. He is an active member of Veterinary National and Internationals Associations such as AVEPA-GMCAE, where he was the past president, SECV, and AEMV.

Jody Nugent-Deal is a registered veterinary technician and a veterinary technician specialist in anaesthesia/analgesia and clinical practice – exotic companion animal. Jody has been with the University of California at Davis Veterinary Medical Teaching Hospital since 1999, working in the Companion Exotics Department for 10 years and currently in the Anesthesia Department, where she is the supervisor. Jody is a founding member of the Academy of Veterinary Technicians in Clinical Practice. She currently serves as Executive Secretary for the group as well as the Appeals Committee Chair. Jody is also an active member of the Academy of Veterinary Technicians in Anesthesia and Analgesia (AVTAA), having served on the Nominations, Credentials and Appeals Committees. Jody is currently a Member-at-Large for the AVTAA. She also serves on the board for the North American Veterinary Anesthesia Society (NAVAS). Jody is passionate about teaching and life-long learning. She currently works as an instructor for the Penn Foster Veterinary Technician Program as well as Veterinary Support Personnel Network (VSPN) and VetMedTeam, teaching anaesthesia and exotic animal medicine. Jody has also lectured throughout North America since 2000 on anaesthesia and various exotic animal medicine topics. She has published numerous articles and book chapters for both canine/feline and exotic animal medicine and anaesthesia topics. Jody has a special interest in anaesthesia, analgesia, pain management and critical care in exotics animals and dogs and cats.

Michelle OBrien is the Veterinary and Wildlife Health Officer for the Wildfowl & Wetlands Trust (WWT) based at Slimbridge in Gloucestershire, UK. After graduating from the Royal Veterinary College in 1999 she spent six and a half years in general and exotics referral practice. She gained the RCVS certificate in Zoological Medicine in 2005 and the ECZM Diploma in Zoo Health Management in 2017. She is now a European Veterinary Specialist in Zoo Health Management and an RCVS recognised Specialist in Zoological Medicine. Michelle has worked at WWT since 2006 and is part of the team that provides veterinary care at seven of the nine WWT centres in the UK. Michelle provides veterinary advice to a number of conservation programmes, including the Great Crane Project and Spoon-billed sandpiper conservation breeding programmes.

Lee Peacock completed her Bachelor of Science (vet) (Hons II) in 2004 and a Bachelor of Veterinary Science (Hons II) in 2005, both from the Faculty of Veterinary Science, University of Sydney, Australia. In 2014 she registered and is currently undertaking a PhD degree at the School of Biological Sciences, Monash University. The theme of her study is entitled 'Avian Disease Ecology: Dynamics of Avian Malaria in Avian Communities of Torres Strait, Australia'. She worked full time in private practice from 2005 to 2013 as a Bird & Exotics Veterinarian in Waterloo, NSW in Australia with birds, wildlife and pocket pets. Since 2013 she has been working as a Locum Veterinarian with birds, wildlife, pocket pets, export checks at the Moonlit Sanctuary Wildlife Conservation Park, Pearcedale VIC; at the Australian Wildlife Health Centre, Zoos Victoria, Healesville Sanctuary, VIC; at the Jetpets Animal Transport, Tullamarine, VIC; at the Jetpets Animal Transport, Mascot, NSW; and at the Bird & Exotics Veterinarian, Waterloo, NSW, all in Australia.

Michael Alan Peirce has spent his career working as a parasitologist with a special interest in haemoparasites of birds and other wildlife and collaborated with researchers globally in studies ranging from those on seabirds in Antarctica to carnivores in Africa. In 1978 he was appointed an Associate of the International Reference Centre for Avian Haematozoa (IRCAH). Following several overseas postings, especially in Africa, he retired from government service in 1994 to establish his own consultancy. After the IRCAH was relocated to the Queensland Museum in Brisbane in 1995 – an association which was productive in expanding the knowledge of Australian wildlife haematozoa, especially avian – he was appointed an Honorary Consultant. A primary focus has been his interest in the haematozoa of African wildlife.

Helene Pendl is the sole proprietor and CEO of Pendl Lab, a company registered in Switzerland. After graduation from Ludwig Maximilian University of Munich, Germany, various avian exotic internships in the US, doctoral thesis in Spain on psittacine haematology, and several years as a small animal, exotic pet, and laboratory animal practitioner, she started her own business on avian and reptile haematology, cytology, pathology and histopathology. Her current working fields include routine diagnostic workups, scientific collaborations and international training programmes. She has published in standard textbooks and journals and has been a regular lecturer at international conferences and postgraduate programmes for more than 25 years.

David Perpiñán is a veterinarian specialised in zoo and exotic animal health and currently working as a freelance consultant in Barcelona, Spain. After graduation from the Autonomous University of Barcelona, he completed an MSc at the same university working on chelonian haematology and later a PhD on viral diseases of ferrets at the University of las Palmas de Gran Canaria, Spain. He has worked in universities, zoos and exotic animal hospitals in Europe and the US for over 15 years, and his current focus includes wildlife conservation and education, exotic animal medicine and zoo animal welfare.

Mary Pinborough qualified as a Biomedical Scientist in 1977, spending many years in the NHS and managing private pathology laboratories. Having had a long interest in reptiles/rescue in 2005, Mary jumped ship and opened her own independent international veterinary diagnostic laboratory for all companion animals specialising in exotic/zoo/wildlife species. She is Technical Services Director of Pinmoore Animal Laboratory Services formed in 2006 with the ethos of providing an exemplary service applying the rigorous standards found in human pathology. The profits of this organisation support a variety of research projects to improve knowledge of a variety of species.

Christal Pollock is a veterinary consultant for Lafeber Company. She completed an internship in Small Animal Medicine and Surgery in private practice and a residency in Avian and Zoological Medicine at the University of Tennessee, College of Veterinary Medicine, Knosville, TN. Christal then served as a Clinical Instructor in the Zoological Medicine Service at Kansas State University, College of Veterinary Medicine. She is Board-Certified in Avian Practice through the American Board of Veterinary Practitioners (ABVP). Christal serves on an ABVP committee and also as Chair for the Association of Avian Veterinarians Membership Committee as well as an Associate Editor for the Journal of Avian Medicine and Surgery. She has written and presented extensively on exotic animal medicine topics. Christal currently works as Editor and Lead Writer for LafeberVet. She wears many other 'hats' for Lafeber Company, including work on product development projects and co-management of the Lafeber Company Veterinary Student Program.

Peter Scott has his own practice, Biotope: specialist veterinary consultancy. A Liverpool veterinary graduate, he obtained his MSc in Aquatic Veterinary Studies at the University of Stirling, UK, based on work with tilapia in Kenya. He is now working with the majority of the UK trout farms, a number of large carp farms, and a 'no-kill' caviar producer. He advises public aquaria and acts as an animal welfare consultant in the media and corporate sectors.

Nico Schoemaker graduated from the Faculty of Veterinary Medicine in Utrecht, the Netherlands, in 1994, which was directly followed by an internship in Companion Animal Medicine and a Residency in Avian Medicine and Surgery at the Department of Clinical Sciences of Companion Animals at Utrecht University. His residency led to certification as an avian specialist in the Netherlands, Europe (Dip. ECAMS) and the US (Dipl. ABVP Avian). In 2003 he defended his PhD entitled 'Hyperadrenocorticism in Ferrets'. In April 2009 he became a founder Diplomate of the Small Mammal Specialty of the ECZM. He is still employed at the Division of Zoological Medicine, Department of Clinical Sciences of Companion Animals, Faculty of Veterinary Medicine, Utrecht University.

Nicole Indra Stacy is a Veterinary Clinical Pathologist specialising in non-domestic species and affiliated with the Aquatic, Amphibian, and Reptile Pathology Programme at the University of Florida, College of Veterinary Medicine, Gainesville, FL. After graduation from Ludwig Maximilian University of Munich, Germany, and a small animal rotating internship, and doctoral thesis in Bern, Switzerland, Nicole pursued a clinical pathology residency at the University of Florida. Her focus has included diagnostics, teaching and basic clinical research in non-domestic species for the past 15 years.

Mark F Stidworthy has been a full-time anatomic pathologist for zoo, exotic, wildlife and aquatic species in the International Zoo Veterinary Group's Pathology Division since 2003. He graduated from the University of Cambridge, UK, in 1996 (with an intercalated degree in pathology) and undertook pathology residency training at Liverpool Veterinary School before obtaining further diagnostic experience and a PhD (on remyelination in the CNS) at Cambridge Veterinary School. Mark is a Fellow of the Royal College of Pathologists, an RCPath examiner for veterinary pathology, and an RCVS Recognised Specialist in Veterinary Pathology (Zoo and Wildlife). Mark was the founder Secretary of the British Society of Veterinary Pathology, and was President of the British Veterinary Zoological Society between 2015 and 2017. He publishes and speaks regularly in the field of zoo, wildlife and aquatic pathology.

Juan Carlos Troiano graduated as a veterinarian in 1981 at the Universidad Nacional del Litoral and in 1986 completed the Herpetology Specialty in the Cátedra de Vertebrados de la Facultad de Ciencias Exactas y Naturales, University of Buenos Aires, Argentina. Since 2002, he has been an Associate Professor at the School of Veterinary Medicine, University of Buenos Aires, Argentina, and Visiting Associate Professor at the School of Veterinary Medicine, Universidad del Salvador, Buenos Aires, Argentina. In 2013 he was awarded the title of Specialist in Wild and Non-Traditional Animal Medicine from the Veterinary School de la Provincia of Buenos Aires. His area of interest includes exotic animal and wildlife clinical pathology, but particularly haematology in reptiles.

Yvonne Rieky Annet van Zeeland is an Associate Professor at the Division of Zoological Medicine, Department of Clinical Sciences of Companion Animals, Faculty of Veterinary Medicine, Utrecht University, the Netherlands. After graduating with merit at the Faculty of Veterinary Medicine, Utrecht University, in 2004, she worked briefly in private practice, and then returned to Utrecht University to successfully complete an internship in Companion Animal Medicine, residency in Avian Medicine and a PhD on feather damaging behaviour in grey parrots. Aside from her recognition as a specialist in avian medicine, she also became a *de facto* Diplomate in Small Mammal Medicine and certified as a parrot behaviour consultant.

IN MEMORIAM

Christine Mary Hawkey BSc, MSc, PhD
(1931–2008)

Dr Hawkey was born in Gosport, Hampshire, a town on the south coast of the United Kingdom, on 5 April 1931. During the Second World War, her family moved to a farm, and this probably contributed to her long-standing interest in animals and wildlife in general. After attending a boarding school and then a school in Portsmouth, she took a degree in zoology at Bedford College, Bedfordshire, England. Dr Hawkey then worked in a pathology laboratory at Paddington General Hospital and St George's Hospital Medical School in London. While working at the St George's Hospital Medical School she obtained her PhD degree. The title of her thesis was 'Extrinsic fibrinolytic activation: The effect in man of plasminogen activation on coagulation and fibrinolytic factors'. Apart from a one-year placement at the Massachusetts General Hospital in Boston, USA, she continued working at St George's Hospital Medical School until 1964 when she was head-hunted for a position as a Research Fellow at the Veterinary Department of the Institute of Zoology, Zoological Society of London (ZSL). She headed the Haematology Unit at the Veterinary Department for the next 27 years until her retirement in 1991. During this time, together with her long-time assistant, Mike Hart, and veterinary personnel based both at Regent's Park and Whipsnade, she undertook the most ambitious haematology study that science has ever seen. This study was undertaken parallel to a work on sex determination in monophormic birds using fibre-optic endoscopy in almost every major zoological collection in the United Kingdom. She was a prolific author, and as a by-product of the above-mentioned study and her tireless work at the ZSL, she published numerous publications in scientific journals, book chapters and the highly sought-after *A Colour Atlas of Comparative Veterinary Haematology* co-authored with Dr Mike Peirce and the late Terry Dennett. She passed away on 23 April 2008 remembered by all those who love her.

(With extracts from the obituary published in the *Veterinary Record*, 12 July 2008 by Edmund Flach.)

Dr Christine M Hawkey in the Haematology Unit within a camp tent during a field trip to Sudan. Dr Hawkey was there as part of a team to assess the efficiency of a government-sponsored programme to study the effect of a new water scheme on the cattle of the Dinka tribe in the Abyei Area of the Ngok Dinka in South Sudan.

Terry Benjamin Dennett
(1938–2018)

Terry was born in Chigwell, Essex, in 1938 and died in Islington, a few days before his 80th birthday. He was widely described as a British activist, community workshop organiser, social historian and photographer. Terry trained originally as a painter, but worked for many years as the Head of the Photography Unit at the Zoological Society of London, based at Regent's Park supporting research personnel of the Institute of Zoology. Terry was involved in and supported actively many projects, including 'The Crisis Project', the 'International Phototherapy Community', was an active member of the Royal Anthropological Institute and was President of the Association for Historical and Fine Art Photography (2006–2011), among others. Together with Jo Spence, Terry founded the Photography Workshop Ltd. in 1974, an organisation with the main objective of, in Terry's own words, 'researching marginalised or hidden histories and traditions of social radicalism'. The Half Moon Gallery later merged with Terry & Jo's Photography Workshop, eventually creating Camerawork Gallery. Terry worked tirelessly for over 25 years for the 'Jo Spence Memorial Archive' perpetuating the legacy of the work of Jo Spence. Terry was a beautiful, creative, caring and very sensitive human being who is greatly missed.

(With extracts from the tribute to Terry Dennett published on the 3 February 2018 in juliawinckler.blogspot.com.)

Terry as seen through the lens of Dr Julia Winckler, an academic and artistic photographer, while looking for old cameras in Eastbourne.

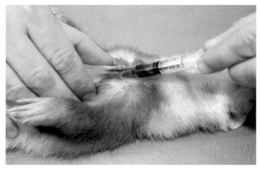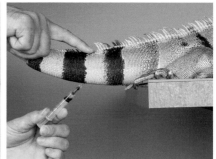

1

GENERAL INTRODUCTION

1.1 Introduction to Haematology

'Blood is the most complex fluid known to man.'

Dr Christine M Hawkey (1931–2008)

Haematology is an important component of clinical laboratory diagnosis and plays a prominent role in differential diagnosis, in monitoring the progress of therapeutic protocols and in providing an educated prognosis. Examination of a stained blood film is an integral part and, many would argue, the most essential assay of any routine haematological examination. In very small animals, where the amount of blood is limited, it may be the only procedure that can be undertaken. The ability to distinguish between normal and abnormal changes in blood cell morphology and to interpret these in terms of their pathological significance is one of the primary tasks of the haematologist.

For haematologists undertaking comparative assessments between different species this task is not easy, because of the large number of species involved. For each of these an understanding is needed of the normal blood cell morphology and how this is influenced by different pathological processes. Although similar principles are probably involved in all animals, in practical terms, it can be extremely difficult to apply them when faced with the problem of assessing the blood film from less common species of mammals, birds, reptiles, amphibians, fish or invertebrates.

Clearly, it is not possible to provide a complete description of normal and abnormal blood cells of all species in one volume, even if this information were available. The blood cells of humans have been extensively illustrated, and there are several existing books devoted to the blood and bone marrow cells of domesticated and laboratory animals, which concentrate mainly on presenting detailed information for individual species. The approach used in this Atlas is somewhat different. We believe that it is valuable to present a comparative study of the morphology of blood cells of different species. By doing this, it is possible to demonstrate some of the basic principles influencing the responses of blood cells to pathological insults and diseases, which would contribute to the understanding of health and diseases in individuals. This eventually should create a situation where interpretation of blood films of sick individuals of any species can be undertaken from an informed background, whether or not the interpreter is familiar with the species in question.

All haematologists would realise that there are many potential drawbacks to this approach. Technical problems created by the difficulty of ensuring standardization of fixation and of staining methods are inevitable, particularly as the materials and methods used by many laboratories can be different and are based on methods used in human haematology. The variations can produce false indications of species similarities and differences. In fact, morphological differences do exist in the normal blood cells of different species, obviously so with regard to the red cells of mammals when compared with those of other species and, perhaps less obviously so, when considering white cell morphology, even among mammals. Many of these differences are illustrated in this Atlas, with the aim of providing some insight into the range of variation which can be considered normal. Inclusion of this normal material should also render the Atlas of interest to medical technologists, comparative physiologists, cell biologists and taxonomists.

The images of this Atlas are originated from blood samples obtained from free-living animals, from specimens housed in zoological collections and from individuals seen by Veterinary Surgeons working in clinical practice from at least 15 different countries. In addition, over 400 slides used to produce the previous publication of the Atlas, were scanned and converted into digital images and are also included in this new publication. The previous publication of this Atlas included over 125 images of blood cells from domestic animals as this was produced as an Atlas of Comparative Veterinary

Haematology. In this new publication, even though the content is about wild and exotic animal haematology, we decided to leave the images from domestic species for comparative purposes. The normal blood cells illustrated in the Atlas are from specimens declared clinically normal after examination from qualified Veterinary Surgeons at the time when the sample was obtained. Although several of the individuals were in the natural environment, there is as yet no evidence that blood cell morphology would be influenced by this fact.

The abnormal cells shown, in most instances, are from clinical cases in which the diagnosis was indicated by clinical examination, response to treatment and, on some occasions, post-mortem and histopathological examination. We have also included examples of cells which are clearly abnormal but where a definitive diagnosis was not made. This has been done to extend the recognition range of abnormal variation and, more importantly, with the hope of stimulating users of the Atlas to ponder upon the possible pathological significance of these cells and/or the changes observed.

Each section of the Atlas is introduced by a description of the comparative morphology, relationships and function of the cells in different species and comments upon possible primary and secondary pathological variations. Within each section, an attempt has been made to demonstrate the range of morphological variation considered to be normal in mammals, birds, reptiles, amphibians, fish and invertebrates, followed by examples of morphological variations associated with different disease processes. Each figure is accompanied by a descriptive legend and, where relevant, by details of cell counts and other haematology changes. As the main objective of the Atlas is to aid with the identification of blood cells in different species, the figures have been grouped together according to the phylogeny of the specimens in which they were identified. Specimens depicted in the Atlas are referred to by their common English names followed by their scientific names to assist in the identification for a more international audience. Unless otherwise stated, blood films were prepared using venous blood stored in commercially available tubes containing ethylenediaminetetraacetic acid (EDTA) and fixed with acetone-free absolute methanol using either May-Grünwald-Giemsa stain or Wright-Giemsa stain, or by using commercially available rapid stain kits such as Diff Quik™. In most instances blood cells were photographed at a magnification of ×100 or ×40.

1.2 History of Haematology

1.2.1 Introduction

Haematology saw its early beginning in ancient Egypt with the practice of draining blood from human patients with the intention of healing a condition or a symptom. This technique was widely employed by physicians throughout the Middle Ages, and it was simply known as bloodletting. During this period, the earliest written record of the use of bloodletting in human medicine can be credited to the Italian physician Prospero Alpini (1553–1617) in his work 'De Medicina Aegyptiorum'.

1.2.2 Invention of the Microscope

The first step towards the development of haematology as a science was made with the construction of the first compound microscope widely attributed to Zacharias Janssen (1585–1632) born in The Hague, The Netherlands. The microscope was made circa 1590 during the Dutch Golden Age, in which Dutch trade, science, military and arts were among the most acclaimed in the world. Historians and scholars firmly believe that his father, Hans Janssen, must have played an important role in the development of the instrument as Zacharias was a teenager in the 1590s.

Galileo Galilei (1564–1642) born in Pisa, Italy, improved upon the Zacharias Janssen design of the microscope. Using lenses with a shorter focal length and turning a telescope around he found he could magnify small objects. His first microscope, produced in 1609, was basically a small telescope with the same two lenses, a biconvex objective and a biconcave eyepiece. Galileo called this piece of equipment the 'occhiolino' or 'little eye'. Giovanni Faber (1574–1629), originally from Bamberg in Bavaria, Germany, is credited with coining the name 'microscope' for the compound microscope. Galileo submitted his invention to the Accademia dei Lincei in 1625.

The English scientist Robert Hooke (1635–1703), born in the Isle of Wight, England, also brought major innovations to the microscope. The Hooke microscopes were two and three lens microscopes made by Christopher Cock from London. Hooke examined the structure of snowflakes, invertebrates and plants. He is attributed with coining the word 'cell' from the Latin 'cella', which means 'small room', because he compared the cells with the small rooms used by monks in monasteries at the time. Hooke described his observations in the book *Micrographia: Or Some Physiological Descriptions of Minute Bodies Made by Magnifying Glasses. With Observations and Inquiries Thereupon* published in January 1665.

Early compound microscopes provided more magnification than single lens microscopes; however, they also distorted more of the image. Dutch scientist Antoine Philips van Leeuwenhoek (1632–1723) designed high-powered single lens microscopes in the 1670s. With these improved microscopes, he was the first to describe spermatozoa from dogs and humans. He also studied yeast, red blood cells, bacteria and protozoa. Van Leeuwenhoek's single lens microscopes could magnify up to 270 times larger than actual size.

In subsequent years, significant advances were not made to improve the microscope until the middle of the 19th century when great improvements contributed to the design and production of high-quality equipment similar to the microscopes we use today (Tables 1.1 and 1.2).

Table 1.1 Haematology through History – Milestones

Discovery of the first compound microscope	Zacharias Janssen, spectacle maker, The Hague, The Netherlands, 1585–1632
Observation of the 'ruddy globules' in blood (red blood cells)	Jan Swammerdam, biologist and microscopist, Amsterdam, The Netherlands, 1637–1680
Detailed description of red blood cells	Antoine Philips van Leeuwenhoek, microbiologist, microscopist, Delft, The Netherlands, 1632–1723
Determination of the importance of number of red cells in blood	William Hewson, anatomist and physiologist, generally accepted as the father of haematology, Hexham, Northumberland, England, 1739–1774
Founder of scientific haematology and its integration into clinical and analytical medicine	Gabriel Andral, pathologist, Paris, France, 1797–1876
Development of the triacid stain	Paul Ehrlich, physician, Strehlen, Upper Silesia, Germany, 1854–1915
Classification of the white blood cells	Paul Ehrlich, physician, Strehlen, Upper Silesia, Germany, 1854–1915
Development of the eosin methylene blue stain	Dimitri Leonidovich Romanowsky, physician, St Petersburg, Russia, 1861–1921
Dawn of the modern haematology era	George Richard Minot, Harvard Medical School, USA, 1885–1950 William Parry Murphy, Harvard Medical School, USA, 1892–1987

Table 1.2 Haematology through History – Inventors

Arneth count or **Arneth** index*	Joseph **Arneth**, physician and haematologist, Burgkunstadt, Germany, 1873–1955
Coplin jar	William Michael Late **Coplin**, physician, USA, 1854–1928
Dacie fluid	Sir John Vivian **Dacie**, haematologist, UK, 1912–2005
Drabkin reagent	David Leon **Drabkin**, biochemist, Pennsylvania, USA, 1899–1980
Giemsa stain	Gustav Giemsa, apothecary and chemist, Hamburg, Germany, 1867–1948
Gram stain	Hans Christian Joachim **Gram**, physician and bacteriologist, Copenhagen, Denmark, 1853–1938
Leishman stain	Sir William Boog **Leishman**, pathologist and army medical officer, Glasgow, Scotland, UK, 1865–1926
May-Grünwald stain	Richard **May**, internist, Munich, Germany, 1863–1936 Ludwig **Grünwald**, internist and otolaryngologist, Austrian born, Germany, 1863–1927
Natt-Herrick diluting fluid	Michael P. **Natt**, 1925–1998, Department of Zoology and Chester A. **Herrick**, 1893–1955, Department of Veterinary Science, University of Wisconsin-Madison, USA
Neubauer haemocytometer	Otto **Neubauer**, physician, Karlsbad, Germany, 1874–1957
Petri dish	Julius Richard **Petri**, bacteriologist, Barmen, Germany, 1852–1921
Romanowsky stains	Dimitri Leonidovich **Romanowsky**, physician, St Petersburg, Russia, 1861–1921
Sørensen buffer	Søren Peder Lauritz **Sørensen**, chemist, Denmark, 1868–1939
Thoma pipette	Richard **Thoma**, pathologist and histologist, Bondor, Germany, 1847–1923
Wright stain	James Homer **Wright**, pathologist, Boston, USA, 1871–1928
Vernier scale	Pierre **Vernier**, mathematician, Ornans, France, 1584–1638

* The Arneth count or Arneth index describes the morphological characteristics of the nucleus of the neutrophil in an attempt to detect disease.
- Neutrophils typically have two or three lobes. In general, older neutrophils have more lobes than younger neutrophils. The Arneth count determines the percentage of neutrophils with one, two, three, four, and five or more lobes.
- Individuals who have a larger percentage of neutrophils with fewer lobes have a *left shift* which can be indicative of disease processes such as infection, malignant tumours, haemolytic crises, myocardial infarction, acidosis and others.
- Individuals with a larger percentage of neutrophils with more lobes have a *right shift* and most commonly have diseases such as vitamin B_{12} or folate deficiency, chronic uraemia, liver disease and others.
- The same principle and count apply to heterophils of birds, reptiles, amphibians and fish.

1.3 Blood Sample Collection

1.3.1 Small Mammals

In small mammals, including the short-tailed chinchilla (*Chinchilla chinchilla*), ferret (*Mustela putorius furo*), European hedgehog (*Erinaceus europaeus*), rabbit (*Oryctolagus cuniculus*), golden or Syrian hamster (*Mesocricetus auratus*) and sugar glider (*Petaurus breviceps*), the cephalic vein can be used for the collection of blood samples. The use of the medial saphenous vein has also been described in the house mouse (*Mus musculus*). Blood samples may be collected from lateral tail veins in the Mongolian gerbil (*Meriones unguiculatus*), the brown rat (*Rattus norvegicus*) and house mouse. Conversely, the dorsal tail vein can be used in the rat, or the ventral tail vein in the sugar glider. A good practice is the placement of a 27- or 25-gauge needle into the blood vessel and, when there is blood in the hub of the needle, to collect the blood sample in a haematocrit tube by capillary action. The cranial vena cava is generally the quickest method to collect the largest amount of blood from the ferret and the sugar glider. The vena cava may also be used as a last resort in the chinchilla, hamster, rabbit, rat and, particularly, the guinea pig (*Cavia porcellus*) and the European hedgehog. Although the ferret heart is located caudally within the long thoracic cavity and the risk of cardiac puncture with caval sticks is minimal, the vena cava can still be lacerated if the ferret moves and all but the most severely debilitated ferrets should also be sedated or anaesthetised. Caval venipuncture is a blind technique which relies on the identification of anatomic landmarks.

The jugular vein tends to lie in a more lateral position in the ferret when compared to the cat. Jugular venipuncture can be difficult in hedgehogs unless they are thin, and guinea pigs have very short necks, making it difficult to obtain blood from their jugular veins as well. It may also be difficult in obese rabbits or does with a large dewlap. Jugular venipuncture may also be attempted in the hamster, rat and mouse, but this site can be quite challenging and requires heavy sedation or general anaesthesia. Blood sample collection from the femoral artery or vein is generally the quickest method that will yield the largest amount of blood in the guinea pig, rat, mouse and hamster. This venipuncture site has been described in the literature in most small mammals except for the ferret, rabbit and gerbil. General anaesthesia is usually required to collect samples in small mammals. Suggested sites for blood collection in small rodents include the lateral saphenous vein, saphenous vein, dorsal and lateral tail veins, ventral tail artery, jugular vein, superficial temporal vein, cranial vena cava, cephalic vein and metatarsal vein. There appears to be a general agreement that rabbit blood vessels are fragile and prone to haematoma formation. This may disrupt the blood flow, leading to thrombosis, ischaemia and sloughing of pinnal tissue. Therefore, the ear is only used as a last resort. The submandibular vein is sometimes used in a laboratory setting and is not recommended in the case of the pet mouse. The orbital and submandibular veins converge behind the mandibular joint to form the jugular vein. Small amounts of blood may be collected from the anaesthetised mouse where these vessels meet. Potential risks include bleeding into the mouth, which can cause aspiration and potential death, or bleeding into the ear canal.

Extracted from: Nugent-Deal J, Pollock C (2010). Venipuncture in small mammals, *LafeverVet*, 20 August 2010.

1.3.2 Large Mammals

Blood samples can be obtained from the ventral tail vein of beavers such as the Eurasian beaver (*Castor fiber*), of the Asian short-clawed otter (*Aonyx cinerea*) and of arboreal prehensile porcupines (*Coendou prehensilis* and *Coendou melanurus*). Conversely, blood samples can be collected from the brush tailed rock wallaby (*Petrogale penicillata*) and the red kangaroo (*Megaleia rufa*) using the jugular vein or, most commonly, the lateral coccygeal vein. The gluteal caudal vein is the preferred site for blood sampling in the seal (*Phoca vitulina*). Alternative routes include the interdigital veins located between fingers on the dorsal and ventral aspects of the hind limbs, the metatarsal vein on the dorsal aspect of the hind limb, the tarsal sinus on the ventral aspect of the hind limb and the extradural vein from the epidural sinus. The collection of blood samples from the interdigital veins is less efficient as the amount of sample that can be obtained is relatively small and it tends to be too stressful to the seals. The brachial plexus of the pectoral flipper is the most commonly used venipuncture site in dugongs (*Dugong dugon*). The dorsal and ventral tail fluke periarterial venous rete is the most widely used site to obtain blood samples from the killer whale (*Orcinus orca*) and the common bottlenose dolphin (*Tursiops truncatus*). The auricular veins are the preferred sites for blood collection in elephants; however, it may be difficult to use such sites in cold conditions unless the ear is hosed down for a few minutes using warm water. Alternative venipuncture sites include the cephalic vein on the proximal medial foreleg, the femoral vein and the saphenous vein on the lower medial hindleg. Blood samples may be obtained from most large mammals from the jugular vein, cephalic vein and the saphenous vein.

1.3.3 Birds

In birds, blood samples are commonly collected using the right jugular vein (*vena jugularis dextra*) as this is generally larger than the left jugular vein in most avian species. Other preferred sites include the cutaneous ulnar or basilic vein (*vena cutanea ulnaris superficialis*) and the medial metatarsal or caudal tibial vein (*vena metatarsalis plantaris superficialis*). The methodology used for the collection of blood samples varies according to the species and the site selected. For example, in long-legged birds such as large bustards, cranes and storks, the jugular or caudal tibial veins are very often used. Blood samples should be collected from the heart or the occipital sinus only if these birds are under anaesthesia and are to be euthanised. It is a poor practice to collect blood samples from clipped nails as cell distribution and cell content is invariably affected.

Blood samples can be obtained from most bird species from 200 g to 4000 g using a basilic vein while the bird is in dorsal recumbency, although most practitioners in the US and Europe dealing with psittacine species prefer jugular venipuncture. In most avian species the optimal area for collecting a blood sample from a basilic vein is along the medial section of the vein. The preferred side is from the right wing if the practitioner is right-handed, while the left wing is the preferred side if the practitioner is left-handed. Venipuncture immediately above the elbow joint is not recommended as haemostasis is difficult to achieve at this site in most cases. The application of digital pressure with the thumb at the proximal humerus helps in raising the vein, making it clearly visible running parallel to the external aspect of the humerus. After separating the feathers and preparing the site with an alcohol swab, the bent needle is gently inserted into the vein at an approximately 45° angle. The sample can now be collected taking the precaution not to exert high negative pressure while withdrawing with the syringe because this will invariably result in the collapse of the vein. While withdrawing the sample, it is recommended to continue maintaining pressure on the proximal humerus to ensure a raised and well-defined vein. The method of approaching the basilic vein dorsally is by entering under the adjacent tendon, which prevents haematoma formation because the underlying tissue exerts pressure on the venipuncture site when the needle is withdrawn. It is essential to avoid sudden movements that can alarm the bird and trigger a struggle as this can easily lacerate the vein and result in a haematoma or, worse, severe haemorrhage.

After collection, the needle should be removed and the blood gently deposited into a 0.5–1.0 ml commercially available paediatric blood storage tube containing the anticoagulant agent ethylenediaminetetraacetic acid (1.5 mg/ml of blood) or lithium heparin (1.8 mg/ml of blood). Squirting the sample through the needle is a poor practice as it may cause severe disruption to the fragile blood cells. For general haematology analysis, EDTA is the anticoagulant of choice as it is not possible to estimate fibrinogen or to count white blood cells accurately in heparinised samples. In some avian species, however, storing blood samples in tubes containing EDTA causes progressive red cell haemolysis and is not recommended; in these cases, it is preferable to use heparinised tubes. This is the case with some species of Corvidae such as the jackdaw (*Corvus monedula*) and raven (*Corvus corax*); Gruidae such as the black-necked crowned crane (*Balearica pavonina*) and grey-necked crowned crane (*Balearica regulorum*); Cracidae such as the black curassow (*Crax alector*); Phasianidae such as the brush turkey (*Alectura lathami*); Bucerotidae such as

the crowned hornbill (*Tockus alboterminatus*); and the ostrich (*Struthio camelus*). Storing blood samples in sodium citrate tubes is recommended when sending samples to a commercial laboratory for processing using laser flow cytometry.

Extracted from: Samour J (2005). Diagnostic value of hematology. In: Harrison GJ, Lightfoot T, editors. *Clinical Avian Medicine*. Spix Publishing, Lake Worth, FL, 587–609.

1.3.4 Reptiles and Amphibians

In lizards and snakes, blood samples can be obtained from the post orbital sinus, the ventral tail vein and the heart. In crocodiles and alligators, the ventral tail vein and the dorsal cervical or supravertebral sinus are the most common anatomical sites from which blood samples are collected. In tortoises, terrapins and turtles, there are numerous sites from where to obtain blood samples, including the supravertebral sinus, jugular vein, heart, sub-carapacial vein, dorsal coccygeal vein, femoral vein and brachial vein. Blood samples collected from tortoises, terrapins and turtles should be stored in tubes containing heparin as anticoagulant as EDTA causes haemolysis. In amphibians, blood samples can be collected from several anatomical sites. For instance, in most species of anurans and salamanders blood samples can be collected from the lingual plexus, facial or maxillary/musculocutaneous vein, ventral abdominal vein, femoral vein and the ventricle of the heart. In some salamanders, samples can also be obtained from the ventral coccygeal vein. Lymphatic vessels are widely found running along the blood vessels.

1.3.5 Fish

The use of general anaesthesia is highly recommended when collecting blood samples from fish in order to eliminate or reduce the stress during handling. If a general anaesthetic is not used, blood sampling should be undertaken in less than 30 seconds. Blood samples can be collected from small fish by severing the caudal peduncle; however, from larger fish, blood samples can be obtained by venipuncture of either the caudal vein (*vena cava caudalis*) or the dorsal aorta vein, by direct puncture of the heart or by using the vascular sinus located caudal and slightly ventral to the dorsal fin. In most fish, blood sampling can be carried out from the caudal vein running along the ventral aspect of the vertebral column, at the caudal peduncle level, through either the ventral or the lateral approach. The venipuncture site for the ventral approach is just behind the anal fin on the underside of the fish. As an average a total volume of 1 ml/kg can be collected from fish. Precaution should always be taken to avoid contamination of the samples by water or any other fluids. A novel method was recently described for serial blood sample collection in farmed Atlantic salmon (*Salmo salar*) by inserting a permanent catheter through the snout and the blood vessels at the level of the gills.

1.3.6 Invertebrates

Haemolymph can be collected from the Mexican red-knee tarantula (*Brachypelma smithi*) from the dorsal midline of the opisthosoma using an insulin syringe fitted with a 30-gauge needle. A different and widely used method for the collection of haemolymph from the great pond snail (*Lymnaea stagnalis*) consists of stimulating gently the foot of the snail using the tip of a micropipette to induce the release of a small quantity of haemolymph. The samples can then be aspirated using the pipette. Haemolymph can be collected from the fruit fly (*Drosophila melanogaster*) using a Nanoject II Auto-Nanoliter Injector™. Thin glass capillaries are used as this allows haemolymph into the needle by capillary action. The needle is inserted into the mesothorax below the wing hinge and kept just under the cuticle. An alternative method for the collection of haemolymph consists of procuring a small group of flies, e.g. 15–20 individuals, euthanising them using CO_2 and removing the head using a scalpel fitted with a fine blade or a pair of dissecting forceps with long fine tips. The bodies of the flies are then transferred into a microcentrifuge tube with a small opening at the bottom. This is placed into a slightly larger centrifuge tube and is centrifuged for 6 minutes at 1500g at 4°C. The haemolymph is then collected from the bottom of the larger centrifuge tube. Haemocytes from caterpillars can be collected by first chilling the larvae on ice for 20 minutes prior to haemocyte extraction. The collection solution contains 60% of Grace's Medium (Biological Industries USA, 100 Sebethe Drive, Cromwell, CT, 06416-1073, USA) supplemented with 10% of Fetal Bovine Serum and 20% of anticoagulant buffer (98 mM NaOH, 186 mM NaCl, 1.7 mM EDTA and 41 mM citric acid, pH 4.5). Approximately 1 µl solution per 10 mg of caterpillar total body mass is injected into each caterpillar under a dissecting microscope. After 30 minutes a small (2–3 mm long) incision is made in the caudal region of the caterpillar using a sterile scalpel. Care has to be taken not to damage the gut or other internal organs. Squeeze the caterpillar gently to collect the buffer/haemolymph mix.

Always remember the rule of thumb in clinical pathology.

'Always label the tubes and not the lids.'

1.4 Blood Sampling

In this section, various contributors depict photographically the methods and routes commonly used in veterinary practice to obtain blood samples for haematology analyses from different wild and exotic animals.

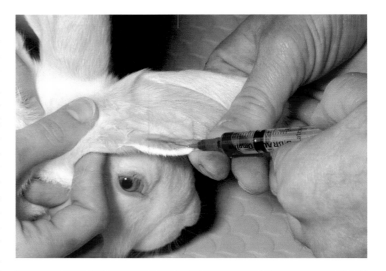

Figure 1.1 Blood collection from the lateral auricular vein in a rabbit (*Oryctolagus cuniculus*). When the vein is held off at the base of the ear, the slightly bent 26-gauge × 1/2″ needle can easily be inserted into the vein. Due to the highly innervated skin, rabbits tend to shake their head at the time of needle insertion. Application of a topical anaesthetic cream or spray may prevent shaking of the head (courtesy of Nico Schoemaker and Yvonne RA van Zeeland).

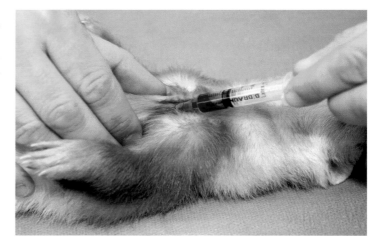

Figure 1.2 Blood collection from the cranial vena cava in a ferret (*Mustela putorius furo*), while it is under sedation with isoflurane in 100% oxygen delivered through a face mask. The 26-gauge × 1/2″ needle is inserted on the left side of the *manubrium* just cranial to the first rib. The syringe is directed towards the contralateral hindleg with a 30° angle in relation to the ferret (courtesy of Nico Schoemaker and Yvonne RA van Zeeland).

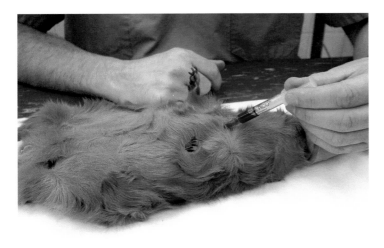

Figure 1.3 Blood collection in a guinea pig (*Cavia porcellus*) is in essence similar to that in a ferret (*Mustela putorius furo*). The thorax of guinea pigs, however, is much shorter, thereby placing the heart much closer to the insertion site. In guinea pigs, the needle should therefore be inserted at an 80° angle in relation to the guinea pig and directed towards the contralateral distal end of the scapula (courtesy of Nico Schoemaker and Yvonne RA van Zeeland).

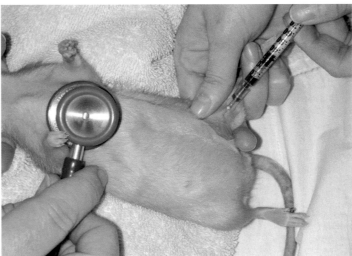

Figure 1.5 Blood sample collection from a femoral vessel in a domestic rat (*Rattus norvegicus*) while under isoflurane anaesthesia in 100% oxygen delivered via face mask. The rat is placed in dorsal recumbency and the limb is positioned at a 90° angle to the long axis of the body. The optimal site for venepuncture is where the body wall meets the upper thigh. A 25-gauge × 3/8″ needle on a 1 ml syringe is directed at an approximately 45° angle to the femur, while using gentle negative pressure. If the blood collected is bright red, the femoral artery has likely been cannulated. Apply firm, gentle direct pressure to the venepuncture site for at least 2–5 minutes (courtesy of Jody Nugent-Deal and Christal Pollock).

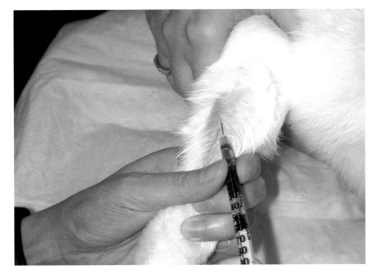

Figure 1.4 Blood sample collection from the lateral saphenous vein in a domestic rabbit (*Oryctolagus cuniculus*). The rabbit is manually restrained in sternal recumbency. This hind end of this conscious patient is gently slid towards the edge of the table, allowing the hind limb to drop off the table. This allows access to the limb without placing any torque on the hindquarter that would increase the risk of injury. The upper leg is grasped just above the stifle to engorge the vessel and prevent the limb from moving (courtesy of Jody Nugent-Deal and Christal Pollock).

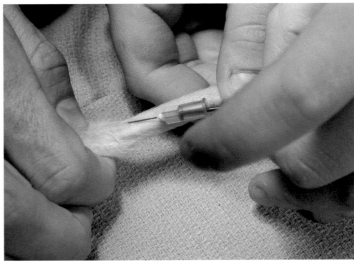

Figure 1.6 Blood sample collection from the lateral tail vein in a conscious rat (*Rattus norvegicus*) using a 25-gauge × 3/8″ needle. The tail is warmed beforehand and the needle is used to cannulate the vessel so that a small sample can be collected using a haematocrit capillary tube. Avoid squeezing or milking the tail during sampling, and provide appropriate haemostasis afterwards (courtesy of Jody Nugent-Deal and Christal Pollock).

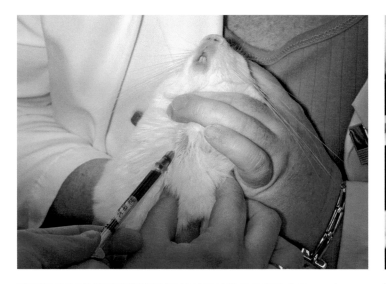

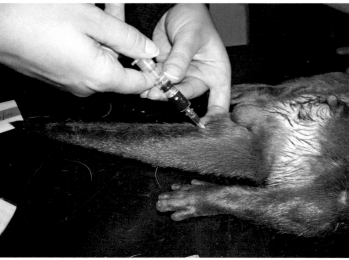

Figure 1.7 Blood sample collection from the jugular vein of a conscious chinchilla (*Chinchilla lanigera*). The chinchilla is manually restrained in sternal recumbency at the edge of an examination table with the head extended upwards and the legs pulled down towards the body. Although jugular venepuncture may be performed in conscious chinchillas, it may be best for patient and phlebotomist to use chemical restraint if the patient is stressed or uncooperative (courtesy of Jody Nugent-Deal and Christal Pollock).

Figure 1.9 Blood sample collection from an anaesthetised Asian short-clawed otter (*Aonyx cinerea*). The otter is chemically restrained using medetomidine/ketamine IM and maintained on inhalant isoflurane in 100% oxygen prior to blood sampling. The otter is positioned in dorsal recumbency on the examination table and blood taken from the ventral coccygeal blood vessel, approximately 2 cm below the base of the tail. A 3 ml syringe and 21-gauge needle are used and blood is stored in EDTA tubes (courtesy of Michelle O'Brien).

Figure 1.8 Blood sample collection from an anaesthetised European beaver (*Castor fiber*). The beaver is chemically restrained using inhalant isoflurane in 100% oxygen prior to blood sampling. The beaver is positioned in dorsal recumbency on the examination table and blood taken from the ventral coccygeal blood vessels, just below the base of the tail. A 5 ml syringe and 21-gauge needle are used and blood is stored in EDTA tubes (courtesy of Michelle O'Brien).

Figure 1.10 Blood sampling from a conscious lion (*Panthera leo*) using the lateral tail vein approximately 8 cm from the tail base. A 21-gauge butterfly needle and 5 ml syringe are used and blood is stored in both EDTA and serum separator tubes. The lion is not sedated but is under behavioural restraint in a small cage to limit movement and being fed treats during the procedure (courtesy of Scott Citino).

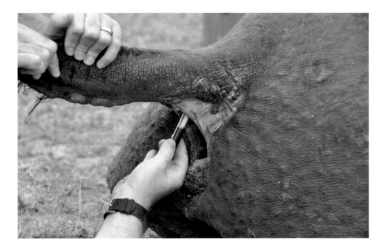

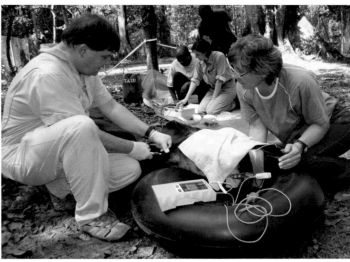

Figure 1.11 Blood sample collection from a free-ranging, anaesthetised Nile (River) hippopotamus (*Hippopotamus amphibius*). The hippopotamus was chemically restrained using butorphanol, azaperone and medetomidine by remote darting from a helicopter. The hippopotamus is in sternal recumbency and blood is taken from the ventral coccygeal vein using a 21-gauge Venoject system (Terumo™) and a serum separator vacutainer tube. This is just outside the anus. There is a flat triangular area on the ventral tail there that allows access to both the ventral coccygeal vein and artery. On many adult hippopotamuses, this is the only anatomical site from where blood samples can be easily obtained (courtesy of Scott Citino).

Figure 1.13 Blood sample collection from a captive okapi (*Okapia johnstoni*) anaesthetised with carfentanil and xylazine delivered by plastic dart. The okapi is in lateral recumbency with head supported by a tire inner tube and blood is being taken from the jugular vein using a 20-gauge Venoject system (Terumo™) and EDTA, sodium heparin, serum separator and royal blue vacutainer tubes. A pulse oximeter is attached to the tongue for haemoglobin oxygen saturation readings and a capnograph cannula is sampling breaths from the nose for end-tidal CO_2 readings. Tire inner tubes are usually used as padding and support for various parts of the body of large ungulates, the head in particular (courtesy of Scott Citino).

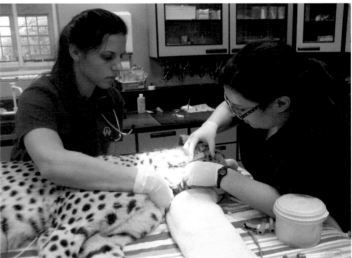

Figure 1.12 Blood sampling from a tiger (*Panthera tigris*) chemically restrained with medetomidine, midazolam and ketamine by blow dart. The tiger is in right lateral recumbency and blood is taken from the medial saphenous vein using a 20-gauge needle and 12 ml syringe, while an assistant holds off the vein, and blood is stored in EDTA and serum separator tubes (courtesy of Scott Citino).

Figure 1.14 Blood sampling from a captive cheetah (*Acinonyx jubatus*) chemically restrained with medetomidine, butorphanol and midazolam and maintained on inhalant isoflurane in 100% oxygen. The cheetah is positioned in left lateral recumbency and blood is being collected from the jugular vein using a Venoject system (Terumo™) and vacutainer tubes (courtesy of Scott Citino).

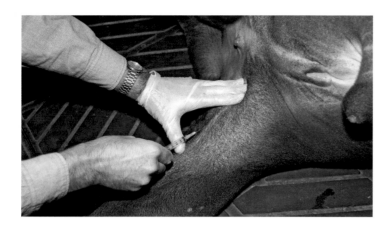

Figure 1.15 Blood sample collection from the saphenous vein in a conscious Malayan tapir (*Tapirus indicus*) using a 21-gauge × 1/2″ needle with a 3 ml syringe. A 25-gauge × 5/8″ needle can also be used. In this case the female is made to lay down on her side; it is also possible in the same vessel in a standing tapir (courtesy of Linda B van Sonsbeek).

Figure 1.17 Blood sample collection from the dorsal digital vein in a conscious black rhinoceros (*Diceros bicornis michaeli*) using a 23-gauge × 0.8 mm butterfly needle with a 30 cm tubing connected to a vacuum tube by using a luer lock adapter. Preferably the temperature in the surrounding should be 23°C or warm water should be used for dilating the blood vessels. This vein can be accessed in both front and hindlegs. In addition, the same vessels can be used in white and Indian rhinoceros to obtain blood samples, although the vessels in the leg of the Indian rhino are very difficult to see and palpate. Ultrasound guidance can be used to locate the blood vessel (courtesy of Linda B van Sonsbeek).

Figure 1.16 Blood sample collection from the radial vein in a conscious black rhinoceros (*Diceros bicornis michaeli*) using a 23-gauge × 0.8 mm butterfly needle with a 30 cm tubing connected to a vacuum tube by using a luer lock adapter. The collection tube in the image is a serum tube with gel. Preferably the temperature in the surrounding should be around 23°C or warm water can be used for dilating the blood vessels (courtesy of Linda B van Sonsbeek).

Figure 1.18 Blood sample collection from the saphenous vein in a conscious Asian elephant (*Elephas maximus*) using a 23-gauge × 3/4″ butterfly needle with a 10 cm tubing (no luer lock adapter is needed). The collection tube in the image is a serum tube with gel. The ear vein can also be used or the same vein more proximal in the hindleg (courtesy of Linda B van Sonsbeek).

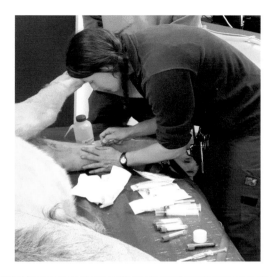

Figure 1.19 Blood sample collection from the saphenous vein in an anaesthetised Asian lion (*Panthera leo persica*) using a 20-gauge × 1½″ needle for collection using a vacuum tube. The collection tube in the image is a serum tube with gel. A tourniquet can be used for dilating the blood vessel to ease the blood sampling procedure (courtesy of Linda B van Sonsbeek).

Figure 1.21 Gloves are used in an etorphine anaesthetised plains zebra (*Equus quagga*, formerly *E. burchellii*) to protect against potentiated opiate exposure. A jugular blood sample is collected using a 1.5″ × 18-gauge needle and a 10 ml syringe. Zebra skin is very tough and best penetrated perpendicular to the jugular vein. The blindfold is commonly used to avoid stress (courtesy of Brett Gardner).

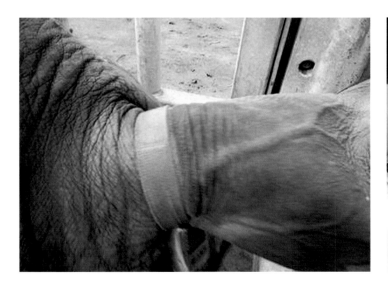

Figure 1.20 Blood sample collection from the auricular vein in a conscious black rhinoceros (*Diceros bicornis michaeli*) using a tourniquet and 23-gauge × 0.8 mm butterfly needle with a 30 cm tubing connected to a vacuum container using a luer lock adapter. Using this method, the animal tends to react more than when using a radial vein. For dilating the veins, the use of a tourniquet fitted at the base of the ear is recommended (courtesy of Linda B van Sonsbeek).

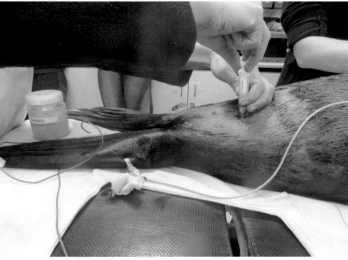

Figure 1.22 The caudal gluteal vein can be used to collect large volumes of blood in *Otariid* seals under manual restraint and under sedation, such as this Cape fur seal (*Arctocephalus pusillus*). The animal needs to be in ventral recumbency. A 1.5″ × 21-gauge needle is used, attached to a 5 ml syringe and inserted 90° to the skin. The point of entry is lateral to the sacral vertebrae, at the junction of the cranial and middle thirds of the distance between the base of the tail and the femoral trochanters (courtesy of Brett Gardner).

Figure 1.23 Blood sample collection from the right jugular vein in a conscious alpaca (*Lama paca*) using a 20-gauge × 1″ needle under manual restraint. The head is slightly tilted to the left to help expose the vessel. Site preferred is in the caudoventral portion of the neck around the processes of the fifth and sixth cervical vertebrae to prevent arterial blood sampling (courtesy of Margarita W Colburn).

Figure 1.24 Preparation for blood collection from an ear vein of a conscious Asian elephant (*Elephas maximus*), under trained behaviour and protected contact. The elephant has been trained to station close to and parallel to the enclosure barrier, with her trunk down for safety purposes. Her trainer maintains this behaviour throughout the procedure by voice commands and assurance and food rewards. She has also been conditioned to have her ear held and abducted to expose the network of vessels on its caudal aspect. A second operator holds the ear in this position and hoses it with warm water for 1–2 minutes (or longer at cold ambient temperatures). This assists with vasodilation, facilitating vein visualisation (courtesy of Helen McCracken).

Figure 1.25 Following warming of the ear, the vessels are palpated to distinguish arteries from veins, with the objective of selecting a vein for venepuncture. Veins are thin walled and easily compressed by digital pressure; arteries are palpable as firm cords that roll under the finger and cannot be compressed. Ear arteries may be cannulated for blood collection for blood gas measurement and direct blood pressure monitoring during anaesthetic procedures. The veins are used for routine blood collection. The selected vein is occluded proximal to the venepuncture site, then blood collected using a 21-gauge × 3/4″ butterfly needle and 3–10 ml syringe, depending on the volume required. Blood is stored in EDTA and serum separator tubes for CBC and chemistry. The elephant in this image has weekly blood collections for the purpose of progesterone monitoring for management of assisted reproduction. For this purpose, blood is collected into plain tubes without gel. Alternate ears are used each week to reduce the likelihood of scarring of veins (courtesy of Helen McCracken).

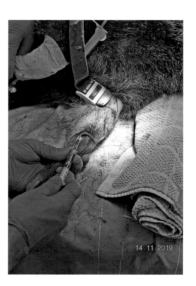

Figure 1.26 Blood collection from the cephalic vein of an anaesthetised collared peccary (*Pecari tajacu*), using a 21-gauge × 3/4″ butterfly needle and 3 ml syringe. The cephalic vein of this species is located in a more medial position on the antebrachium than in many other species and is therefore best accessed from the medial aspect of the extended undermost limb, with the animal in lateral recumbency. Placement of a tourniquet above the elbow assists with distension and therefore visualisation of the vein. Clipping off the very stiff, bristle-like hair must be done with care to avoid causing skin surface capillary ooze that can obscure the vein. Blood collected is stored in EDTA and serum separator tubes (courtesy of Helen McCracken).

Figure 1.27 Blood collection from the jugular vein of a conscious Rothschild's giraffe (*Giraffa camelopardalis rothschildi*) using an 18-gauge × 1.5″ needle, 40 cm extension set and 10 ml syringe. The giraffe was conditioned to enter and station in a custom designed chute with an elevated platform to enable the operator access to his head and neck. The giraffe was encouraged to move his head and neck towards the operators using a bucket of food, the quantity sufficient to keep him eating until the venepuncture was completed. Once he was engaged in eating, the vein was occluded with a thumb over the jugular groove proximal to the intended venepuncture site, and the needle, with the extension set attached, was inserted into the distended vein, perpendicular to the long axis of the neck and directed towards the vertebral bodies at an angle of 30–45°. When blood flow commenced, the syringe was attached and blood collected and stored in EDTA and serum separator tubes (courtesy of Helen McCracken).

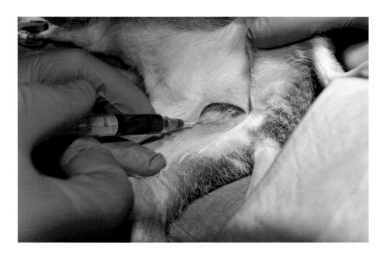

Figure 1.29 The femoral vein is the preferred site for blood collection from many small to medium marsupial species as it will deliver a volume adequate for diagnostic work. Anaesthesia is required for this technique. In this image, blood is collected from the femoral vein of an anaesthetised female Eastern barred bandicoot (*Perameles gunnii*), using a 25-gauge × 5/8″ needle and 3 ml syringe. In the female bandicoot, the vein is most easily visualised inside the pouch. In the male it is generally visible through the skin, or one may locate it by palpating the pulse in the adjacent artery. The needle is inserted at a 30–45° angle to the skin. Due to the pulse of the adjacent artery, it is not necessary to have the vein occluded proximally in order to achieve adequate blood flow. Blood is collected into paediatric EDTA and serum separator tubes (courtesy of Helen McCracken).

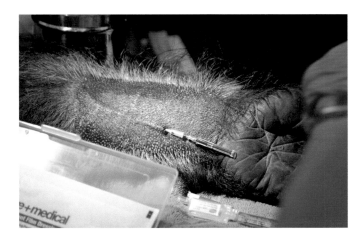

Figure 1.28 Placement of an intravenous catheter in the median antebrachial vein of an anaesthetised Western lowland gorilla (*Gorilla gorilla gorilla*) using a 21-gauge × 1.25″ catheter. Blood may be collected via the catheter, into EDTA and serum separator tubes. The gorilla is in dorsal recumbency, with the forearm supinated. This vein is accessed on the medial aspect of the distal antebrachium. It is best visualised by clipping the site and placement of a standard human tourniquet immediately proximal to the elbow. Placement of the catheter is aided by first making a short longitudinal skin incision over the vein because the skin is thick and may otherwise drag on the catheter, impeding placement (courtesy of Helen McCracken).

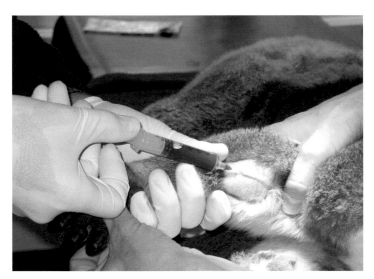

Figure 1.30 Blood collection from the cephalic vein of an anaesthetised Southern koala (*Phascolarctos cinereus victor*), using a 21-gauge × 1″ needle and 5 ml syringe. Blood is stored in EDTA and serum separator tubes. The koala is positioned in lateral recumbency and the vein accessed on the anterior aspect of the elbow or proximal antebrachium of the uppermost forelimb. The vein is best visualised after clipping of the site and occlusion by the thumb of an assistant proximal to the selected venepuncture site. This venepuncture site is also commonly used in conscious koalas, either restrained in a sack with the forearm extended out of the bag or with minimal restraint for quiet, captive individuals while sitting in a low tree fork (courtesy of Helen McCracken).

Figure 1.31 Voluntary blood sampling in a bottlenose dolphin (*Tursiops truncatus*) from the central vein of the fluke. Collection can be done with either a dorsal or a ventral approach, although inserting the needle from the ventral aspect of the fluke is more common. This is practically the same for all cetaceans. Commonly, 21–23-gauge butterfly needles are used depending on the age of the animal. Training cetaceans for voluntary blood sampling reduces stress and unnecessary and dangerous physical restrain (courtesy of Guillermo Sanchez Contreras).

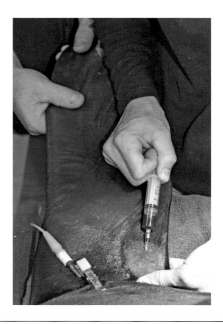

Figure 1.33 Blood collection from the brachial vein of an anaesthetised Australian fur seal (*Arctocephalus pusillus doriferus*) using a 22-gauge × 1″ needle and 5 ml syringe. Blood is stored in EDTA and serum separator tubes. For this procedure, the seal is positioned in lateral or dorsal recumbency with the forelimb abducted from the body and held in full extension, exposing the site of the vein close to the axilla on the medial aspect of the limb. The vein runs close and parallel to the caudal aspect of the flipper, and is best visualised after clipping of the site and occlusion by a thumb proximal to the selected venepuncture site. This venepuncture site may also be used for blood collection from conscious seals under trained behaviour (courtesy of Helen McCracken).

Figure 1.32 Voluntary blood sampling in a South American sea lion (*Otaria flavescens*) from the interdigital veins of the hind flippers. Collection can be done from any of the interdigital veins in either flipper. This is practically the same for all sea lions, fur seals and walruses. Commonly, 23-gauge butterfly needles are recommended although thinner needles may be necessary in smaller animals. Training pinnipeds for voluntary blood sampling reduces stress and unnecessary and dangerous physical restrain (courtesy of Guillermo Sanchez Contreras).

Figure 1.34 Blood sample collection from an Andean condor (*Vultur gryphus*) using the medial metatarsal or caudal tibial vein (*vena metatarsalis plantaris superficialis*). Condors in common with other vultures have very thick skin covering the tarsometatarsal area. A butterfly 23-gauge needle connected to a 3 ml syringe was used in this case (courtesy of Shady Heredia).

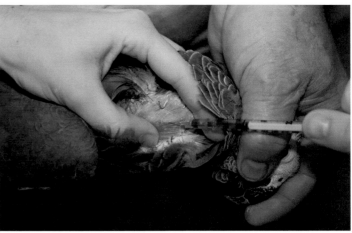

Figure 1.35 Blood sample collection from the medial metatarsal vein in a racing pigeon (*Columba livia*). To enable entering this superficial vein as parallel as possible, it helps to bend the tip of the 26-gauge × 1/2″ needle. After blood has been collected a tape needs to be placed around the metatarsus and kept in place for 72 hours to assure sufficient closure of the puncture site (courtesy of Nico Schoemaker and Yvonne RA van Zeeland).

Figure 1.37 The right jugular vein is the most convenient location to collect blood from a parrot. Especially in grey parrots (*Psittacus erithacus*) this vein is very large. The left jugular vein is much smaller. Holding off the vein at the thoracic inlet will enhance visualisation of the vessel. The approach to the vessel is easier in the direction of the heart, as the broad body size may hinder parallel insertion of the needle, although some veterinarians do prefer this approach. In the photo the approach towards the heart is shown (courtesy of Nico Schoemaker and Yvonne RA van Zeeland).

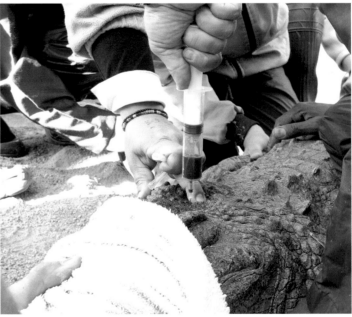

Figure 1.36 Blood sample collection from a juvenile gyrfalcon (*Falco rusticolus*) from the cutaneous ulnar or basilic vein (*vena cutanea ulnaris superficialis*). The feathers around the area are wet with surgical spirit and pressure on the proximal area is applied using the index finger to dilate the vein. The needle is bent prior to venepuncture to ease the blood collection procedure. Negative pressure should be avoided as the vein would invariably collapse (courtesy of Jaime Samour).

Figure 1.38 In medium to large size crocodilians the most reliable site for venepuncture is the supravertebral vein. It is accessible under manual restraint with qualified handlers and with the animal's mouth taped shut to prevent accidental injury. This Nile crocodile (*Crocodylus niloticus*) has its eye covered to assist with keeping the animal calm. A 2″ needle is attached to a 3–20 ml syringe depending on the volume of blood required. The needle is inserted 90° at the caudal extent of the occiput, exactly on the midline, in the most obvious depression, typically where the disinfectant pools. Mild ventroflexion of the head will facilitate entry into the supravertebral vein. The needle must be advanced slowly with adequate restraint of the animal to prevent accidental injury of the underlying spinal cord (courtesy of Brett Gardner).

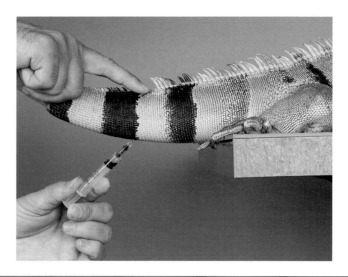

Figure 1.39 Blood sample collection from the ventral coccygeal vein in a green iguana (*Iguana iguana*). It is recommended to use a 23-gauge × 1″ needle fitted to a 1 ml syringe to obtain blood samples from adult individuals. The needle is inserted in the midline pointing slightly cranial until the ventral surface of the vertebrae is reached (courtesy of Nico Schoemaker and Yvonne RA van Zeeland).

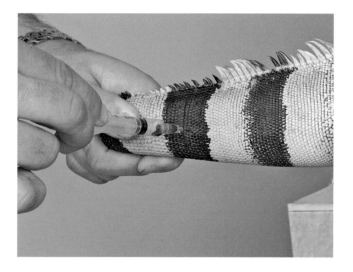

Figure 1.40. Blood sample collection from the ventral coccygeal vein in a green iguana (*Iguana iguana*) using the lateral approach. When collecting blood, the needle is inserted in a horizontal plane, just ventral to the transverse process up to the middle of the tail (courtesy of Nico Schoemaker and Yvonne RA van Zeeland).

Figure 1.41 The ventral coccygeal vein may be used for blood collection from crocodilian species under both manual and chemical restraint. In this image, blood is collected from an anaesthetised Philippines crocodile (*Crocodylus mindorensis*), using a 19-gauge × 2″ needle and 5 ml syringe. The animal may be restrained in ventral recumbency, as in this case, or vertically with the tail down, so that the operator has access to the ventral aspect of the tail. The vein is situated on the midline of the tail, immediately ventral to the coccygeal vertebrae, and is at its greatest diameter and therefore most easily accessed in the cranial half of the tail. On contacting the bone, the needle will generally be in the vein and blood will start dripping, as in this case. Occasionally, one needs to pull back on the needle a little to start the blood flow. Blood is collected into lithium heparin and serum separator tubes (courtesy of Helen McCracken).

Figure 1.42 The dorsal cervical sinus (external jugular vein) is easily accessed under manual restraint in sea turtles. This juvenile green sea turtle (*Chelonia mydas*) is held with the caudal aspect of the body slightly elevated and the neck extended to facilitate blood sampling. A 1″ × 23-gauge needle is used with a 3 ml syringe. The vessel is located along the length of the dorsolateral neck on both the right and left side, at approximately the one and eleven o'clock positions, as indicated by the strap-like biventer cervicis muscle. It can be accessed anywhere along its length (courtesy of Brett Gardner).

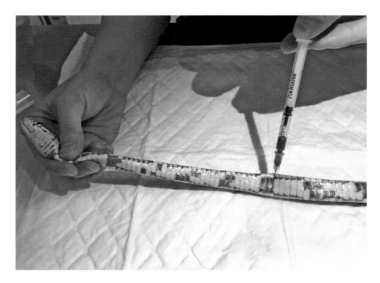

Figure 1.45 Blood sample collection from the right jugular vein in a Senegal chameleon (*Chamaeleo senegalensis*). The jugular vein is located and venepuncture is achieved by inserting a 25–27-gauge needle attached to a 1 ml syringe at a 30° angle (courtesy of Xavier Valls Badia).

Figure 1.43 Blood sample collection directly from the heart in a corn snake (*Pantherophis guttatus*). The snake is placed in dorsal recumbency and the heart located through the heart beat. A 26–27-gauge needle fitted with a disposable 1 ml syringe is used. The needle is introduced between two scales in a perpendicular direction of the longitudinal axis of the snake. After collecting the blood sample, the needle is removed and blood smears are made directly from the syringe without the need of using any anticoagulant in neither the needle nor the syringe in order to avoid artefacts (courtesy of Xavier Valls Badia).

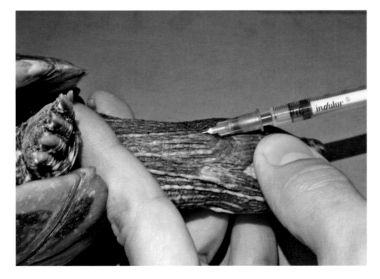

Figure 1.46 Blood sample collection from the dorsal coccygeal vein in a Russian tortoise (*Agrionemys horsfieldii*). The disadvantage of using this route is that the blood sample tends to get contaminated with lymph resulting in misleading blood cell counting and haematocrit estimation. 1 ml syringes fitted with 25–27-gauge × 5/8″ needles are commonly used (courtesy of Xavier Valls Badia).

Figure 1.44 Blood sample collection from the dorsal right jugular vein in a red-eared slider (*Trachemys scripta elegans*). The head is grabbed with one hand holding it with the thumb and ring fingers, while using the middle and index fingers to apply pressure at the base of the neck to dilate the vein and ease the blood sampling procedure. With the other hand venepuncture is carried out at an angle of 30° using a 1 ml syringe fitted with 25–27-gauge × 5/8″ needles (courtesy of Xavier Valls Badia).

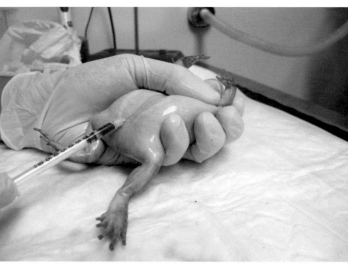

Figure 1.47 Blood sample collection from a Dumeril's boa (*Acrantophis dumerili*) using the ventral coccygeal vein. Precaution has to be taken in males to avoid venepuncture close to the cloaca as to avoid injuring the hemipenes. This is the ideal venepuncture site in Serpentes, although finding the vein offers a certain degree of difficulty. In this case, a 25-gauge × 1″ needle fitted to a 1 ml syringe was used introducing the needle between two scales at an angle. Different needles and syringes are used for larges snakes (courtesy of Xavier Valls Badia).

Figure 1.49 Blood sample collection from an American bull frog (*Lithobates catesbeianus*) using the ventral abdominal vein. It is highly recommended to manipulate all amphibians using wet latex or rubber gloves to avoid injury to the fragile skin. In this case, the blood sample was collected using a 0.5 ml disposable syringe fitted with a 29-gauge × 1/2″ needle (courtesy of Xavier Valls Badia).

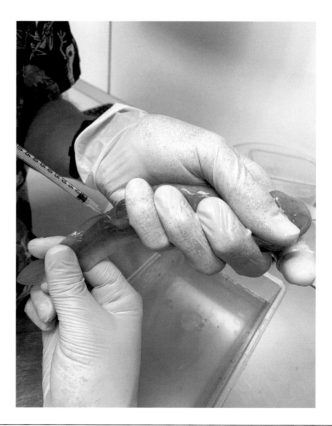

Figure 1.48 Blood sample collection by cardiocentesis from a decalcified Home's hinge-back tortoise (*Kinixys homeana*). In weak individuals the carapace is soft and it is possible to obtain blood samples directly from the heart. This route is used mainly in moribund individuals. Diagnosis is relevant in such individuals as they tend to be kept in groups becoming necessary to discard infectious conditions that could affect the group. In this case, a 25-gauge × 5/8 needle attached to a 1 ml syringe was used (courtesy of Xavier Valls Badia).

Figure 1.50 Blood sample collection from an albino axolotl (*Ambystoma mexicanum*). The specimen is manipulated very carefully with a container with water to avoid injury in case it falls down. A 0.5 ml syringe fitted with a 29-gauge × 1/2″ needle was used in this case (courtesy of Xavier Valls Badia).

Figure 1.51 Blood sample collection from a spiny-tailed lizard (*Uromastyx* spp) utilising the ventral coccygeal vein. For venepuncture the needle is introduced in the medial aspect of the tail between two scales at a 70–90° angle. In this case, a 1 ml syringe fitted with a 27-gauge × 5/8″ needle was used (courtesy of Xavier Valls Badia).

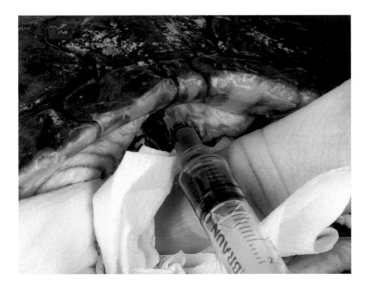

Figure 1.52 Blood collection from the subcarapacial venous plexus in a red-eared slider (*Trachemys scripta elegans*). To allow access to the insertion site, the head is pushed backwards. A 23-gauge × 1″ needle is slightly bent upwards and inserted just caudal to the middle scute of the carapace. Instead of drawing blood, it is common to draw up lymph fluid as the lymph vessels lie just adjacent to the subcarapacial venous plexus (courtesy of Nico Schoemaker and Yvonne RA van Zeeland).

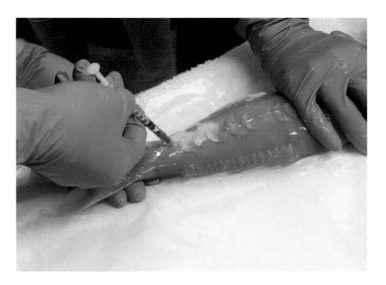

Figure 1.53 The caudal aspect of the lateral vein is best utilised in sedated or anaesthetised fish. A 25-gauge needle attached, to a 1 ml syringe is used in a lateral approach in a koi carp (*Cyprinus carpio*). The needle is inserted parallel and ventral to the lateral line. It is advanced at 45° in a craniomedial direction towards the ventral aspect of the vertebra. Gentle negative pressure, sedation and light post-venepuncture digital pressure will prevent haemolysis and local bruising. In most cases, it is advisable to heparinise the needle and syringe prior to collection (courtesy of Brett Gardner).

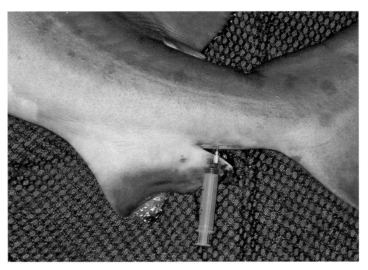

Figure 1.54 Blood collection, midline from the caudal vein of a sand tiger shark (*Carcharias taurus*) using a ventral approach and some tail flexion. Elasmobranch skin is difficult to penetrate; therefore a 14-gauge × 51 mm needle was used in this case. For a smaller individual a scalpel blade or 14-gauge can be used to create an entry point, then switch to a needle size more suited to the vessel. This was performed under ketamine/medetomidine anaesthesia. Samples can often be obtained from manageable individuals under 'tonic immobilization' (courtesy of Peter Scott).

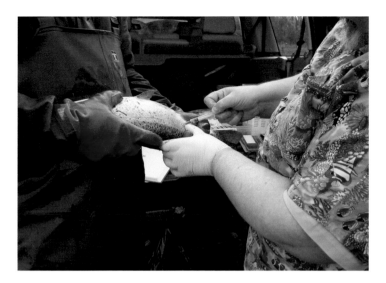

Figure 1.55 Blood collection, from the caudal vein of a rainbow trout (*Oncorhynchus mykiss*) using a lateral approach, which is often less deep that the ventral approach. The vessel lies in contact with the ventral surface of the vertebral column. A 2 ml syringe and 23-gauge × 25 mm needle was used. Heparin was used as an anticoagulant in this case. Depending on fish size and time taken, it may be helpful to heparinase the syringe prior blood sampling (courtesy of Peter Scott).

Figure 1.57 Sample collection for an anaesthetised burgundy goliath bird eater (*Theraphosa stirmi*). The theraphosid was anaesthetised with cotton wool impregnated with isoflurane gas in a ventilated scavenged environment. The theraphosid was placed in ventral recumbency and the prosoma secured with gentle pressure. Haemolymph was collected through puncture of the cranial 1/3 opisthosomal midline with a 25-gauge needle and 1 ml syringe. Cellularity of samples was high as gauge of needle was wide enough to allow the relatively large invertebrate cells to be collected (courtesy of Benjamin M Kennedy).

Figure 1.56 Sample collection from an anaesthetised Chilean rose tarantula (*Grammostola rosea*). The theraphosid was anaesthetised with isoflurane gas in a small animal induction chamber. The theraphosid is placed in ventral recumbency and held in place with a finger or cardboard divided over the prosoma (thorax). Haemolymph, in this case, was collected through puncture of the cranial 1/3 opisthosomal midline with a 29-gauge and 1 ml syringe. Cellularity of the sample was low and haemolymph could be biochemically analysed (courtesy of Benjamin M Kennedy).

Section A

Basic Wild and Exotic Animal Haematology

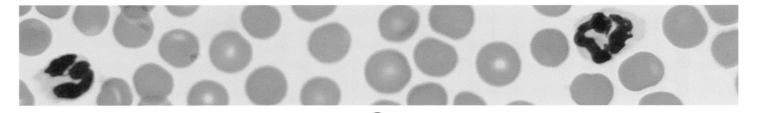

2
HAEMATOPOIESIS

In mammals, birds and reptiles, erythropoiesis takes place in the bone marrow. In amphibians, however, the process of erythropoiesis is more complex. For instance, in juvenile amphibians, erythropoiesis takes place in the liver and kidney, while in adult aquatic and terrestrial frogs erythropoiesis takes place in the bone marrow, spleen and the liver. In a wide number of elasmobranch fish, erythropoiesis takes place in the organ of Leydig, a white mass located in the dorsal and ventral wall of the oesophagus, the epigonal organ, associated with the gonads, the spleen and the thymus. In more primitive holocephalans, the site of granulocytopoiesis is found in the tissues within the cranium. In teleost fish, the main erythropoietic organ is the head kidney, with minor sites including the spleen, liver and thymus. In cases of anaemia and marrow dysfunction, however, it is not uncommon to find immature red blood cells in the circulating blood and, therefore, the recognition of normal and abnormal red blood cell precursors is of diagnostic importance. Little information is available on the process of red blood cell maturation in birds and even less for reptiles. On the assumption that, apart from differences related to loss of the cell nucleus in mammals, the process is essentially similar in all three groups, a simple classification scheme based on the developmental states involved in erythropoiesis in mammals can be used to identify and compare immature red blood cells in the other Classes of vertebrates. In mammals, the earliest recognisable cell of the red blood series is the proerythroblast, which gives rise to a series of nucleated cells, the erythroblasts. These subsequently undergo the typical progressive changes associated with increasing maturity, including reduction in size of both cytoplasm and nucleus, loss of nucleoli, decreasing cytoplasm basophilia and assumption of functional characteristics. In the case of red blood cells, the latter involves the synthesis of haemoglobin and the acquisition of a shape suitable for effective gaseous exchange and for circulation through the vasculature. In mammals, it involves also the loss of the nucleus. The classification scheme derived from these

observations is inevitably an oversimplification but provides a means of defining and, perhaps, of comparing the stage of immaturity of early red blood cells in the circulation of mammals, birds, reptiles, amphibians and fish.

Under normal circumstances, the production rate of new red blood cells is equal to the destruction rate of outworn cells and this is determined by the red blood cell life span. There are marked species differences in the red blood cell life span, which, in general, is directly related to metabolic rate and, therefore, to body weight. Thus, red blood cells do not survive as long and erythropoiesis is more active in animals of a small size. There appears to be a direct relationship between normal red blood cells turnover rate and the rate at which an animal can replace red blood cells lost through haemorrhage or haemolysis. This process is more rapid in small animals than in larger ones. These points are important when considering indicators of active erythropoiesis which might be expected to show on a stained blood film or in a preparation stained supravitally for reticulocytes. In small mammals, it is normal to find a moderate number of polychromatic red blood cells in the blood film and the reticulocyte count is relatively high. Such animals respond rapidly and dramatically to red blood cell loss. In contrast, in most large mammals, polychromatic red blood cells are rarely present under normal circumstances and the reticulocyte count is low. In some species, including horses and other Perissodactyla, red blood cells do not enter the circulation until fully mature and neither reticulocytes nor polychromatic red blood cells are seen in samples of peripheral blood, even during recovery from acute blood loss. In mammals, reticulocytes and polychromatocytes, particularly those produced in response to severe blood loss, are larger than mature red blood cells. The cells that enter the bloodstream under these circumstances are larger than normal in horses and other species in which circulating reticulocytes do not occur. In these species, a finding of macrocytosis is a useful indication of increased erythropoiesis.

In birds, reptiles, amphibians and fish, interpretation of samples stained for reticulocytes is somewhat difficult because reticular material is present in virtually all of the red blood cells. In the earliest cells, this material forms a band of particles encircling the nucleus. During maturation, the particles are gradually reduced in number and become dispersed across the cytoplasm but some may persist throughout the lifespan of the cell. There are no firm data on the amount of reticulum present in the cells as they leave the marrow, and it is likely that this will vary in different species. Several methods have been used for defining reticulocytes in birds, reptiles, amphibians and fish, including those cells which contain a complete perinuclear ring of reticular particles or those which contain five or more particles. At the present state of knowledge, evidence of active erythropoiesis in birds, reptiles, amphibians and fish is probably best based on the number of polychromatic cells present and the appearance and size of the nucleus and cytoplasm in Romanowsky (named after Dmitri Leonidovich Romanowsky, 1861–1921) stained blood films. Although systematic comparative studies have not yet been undertaken, there is subjective evidence that, as in mammals, small birds normally show greater erythropoietic activity than larger ones. There is evidence suggesting that the presence of immature erythrocytes increases during the periods of ecdysis and is more prevalent in young growing reptiles.

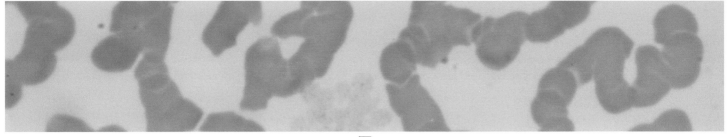

3
NORMAL AND ABNORMAL RED BLOOD CELLS

In all vertebrates except mammals, the red blood cells are ovoid and nucleated. The red blood cells of birds appear elliptical in outline and have elliptical nuclei, whereas those of reptiles and amphibians have more rounded ends and the nuclei are usually round. The red blood cells in fish in common with birds are also ellipsoidal in shape, with a centrally situated, elongated biconvex nucleus. Mammalian red blood cells are anucleated and, with the exception of the Camelidae, take the form of biconcave discs (Table 3.1). Camelidae, including camels, llamas, vicuñas, alpacas and guanacos, have flat, elliptical, anucleated red blood cells. All of these morphological differences are strikingly obvious on stained blood films.

Also obvious is the high degree of size variation in the red blood cells of the main vertebrate groups. As a general rule, avian red blood cells are larger than those of mammals, and reptilian red blood cells are larger than those in birds. The largest red blood cells occur in amphibians, with the record held by the three-toe salamander (*Amphiuma tridactylum*) of 70 μm × 40 μm. In mammals, the smallest red blood cells are found in the lesser mouse-deer or lesser Malay chevrotain (*Tragulus kanchil*) (1.4 μm–3.3 μm), domestic goats (*Capra aegagrus hircus*) (2.5 μm–3.9 μm) and in domestic sheep (*Ovis aries*) (4 μm to 6 μm). There is much interspecies variation, particularly in mammals where the mean cell diameter ranges from 1.4 μm to 3.3 μm in the lesser mouse-deer to 9.0 μm to 9.6 μm in the African elephant (*Loxodonta africana*). As a general rule, Elasmobranchii fish (a subclass of Chondrichthyes or cartilaginous fish, including sharks (superorder Selachii) and rays, skates, and sawfish), have larger red blood cells than teleost fish (the largest infraclass in the class Actinopterygii, the ray-finned fish, making up to 96% of all living fish species). Deep sea teleosts have larger size red blood cells than normal teleosts.

Because animals with small red blood cells have high cell counts and vice versa, the normal packed cell volume (PCV/haematocrit) is relatively constant in mammals and birds. In mammals, there is some indication that red blood cell size is related to body size within a zoological family, but this rule can be modified by environmental factors. Small red blood cells are found in species which normally live at high altitudes (e.g. some species of sheep and goats) where oxygen availability is low. These small cells provide an increased red blood cell surface area for enhanced oxygen uptake by diffusion. Diving mammals (seals and cetaceans), which are exposed to long periods without access to oxygen while submerged, have relatively thick red blood cells which act as a slow-release oxygen store. Apart from minor age-related differences, there are normally no significant intraspecies differences in cells size, and in most healthy individuals, red cell size is fairly uniform.

The circulatory system of insects and other arthropods (spiders and crustaceans) is an open system and very different, in both structure and function, from the closed circulatory system found in vertebrates. This system is integrated mainly by the haemolymph and haemocytes. The haemolymph is a blood-like fluid circulating in the interior of the body and in direct contact with the tissues. The haemocytes are suspended in the haemolymph and play a role in the arthropod immune system and are phagocytes of invertebrates. In the common fruit fly or vinegar fly (*Drosophila melanogaster*), haemocytes can be divided into two groups, namely, embryonic and larval. Embryonic haemocytes are derived from head mesoderm and enter the haemolymph as circulating cells. Larval haemocytes, on the other hand, are responsible for tissue remodelling during development. Specifically, they are released during the pupa stage in order to prepare the fly for the transition to an adult stage. There are four types of haemocytes found in fruit flies, including secretory, plasmatocytes, crystal cells and lamellocytes. The haemolymph is composed of water, inorganic salts (mostly sodium, potassium, calcium, chlorine and magnesium) and organic compounds (mostly carbohydrates, proteins and lipids). The haemolymph of lower arthropods, including most

Table 3.1 Normal Blood Cells in Mammals, Birds, Reptiles, Amphibians and Fish

Cell	Mammals	Birds	Reptiles	Amphibians	Fish
Erythrocytes	Anucleated, biconcave disc (oval in Camelidae)	Nucleated, oval	Nucleated, oval	Nucleated, oval	Nucleated, oval
Haemostatic cells	Platelets	Thrombocytes	Thrombocytes	Thrombocytes	Thrombocytes
Leucocytes (granulocytes)	Neutrophils Eosinophils Basophils	Heterophils Eosinophils Basophils	Heterophils Eosinophils Basophils	Heterophils Eosinophils Basophils	Neutrophils Eosinophils Basophils
Mononuclear cells (agranulocytes)	Lymphocytes Monocytes	Lymphocytes Monocytes	Lymphocytes Monocytes Azurophils	Lymphocytes Monocytes	Lymphocytes Monocytes Macrophages

insects, is not used for oxygen transport because such animals breathe through other means.

3.1 Reversible Changes in Red Blood Cell Shape

Apart from the basic differences in the shape of mammalian, avian, reptilian, amphibian and fish red blood cells already described, and the variation shown by Camelidae red blood cells, reversible shape changes occur normally in the cells of some non-human mammals which can alter the appearance of the cells on stained blood films. These shape changes are usually associated with the presence of haemoglobin variants which polymerise or crystallise under the conditions prevailing in blood samples being processed for testing and are probably correctly considered as *in vitro* phenomena. As yet, no pathological consequences of these phenomena have been described. The most striking is the sickling tendency seen in most species of deer and in some species of sheep, goats, antelopes and small carnivores (Table 3.2). In these animals, variants of adult haemoglobin occur which form insoluble, elongated polymers when in the oxygenated state. The presence of these polymers inside the cells leads to deformation of the cell into sickle, spindle, holly leaf, triangular and other bizarre shapes, resembling those seen in human patients with sickle cell disease. The time taken for haemoglobin to polymerise in this way is longer than the length of time taken for the cells to pass through arteries, providing one explanation for the fact that the shape changes do not occur *in vivo*. They are seen, however, in blood exposed for more than a few seconds to atmospheric oxygen as, for example, when a sample is manually or mechanically mixed or spread on a microscopic slide during preparation of a blood film.

A similar, less extensively studied shape-change phenomenon, associated with *in vitro* intracellular haemoglobin crystallisation, is seen in some other species of mammals. In these, the cells take the shape of the haemoglobin crystals. Examples of this phenomenon include the finding of square red blood cells in the mara (*Dolichotis patagonum*

and *D. salinicola*), triangular red blood cells in the markhor (*Capra falconeri*) and matchstick red blood cells in the white rhinoceros (*Ceratotherium simum*) and blackbuck (*Antilope cervicapra*). Haemoglobin crystallisation may explain the occurrence of individual red blood cells which appear to break down into a number of small sickle cells in the capybara (*Hydrochoerus hydrochaeris*).

3.2 Inclusion Bodies

Apart from intra-erythrocytic parasitism, which is dealt with in the section about blood parasites, Howell-Jolly bodies (named after William Henry Howell, 1860–1945, and Justin Marie Jolly, 1870–1953), Heinz bodies (named after Robert Heinz, 1865–1924) or small diffuse basophilic granules (basophilic stippling) are sometimes observed in the red blood cell cytoplasm. Of these, Howell-Jolly bodies are indicative of diminished splenic function or abnormal nuclear division in man but occur with some frequency under normal conditions in carnivorous, rodents, marsupials and small primates. They are single, eccentrically placed spherical structures containing DNA, which stain dark with Romanowsky stains.

Basophilic stippling is manifested by the presence of a number of small, irregular granules, distributed throughout the cytoplasm and has been noted in mammals and in birds. The granules consist of aggregates of ribosomes and stain purple or blue with Romanowsky stains. Their presence is generally considered to be pathological and may be associated with lead toxicosis or hypochromic anaemia. Basophilic stippling is sometimes seen in artiodactyls during the first few weeks of life, when replacement of foetal red blood cell by adult cells is in progress. In these circumstances it appears to be without pathological significance.

Heinz bodies are not seen in Romanowsky stained films, but in samples stained supravitally with new methylene blue or methyl violet, they appear as one or more pale blue or violet structures of irregular size and shape, often associated with the cell membrane. They can also be seen in unfixed, unstained blood films as refractile structures within the red

Table 3.2 Mammals in which Non-Pathological Reversible Red Cell Sickling Has Been Recorded _In Vitro_

Deer	Antelope	Sheep and Goats	Carnivores
Indian muntjac (_Muntiacus muntjac_)	Nyala (_Tragelaphus angasi_)	Domestic sheep (_Ovis aries_)	Slender mongoose (_Galerella sanguinea_)
Reeve's muntjac (_Muntiacus reevesi_)	Bongo (_Tragelaphus eurycerus_)	Soay sheep (_Ovis aries_)	Blotched genet (_Genetta maculata_)
Fallow deer (_Dama dama_)		Barbary sheep (_Ammotragus lervia_)	Spotted genet (_Genetta genetta_)
Persian deer (_Dama dama mesopotamica_)		Mouflon (_Ovis orientalis orientalis_)	
Axis deer (_Axis axis_),		Domestic goat (_Capra aegagrus hircus_)	
Hog deer (_Hyelaphus_ porcinus)		Markhor (_Capra falconeri_)	
Timor deer (_Rusa timorensis_)			
Swamp deer (_Rucervus duvaucelii_)			
Sika deer (_Cervus nippon_)			
Red deer (Cervus elaphus)			
Père David's deer (_Elaphurus davidianus_)			
Mule deer (_Odocoileus hemionus_)			
White tailed deer (_Odocoileus virginianus_)			
Pudu (_Pudu mephistophiles_)			
Roe deer (_Capreolus capreolus_)			

blood cells, and under these circumstances they have been termed 'erythrocyte refractile bodies' (ER bodies). Heinz bodies consist of denatured haemoglobin and, in man, they are indicative of splenic dysfunction, toxic changes produced by oxidant drugs or chemicals, defects in the enzymes which protect haemoglobin from excessive oxidation or of the presence of unstable haemoglobin variants. In other mammals, their presence has been associated with phenothiazine treatment, brassica toxicosis, onion toxicosis (several domesticated species), maple leaf toxicosis (horses), copper toxicosis (sheep), exposure to methylene blue dye (cats), selenium deficiency (cattle), prolonged corticosteroid therapy (dogs) and wasting marmoset syndrome (marmosets and tamarins). They have also been described in birds which have ingested oil from contaminated plumage. Heinz bodies have been associated with the consumption of acetaminophen, garlic, and onions by cats, dogs, and a number of primates. Thiosulfate compounds in the flesh of onions have been identified as the cause. Propylene glycol used to be a common ingredient in soft, moist cat food. The Food and Drug Administration (FDA) acknowledged that propylene glycol caused Heinz body formation in the red blood cells of cats. Subsequently, in 1996 the Center of Veterinary Medicine of the FDA amended regulations to specifically ban incorporating propylene glycol in cat food. In some species, such as cats and common marmosets, Heinz bodies are not always associated with ill health, and in others, for example, the white rhinoceros, they can be produced artefactually during incubation of the red cells with new methylene blue. The haemoglobin of these animals may have a lowered stability threshold.

3.3 Rouleaux Formation

Rouleaux formation can be identified in the thicker areas of blood films and in wet preparations of whole blood where it appears as a face-to-face superimposition of the cells into stacks of various lengths and conformations. Rouleaux formation is confined to discoid red blood cells and does not occur in camelids, birds, reptiles, amphibians and fish. In man, in some other primates and in some carnivores and deer, a high degree of rouleaux formation is indicative of the presence of an inflammatory condition. Among other mammals, the clinical interpretation of the presence or absence of rouleaux on a blood film is complicated by species differences. For instance, in horses and other Perissodactyla, marked rouleaux is normal in healthy animals, whereas, in most rodents and many Artiodactyla, it is rarely observed, even in individuals suffering from severe inflammation. The significance of red blood cell rouleaux depends, therefore, upon the species under consideration. Care must be taken to distinguish between rouleaux and agglutination. The latter is characterised by a close, irregular juxtaposition of red blood cells into clumps of various sizes, indicating an abnormal immune response, and is never seen under normal circumstances. In some species, the presence of intra-erythrocytic haemoglobin crystals can disrupt the spreading of a blood film in such a way that the red blood cells appear to be agglutinated.

3.4 Artefacts

The biconcave discoid red blood cells of many species of mammals show a tendency for crenation in blood films prepared from samples collected in EDTA. This is considered to be an artefact arising from the general use of EDTA at a concentration which may not be isotonic with the cells of the species. Care is necessary to distinguish crenated cells, which have no diagnostic significance, with burr cells or echinocytes, which indicate clinical abnormality. It has been noted that exposure to EDTA causes the red blood cells of some species of birds and reptiles to haemolyse. Affected species include crowned cranes (*Balearica regulorum* and *B. pavonina*), ostrich (*Struthio camelus* and *S. molybdophanes*), kookaburras (*Dacelo* spp.), most Corvidae, most reptiles, most amphibians and most fish. Blood films from these animals should be prepared from non-anticoagulated blood or from heparinised samples. The presence of heparin imparts a blue tinge to blood films treated with Romanowsky stains and may cause clumping of the platelets, thrombocytes and white blood cells.

In all animals, the breakdown of a small proportion of red blood cells is inevitable during film preparation, and, whereas this is usually unnoticed when dealing with mammalian blood, the nuclei released when avian and reptilian red blood cells become disrupted remain visible and can cause confusion. They usually appear as amorphous, pale purple masses with an indistinct outline and are larger in size than the nuclei of intact red blood cells. Incorrect identification as lymphocytes or thrombocytes can be avoided on the strength of their lack of structure, indefinite outline and absence of cytoplasm (Tables 3.3 and 3.4).

3.5 The Blood Film

3.5.1 Red Blood Cells

Background information about normal red blood cell size, shape and staining characteristics is required for the species in question in order to assess the possible pathological significance of these parameters in a given clinical case. In normal adult animals, the cells can be expected to appear as regular discs or elliptocytes, either nucleated or not according to the class of animals they come from. Since the mean cell haemoglobin concentration (MCHC) is similar in all mammals and birds, mammalian and avian red blood cells normally should appear fully haemoglobinised. However, the significance of observations relating to the degree of polychromasia, the

Table 3.3   A Simple Classification Scheme for Normal Erythropoiesis in Mammals

Mammals

Proerythroblasts	Large round cell, large central or excentric nucleus occupying most of the cell. Nuclear chromatin finely stippled, nucleoli or nucleolar spaces present. Cytoplasm strongly basophilic.
Basophilic erythroblasts	Smaller round cell, nuclear chromatin coarsely granular, no nucleoli. Cytoplasm basophilic
Polychromatic erythroblasts	Smaller round cell, smaller nucleus in relation to cytoplasm, irregular clumping of chromatin. Cytoplasm grey or slightly eosinophilic
Orthochromatic erythroblasts	Cytoplasm fully eosinophilic or with basophilic tinge. Nucleus reduced in size, uniformly basophilic or pyknotic
Reticulocytes	Slightly smaller cell, no nucleus. Cytoplasm fully eosinophilic or with basophilic tinge. In Camelidae may be slightly oval. Reticulocytes stains reveal presence of one or more granules or strands
Mature erythrocytes	Smaller, fully eosinophilic round cell with central pallor indicating biconcavity. In Camelidae, fully eosinophilic, oval cell with no central pallor

Table 3.4   A Simple Classification Scheme for Normal Erythropoiesis in Birds, Amphibians, Reptiles and Fish

Birds, Reptiles, Amphibians and Fish	
Proerythroblasts	Large round cells or amoeboid cell. Nuclear chromatin forming coarse open network with marked clumping. Large nucleolus. Copious basophilic cytoplasm with mitochondrial spaces
Basophilic erythroblasts	Smaller round cell, nuclear chromatin clumped, nucleolus smaller but still visible. Cytoplasm basophilic, no mitochondrial spaces
Early polychromatic erythroblasts	Smaller round cell, nucleus relatively small with clumped chromatin, no nucleolus. Cytoplasm grey or slightly eosinophilic
Late polychromatic erythroblasts	Smaller round or slightly oval cell, round or slightly oval nucleus with irregular chromatin clumps. Cytoplasm grey or slightly grey to pale eosinophilic
Orthochromatic erythroblasts	Cytoplasm fully eosinophilic. Nucleus larger than that of mature cell, with irregular chromatin clumping. Reticulocyte stains reveal extensive cytoplasmic granules forming a perinuclear band. Granules become progressively reduced in number and dispersed through the cytoplasm
Mature erythrocytes	Oval cell, cytoplasm uniformly eosinophilic. Nucleus oval, elongated or, in reptiles, sometimes round

Table 3.5   Significance of Some Red Cell Morphological Changes

Observation	Indication	Mammals	Birds, Reptiles, Amphibians, Fish
Hypochromia	Anaemia, mineral (iron) deficiency	Significant	Significant
Poikilocytosis	Metabolic defect, increased erythropoiesis	Significant	Significant
Anisocytosis	Metabolic defect, increased erythropoiesis	Significant if no tendency for Hb polymerisation	Significant
Target cells (Leptocytes)	Liver disease, hypochromic anaemia	Significant	Not described
Stomatocytes	Liver disease, haemolytic anaemia	Significant	Not described
Spherocytes	Haemolytic anaemia	Significant	Not described
Burr cells (Echinocytes)	Uraemia, hyperthyroidism, hypertonicity	Significant	Not described
Schistocytes	Disseminated intravascular coagulation, micro-angiopathic anaemia, severe sepsis	Significant	Not described
Howell-Jolly bodies	Hyposplenism, abnormal nuclear division	Significant in some species	Probably significant
Heinz bodies	Exposure o oxidants, unstable Hb, enzyme defects, hyposplenism	Significant in some species	Significant
Basophilic stippling	Iron deficiency, lead toxicosis	Significant in adults	Significant

size of the cells and, in many instances, their shape, can depend upon the characteristics of the species under examination. For example, poikilocytosis is extremely common in blood samples from deer and other mammals with a non-pathological sickling tendency and in those with enhanced haemoglobin crystallisation, but is rare in Camelidae under any circumstances. Overt macro or microcytosis cannot be defined without knowledge of the normal MCV for the species. The possible significance of Howell-Jolly bodies, Heinz bodies and rouleaux formation must be considered according to the species. Care must be taken to distinguish between true burr cells (echinocytes) and the crenated cells produced artefactually in EDTA samples from many mammals. As a general rule, the morphological observations most consistently indicative of underlying pathology are hypochromia, spherocytes, target cells, agglutination, the presence in the blood of large numbers of immature cells and evidence of abnormal cell division (Table 3.5).

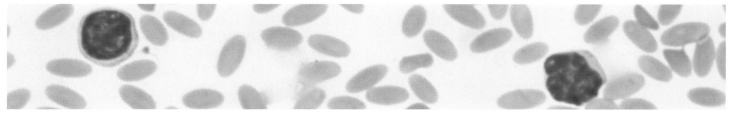

4

NORMAL AND ABNORMAL WHITE BLOOD CELLS

4.1 Granulocytes

The classification of white blood cells and the identification of morphological abnormalities on stained blood films depends not only upon observation of differences in the form and size of the nucleus and cytoplasm but also upon the staining characteristics of the cell constituents. The stains most widely used for this purpose are Romanowsky stains. These contain azure complexes that react with acidic groupings, including those of nucleic acids and proteins of the cell nuclei and primitive cytoplasm, as eosin Y, which has an affinity for the basic groupings on haemoglobin and other molecules. Several different varieties of Romanowsky stains are available, each of which produces a slightly different result. All are suitable for comparative work. The choice is a matter of personal preference. It should be recognised, however, that these stains are difficult to standardise, even when used on a single species. When applied to a range of species, the variations seen may result from uncontrolled staining differences as well as from true species diversity. In addition, the morphology of the cells and the way in which they take up the stains can be modified by technical factors such as the anticoagulant used, improper drying of the cells on the slide, prolonged storage or inadequate fixation.

White cells with granular cytoplasm occur in the blood of all mammals, birds, reptiles, amphibians and fish. Mammalian granulocytes are classified as neutrophils, eosinophils and basophils based on the strength of the reaction of their cytoplasmic granules with Romanowsky stains. This classification is largely descriptive and originates from observations on the reaction of human white blood cells with Romanowsky stains. In normal human blood, the neutrophil granules are relatively small and stain weakly with the azure complexes. The basophil granules are larger, spherical, completely fill the cytoplasm and react strongly with the azure complexes. The eosinophil granules are round, numerous and have an affinity for eosin changes in the staining characteristics, and distribution of the granules can be of diagnostic significance.

Among other mammals, the granulocyte nuclei are polymorphic in all species, although there is much species variation in the degree of lobulation and in the number of lobes normally present. There is also species variation in the size, shape, distribution and staining characteristics of the cytoplasmic granules, particularly those of the neutrophils. Neutrophil granules can be basophilic as, for example, in chimpanzees and some artiodactyls, or strongly eosinophilic, as in rabbits and some rodents. In some species, the cytoplasm of these cells appears agranular or contains only faint pink particles. In these animals, the cells are classified as neutrophil granulocytes based on the strength of their polymorphic (lobed) nuclei and the absence of specific eosinophilic or basophilic granules. The granules of the eosinophils and basophils can also show species variation in size, number, shape and staining intensity. The cytoplasm of eosinophils, if visible, usually appears pale blue in some species.

The same three types of granulocytes can be distinguished in the blood of birds, reptiles, amphibians and fish. And although there are differences in morphology and terminology, it is probable that the cells are functionally similar to those of mammals. The terminological differences stem from the fact that the cells of birds, reptiles, amphibians and fish, which are generally assumed to be homologous with mammalian neutrophils, have cytoplasm containing a large number of strongly eosinophilic, usually spiculate or oval granules. The term 'neutrophil', therefore, is not appropriate and these cells are generally known as heterophils. The eosinophils in reptiles are large cells with spherical eosinophilic granules. However, the granules of the eosinophils in the American or green iguana (*Iguana iguana*), stain blue-green in colour with Romanowsky stains. Chelonians and crocodilians tend to have large basophils, while lizards tend to have smaller basophils. In some amphibian species, the basophil is the most predominant white blood cell.

In reptiles, a further cell type, the azurophil, is found. These are cells frequently observed in the blood films of crocodiles, lizards and snakes, but seldom found in those of turtles and tortoises. Although azurophils are considered by some haematologists to represent neutrophil granulocytes, these cells are included in this Atlas in the same section as mononuclear cells.

The heterophil of birds show little interspecies variation and characteristically have a lobed, usually bilobed, nucleus and spiculate or oval granules, which stain brick red with Romanowsky stains and usually fill the cytoplasm. If visible, the cytoplasm appears colourless or faintly pink. In contrast to those of birds, reptilian heterophils show considerable morphological diversity. In the Serpentes (snakes), Crocodilia (crocodiles and alligators) and Testudines (tortoise, turtles and terrapins), they are large cells with an unlobed, round or oval, often excentric nucleus and spiculate granules which fill the cytoplasm. The granules stain bright orange-red in lizards, snakes and chelonians and brick red in crocodilians. In some species of Saurian (lizards), the heterophil nucleus is round, while in others, it is lobed and resembles that of birds.

Neutrophil/heterophil granulocytes are the most numerous of the granulocytes found in mammals, birds and reptiles. Their primary function is killing bacteria, involving chemotaxis, opsonisation, ingestion and lysis. An increase in the number of these cells in the circulating blood occurs in response to bacterial and fungal infections, tissue damage, some metabolic diseases, myeloid leukaemias and stress. There are marked species differences in the quantitative response detectable in the peripheral blood, and morphological variation can often be more informative than absolute cell counts.

In birds and reptiles, the presence of eosinophilic granules in the heterophils gives rise to problems in distinguishing between these cells and the true eosinophils. Further problems can result from inadequate fixation that results in degranulation of the heterophils and basophils. In practice, avian and reptilian heterophils and eosinophils can usually be differentiated by the appearance of their granules. Heterophil granules are usually spiculate or oval and, in birds, stain with a brick-red colour, whereas the granules of true eosinophils are usually round and stain clear, bright red. If visible, the cytoplasm of the heterophils is colourless or pale pink and that of the eosinophils is blue. Marrow biopsy for diagnostic purposes is not often feasible in such animals but, since immature granulocytes are rarely found in the bloodstream of healthy individuals, their presence on a peripheral blood film is a valuable indication of increased demand or marrow dysfunction, and it is important to recognise these cells when they occur. Granulopoiesis has been studied extensively in man and in some other mammals, although less is known

about the subject in birds and reptiles, as it appears that the process is essentially similar in the three groups. The earliest recognisable cell of the granulocyte series is the myeloblast (granuloblast), which gives rise to promyelocytes (progranulocyte), myelocytes, metamyelocytes and mature granulocytes in sequence. Succeeding maturation stages are differentiated on the basis of a progressive decrease in cell size and cytoplasmic basophilia, condensation and, in mammals and birds, segmentation of the nucleus and the appearance of an increasing complement of specific cytoplasmic granules. Difficulties can arise from the fact that there is a transient appearance of basophilic granules in avian and reptilian heterophil myelocytes and metamyelocytes, which leads to confusion between these cells and mature basophils. Additional problems in distinguishing between heterophils and eosinophils can arise from the fact that the eosinophilic granules in immature avian heterophils are spherical as opposed to rod-shaped. Thus, there is a possibility for the confusion of immature cells of the heterophil series with both basophils and eosinophils.

In reptiles and amphibians, the largest mean diameter of heterophils has been identified in urodeles, and the smallest mean diameter has been identified in turtles. In goldfish, the number of neutrophils is relatively low, accounting for 5.1% of the total white blood cell count. In the peripheral bloodstream of goldfish, the neutrophils show two- or three-lobed nuclei, but very often the nuclei are kidney-shaped.

4.2 The Blood Film

The most usefully observed morphological changes affecting neutrophils and heterophils are those that reflect immaturity or toxicity. The presence in the circulating blood of immature granulocytes is a valuable indicator of myeloid hyperplasia associated with either malignancy of infections or other inflammatory conditions. In some species, the neutrophil's heterophil response to inflammation is qualitative rather than quantitative and, in these animals, the finding of cells showing signs of immaturity is a more reliable index of increased demand, for example, in bacterial infection, than the total white cell count. Information about the stage of maturity of the neutrophils is also important for distinguishing between disease and stress-induced neutrophilia. Immature granulocytes can be identified from their larger size, cytoplasmic basophilia and lack of specific granules. In animals in which the neutrophils/heterophils are normally lobed, immaturity is further characterised by a decrease in the number of or the absence of nuclear lobes (left shift). In reptiles in which the heterophils nucleus is normally round, lobulation can be an indicator of infection. The presence of increased basophilia, toxic granules, Döhle bodies (name after Karl Gottfried Paul Döhle, 1855–1928) or cytoplasmic vacuoles in the cytoplasm can also be of diagnostic significance. Some of these criteria, for example, the finding of metamyelocytes and their precursor or cells containing Döhle bodies or cytoplasmic vacuoles, should be applicable to all species. Pathological variations in nuclear lobulation and cytoplasmic granules may not be recognised without reference to normal cells of the species concerned.

4.3 Normal and Abnormal White Blood Cells – Lymphocytes, Monocytes and Azurophils

4.3.1 Lymphocytes

Typical lymphocytes are regular cells with round, central or slightly excentric, round nuclei and a varying amount of clear, pale blue cytoplasm. They are present in the blood of all mammals, birds, reptiles, amphibians and fish. In some species, the cytoplasm of a proportion of the cells contains a few azurophilic or basophilic granules. In blood films made from avian or reptilian blood, the lymphocytes often appear to be deformed by close juxtaposition of red blood cells. In birds and particularly in reptiles, it can be difficult to distinguish between lymphocytes and thrombocytes. This is usually easier in samples in which partial activation of the thrombocytes has occurred. Some practice is needed to avoid misidentification of disrupted avian and reptilian red blood cells as lymphocytes.

Lymphocytes are concerned with immunological reactions and, in mammals and birds, can be divided into two minor groups with different functions and membrane properties. These are the B lymphocytes, which produce immunoglobulins, and T lymphocytes, which are responsible for cell-mediated immunity. B and T lymphocytes cannot be differentiated without the use of specialised techniques, which are beyond the scope of this Atlas.

Lymphopoiesis takes place in the spleen and thymus and, additionally, in the tonsillar glands of reptiles, the bursa of Fabricius in birds, and the lymph nodes, Peyer's patches and to some extent in the bone marrow in mammals. Immature lymphocytes are distinguished from the mature form by their larger size, increased cytoplasmic basophilia and the presence of nucleoli. In juvenile animals of many species, the lymphocyte is the predominant white blood cell in the bloodstream and in some species, this situation persists throughout life. A pathological increase in lymphocytes occurs in some viral infections and lymphoproliferative disorders. Lymphopenia can be associated with stress.

Small and large lymphocytes appear to be the dominant white blood cell in reptiles and amphibians. The largest mean diameter of small lymphocytes has been observed in urodeles and the smallest mean diameter has been observed in lizards. Large and small lymphocytes have been identified in several fish species. In the goldfish (*Carassius auratus*), the lymphocyte is the predominant white blood cell accounting for up to 92.5% of the total white blood cell count.

4.3.2 Monocytes

Monocytes are usually larger than lymphocytes and can be distinguished by the presence of a relatively pale staining, kidney-shaped nucleus and abundant, slightly opaque, blue-grey cytoplasm that appears faintly granular, giving a lace-like structural appearance. They are found in the blood of all mammals, birds, reptiles, amphibians and fish. In most species, monocytes are relatively uncommon but in elephants, cells with distinctly bilobed nuclei and blue-grey cytoplasm, which have the cytochemical characteristics of monocytes, occur in large numbers.

Monocytes are derived from promonocytes in the bone marrow and circulate in the bloodstream before entering the tissues as mature macrophages. Their main function depends upon their phagocytic activity against invading organisms, necrotic cells and cell debris and on their ability to concentrate antibodies for presentation to the lymphocytes. Monocytosis occurs in some bacterial infections, including tuberculosis and brucellosis, some protozoal viral and fungal infections, malignant conditions and collagen diseases, and also during recovery from acute infections. In many mammalian species, the monocyte count is higher in juveniles than in adults. In reptiles and amphibians, the largest mean diameter has been estimated in urodeles and the smallest mean diameter has been estimated in turtles.

4.3.3 Azurophils

Azurophils occur in the blood of reptiles where they are found in low numbers in normal lizards, crocodilians and chelonians and in a higher number in snakes. They are mononuclear cells, the shape of which can vary from round, lymphocyte-like cells to larger cells that have a monocytoid appearance. These different forms may represent stages in cell maturation. An outstanding feature of azurophils is the metachromatic reaction of their cytoplasm with Romanowsky stains. This makes the cells relatively easy to recognise in routine blood films.

Information is lacking about the origin and function of azurophils and these cells have been variously considered as allied to the granulocytes or monocyte series. There is little doubt that they play some part in the inflammatory response, particularly in snakes, and azurophilia and/or toxic changes in cell morphology are usually indicative of infection.

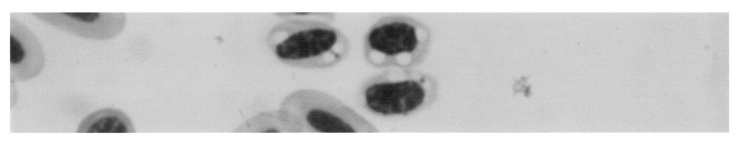

5

NORMAL AND ABNORMAL PLATELETS AND THROMBOCYTES

5.1 Platelets and Thrombocytes

By definition, the haemostatic blood cells of all vertebrate animals are correctly described as thrombocytes. In mammals, however, these cells are often referred to as platelets in order to take into account the striking morphological differences which distinguish them from the thrombocytes of other vertebrates. Strictly speaking, platelets are not true cells as they do not contain a nucleus. They are the smallest elements in the circulating blood and are formed from the cytoplasm of megakaryocytes, mainly in the bone marrow, where each megakaryocyte gives rise to several thousand platelets. Interspecies differences in platelet number, size and granularity occur, and, as a general rule, there is an inverse relationship between platelet size and platelet count.

In contrast, the thrombocytes of birds and reptiles are significantly larger than mammalian platelets and are nucleated. They are produced from mononuclear precursors in the bone marrow. Non-activated thrombocytes may be oval, spindle-shape or round with a round or oval, completely dense nucleus and clear, blue-grey cytoplasm in which one or two small basophilic granules may be visible. In some avian and, particularly, in some reptilian species, it can be difficult to distinguish between thrombocytes and lymphocytes, and

without experience, there is sometimes confusion between disrupted red cell nuclei and thrombocytes.

Although the stimuli for adhesion and aggregation may be different, platelets and thrombocytes appear to function similarly in haemostasis. The propensity for these cells to become activated by contact with foreign surfaces is an integral part of their haemostatic function but also gives rise to artefactual changes in their morphology in all but the most carefully collected blood sample. The morphological changes associated with platelet activation include spreading, pseudopodia formation, vacuolation, aggregation and loss of granules. Thrombocyte activation leads to cytoplasmic vacuolation, alterations in outline and aggregation. Although in practice, a small degree of activation often facilitates the correct identification of thrombocytes, it can also obscure clinically significant quantitative and morphological changes.

On films prepared from correctly collected blood samples, thrombocytopenia can be diagnosed from a lack of platelets, and the presence of activated platelets or thrombocytes is an indication of disseminated intravascular coagulation. Thrombocytosis occurs as a response to haemorrhage or bacterial infection. Platelet anisocytosis and the presence of giant forms are often associated with myeloid and lymphoid hyperplasia.

Section B

Atlas of Wild and Exotic Animal Haematology

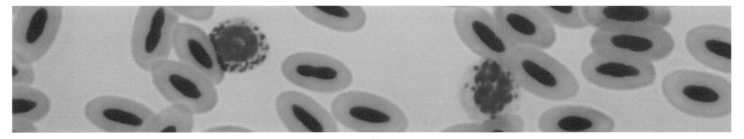

6

NORMAL AND ABNORMAL BLOOD CELLS

6.1 Haematology

6.1.1 Normal Species Variation in Red Blood Cell Size

The blood films were prepared from the blood of healthy adult animals and are shown at the same magnification. The normal range for mean cell volume (MCV) is given for each of the species illustrated.

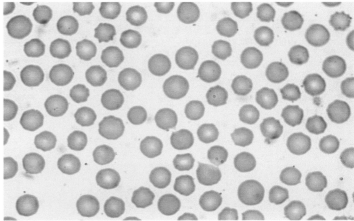

Figure 6.2 Intermediate-sized red blood cells from a domestic cow (*Bos taurus*) (MCV 40–60 fl).

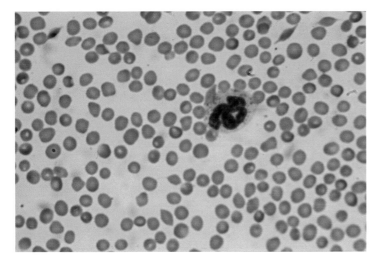

Figure 6.1 Small red blood cells from a domestic goat (*Capra aegagrus hircus*) (MCV 19–24 fl).

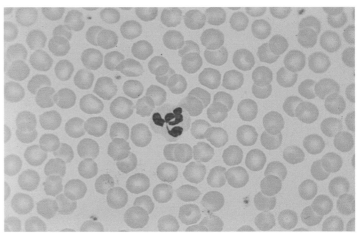

Figure 6.3 Intermediate-sized red blood cells from a domestic dog (*Canis lupus familiaris*) (MCV 70–85 fl).

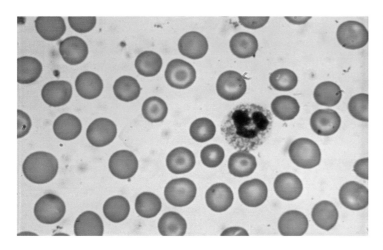

Figure 6.4 Relatively large red blood cells from a Canadian beaver (*Castor canadensis*) (MCV 92–110 fl).

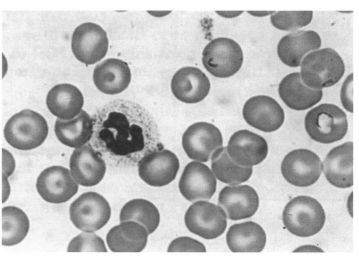

Figure 6.6 Large red blood cells from an Indian elephant (*Elephas maximus indicus*) (MCV 112–130 fl). The red blood cells of elephants often appear as target cells on air-dried blood films. This is an artefact associated with their large size.

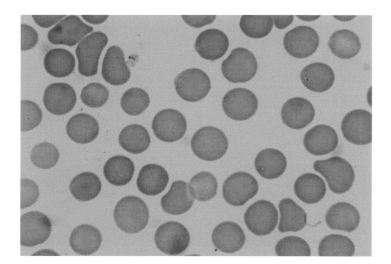

Figure 6.5 Red blood cells from a bottlenose dolphin (*Tursiops truncatus*). They appear smaller than those of the Indian elephant (*Elephas maximus indicus*), although their MCV is similar (MCV 100–120 fl) because, like the red blood cells of other diving mammals, they are thicker than those of terrestrial mammals and probably act as a slow-release oxygen store during diving.

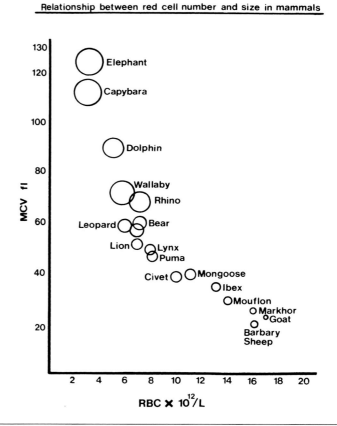

Figure 6.7 The relationship between red blood cell size and number. The cells are drawn according to the mean cell diameter. The positioning occupied by the bottlenose dolphin (*Tursiops truncatus*) is consistent with an increased mean cell average thickness.

8

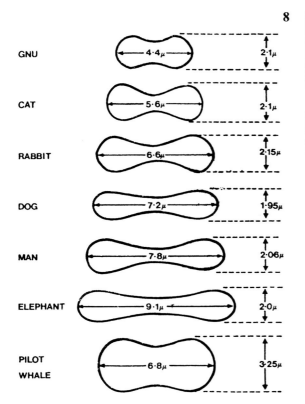

GNU 4·4μ 2·1μ

CAT 5·6μ 2·1μ

RABBIT 6·6μ 2·15μ

DOG 7·2μ 1·95μ

MAN 7·8μ 2·06μ

ELEPHANT 9·1μ 2·0μ

PILOT WHALE 6·8μ 3·25μ

Figure 6.8 Comparison of red blood cell size and thickness in various mammals. The cells of the pilot whale (*Globicephala macrorhynchus*) are thicker than those of other non-diving species.

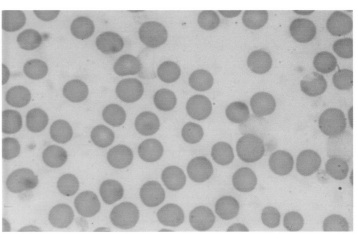

Figure 6.10 Red blood cells from a healthy domestic donkey (*Equus asinus*) (MCV 67–70 fl). These cells are larger than those of horses.

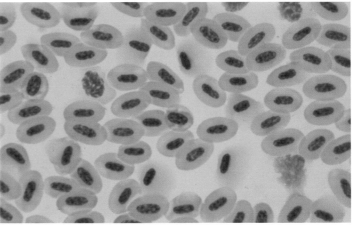

Figure 6.11 Red blood cells from a budgerigar (*Melopsittacus undulatus*) (MCV 99–105 fl). In general, avian red blood cells are large compared with those of most mammals as the size may be directly related to body size.

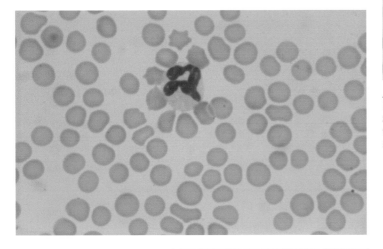

Figure 6.9 Red blood cells and neutrophil from a healthy domestic horse (*Equus caballus*) (MCV 40–56 fl). The neutrophil has a granular cytoplasm.

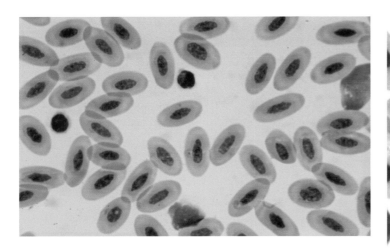

Figure 6.12   Red blood red cells from a common buzzard (*Buteo buteo*) (MCV 150–170 fl).

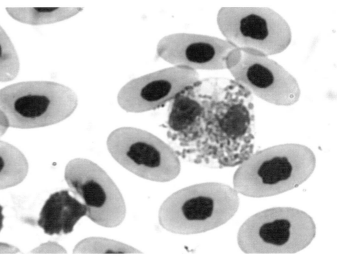

Figure 6.14   Red blood cells from a green iguana (*Iguana iguana*) (MCV 250–290 fl). Reptilian red blood cells are usually larger than those of birds. A normal heterophil is also shown.

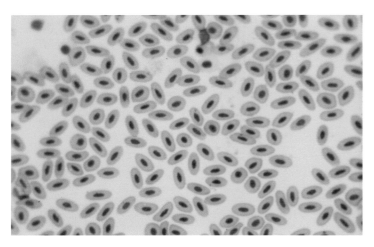

Figure 6.13   Red blood cells from an emu (*Dromaius novaehollandiae*) (MCV 250–280 fl).

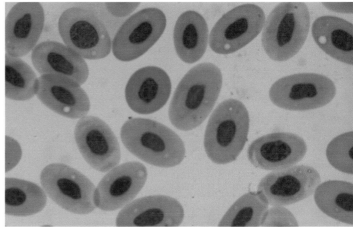

Figure 6.15   Normal red blood cells from an Indian rock python (*Python molurus*) (MCV 290–333 fl).

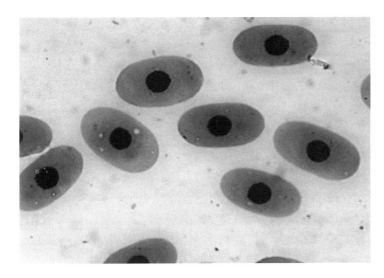

Figure 6.16 Red blood cells from an Aldabra (*Aldabrachelys gigantea*) giant tortoise (MCV 400 fl).

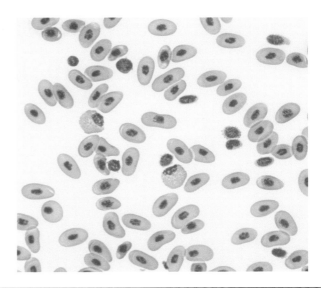

Figure 6.18 Normal red blood cells, white blood cells and thrombocytes from a baw baw frog (*Philoria frosti*), a critically endangered Australian species (MCV 830 fl) (courtesy of Helen McCracken), ×400.

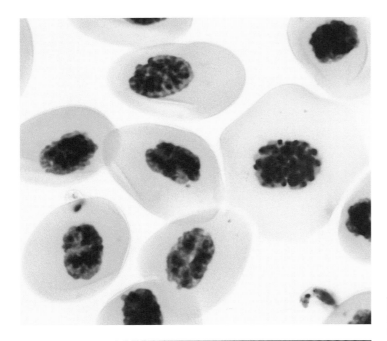

Figure 6.17 Red blood cells of an axolotl (*Ambystoma mexicanum*) showing strong nuclear chromatin clumping (MCV 8,000–12,000 fl) (courtesy of Helen McCracken).

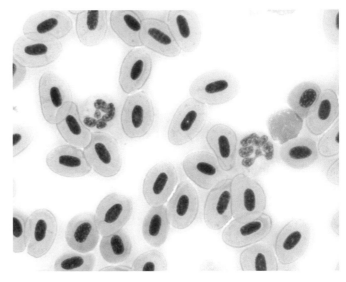

Figure 6.19 Red blood cells of a clinically normal sockeye (*Oncorhynchus nerka*) or red salmon. The name is given as members of this species are red in hue during spawning (MCV 440–550 fl) (courtesy of Nicole Stacy).

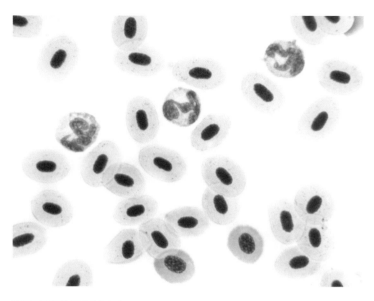

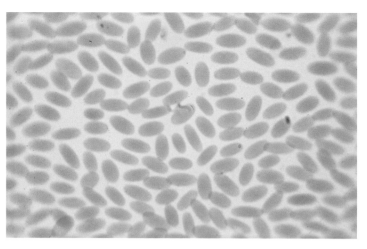

Figure 6.22 Elliptical red blood cells from a Bactrian camel (*Camelus bactrianus*). Elliptocytes occur in all members of the Camelidae. The outline of these cells always appears slightly indistinct on stained blood films.

Figure 6.20 Normal red blood cells from a pink or humpback salmon (*Oncorhynchus gorbuscha*). This is the smallest and most abundant species of the Pacific salmon (MCV 440–550 fl) (courtesy of Nicole Stacy).

6.1.2 Normal Variation in Red Blood Cell Shape

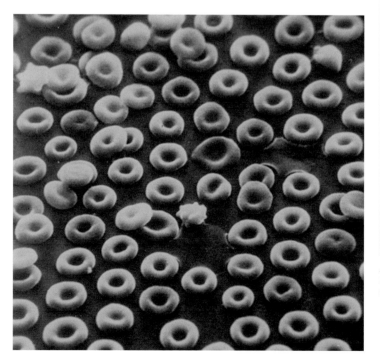

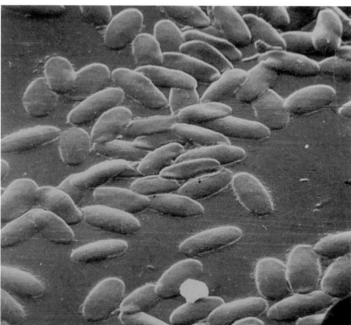

Figure 6.23 SEM image of alpaca (*Vicugna pacos*) elliptocytes showing the absence of a central depression.

Figure 6.21 Scanning electron microscopy (SEM) image of red blood cells from a healthy masked palm civet (*Paguma larvata*) showing the biconcave discoid shape common to most mammals.

Figure 6.24 SEM of North American (*Meleagris gallopavo*), turkey red blood cells, each with a nuclear bulge.

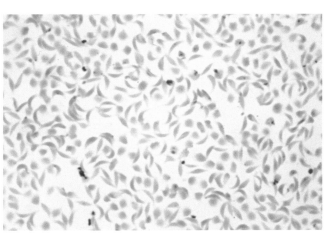

Figure 6.26 Sickled red blood cells in an oxygenated blood sample from a healthy adult red deer (*Cervus elaphus*) (×40). Reversible sickling is shown by the red cells of adult individuals of most species of deer, some bovine and caprine animals and small carnivores. In these animals, sickling is an *in vitro* phenomenon and is not associated with any abnormal clinical signs. The shape change occurs as a result of the presence of haemoglobin variants which form insoluble tactoids when oxygenated. Several different sickling haemoglobins have been described, all of which are different from human HB-S.

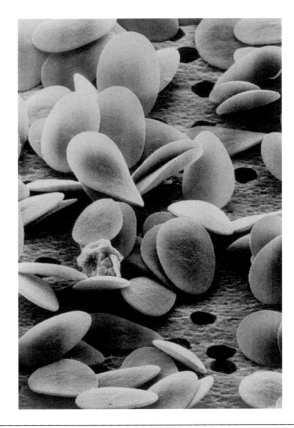

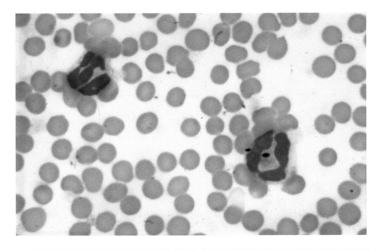

Figure 6.27 Discocytic red blood cells and two neutrophils from a normal reindeer (*Rangifer tarandus*). Red cell sickling has not been found in this species.

Figure 6.25 SEM of red blood cells from a red-eared terrapin (*Trachemys scripta elegans*). Reptilian red blood cells are flatter and appear thinner than those of birds. No nuclear bulge is apparent.

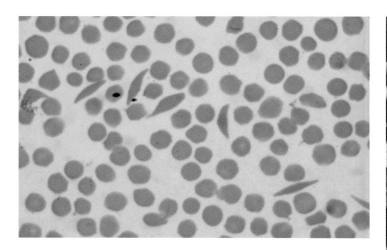

Figure 6.28 Typical appearance of red blood cells in a fresh film prepared from venous blood of a sickling species of deer, in this case an adult Père David deer (*Elaphurus davidianus*). Partial oxygenation of the sample during preparation of the film accounts for the presence of some sickled cells.

Figure 6.30 SEM showing cell deformation in oxygenated blood from a Reeve's muntjac (*Muntiacus reevesi*), a sickling haemoglobin variant. Sickle cells, echinocytes, spur cells and holly-leaf cells are present.

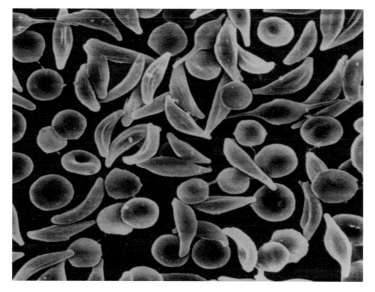

Figure 6.29 SEM of sickled red blood cells in an oxygenated blood sample from a swamp deer (*Rucervus duvaucelii* syn. *Cervus duvaucelii*). Many sickle cells, one spindle-shaped cell and several normal cells are present.

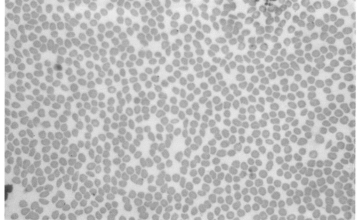

Figure 6.31 Minimal red blood cell deformation in a fresh, unmixed, blood sample from a blotched genet (*Genetta maculata*) (×40).

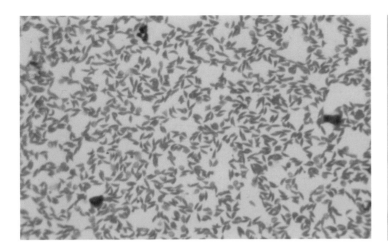

Figure 6.32 All red cells sickled in the same sample as the previous image after exposure to oxygen (×40).

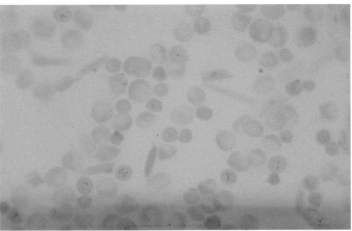

Figure 6.34 Matchstick cells found in the blood of a healthy adult white rhinoceros (*Ceratotherium simum*). These are associated with the presence of a haemoglobin variant which crystallises in the same form. The blood film was stained supravitally with new methylene blue before using the May-Grünwald-Giemsa technique. A few of the cells contain Heinz bodies. This finding is common and without clinical significance.

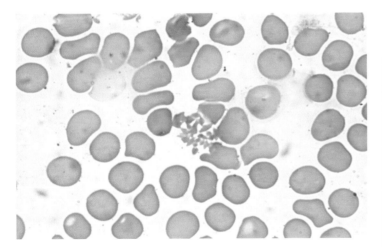

Figure 6.33 Micro sickles from a capybara (*Hydrochoerus hydrochaeris*). Individual red blood cells appear to break up into several small sickle or diamond-shaped fragments.

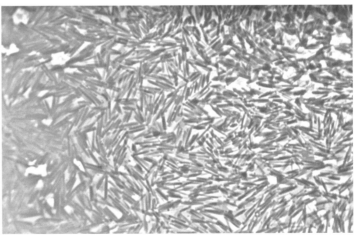

Figure 6.35 Haemoglobin crystals from haemolysed blood of the rhinoceros in the previous image.

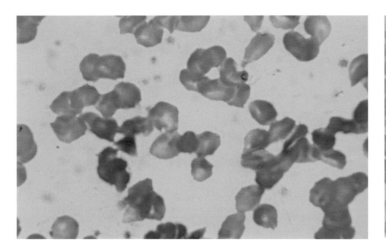

Figure 6.36 Intracellular haemoglobin crystallisation in a capybara (*Hydrochoerus hydrochaeris*). Red blood cells containing crystals interfere with the spreading of the film and this can give the impression of autoagglutination.

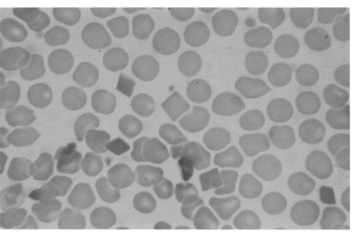

Figure 6.38 Square and diamond-shaped red blood cells due to intracellular haemoglobin crystallisation in a mara (*Dolichotis patagonum*).

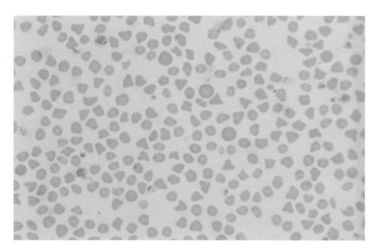

Figure 6.37 Triangular red blood cells caused by intracellular haemoglobin crystallisation in a markhor (*Capra falconeri*). Adult specimens of this species have extremely small red blood cells (MCV 15–19 fl).

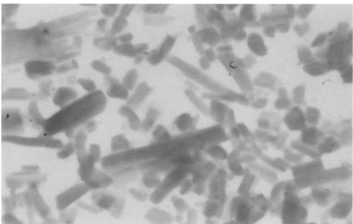

Figure 6.39 Haemoglobin crystals in the blood of a domestic dog (*Canis lupus familiaris*), after prolonged storage in EDTA.

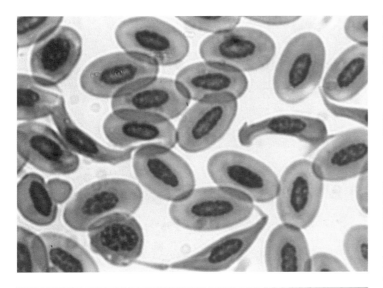

Figure 6.40 Unusual red blood cell morphology in a greater Indian hill mynah bird (*Gracula religiosa intermedia*). This shape change is reversible and its significance has not yet been determined.

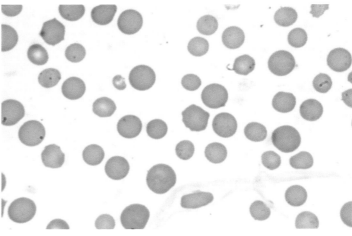

Figure 6.42 Anisocytosis in a normal adult tiger (*Panthera tigris*). The red blood cells of most healthy mammals show a moderate amount of size variation.

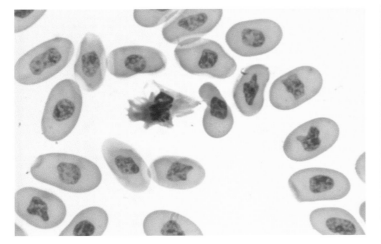

Figure 6.41 Apparent intracellular haemoglobin crystallisation in a taipan snake (*Oxyuranus microlepidotus*).

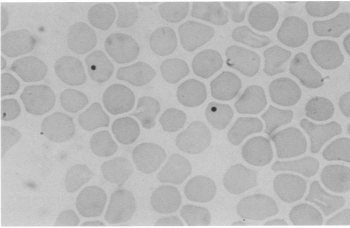

Figure 6.43 Two red blood cells containing Howell-Jolly bodies from a red-necked wallaby (*Macropus rufogriseus*), These represent nuclear remnants and are often found in small numbers in healthy marsupials, felines, rodents and small primates.

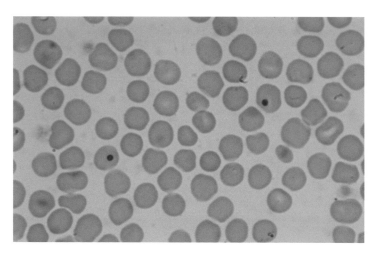

Figure 6.44 Two red blood cells containing Howell-Jolly bodies from a brown capuchin monkey (*Cebus apella*).

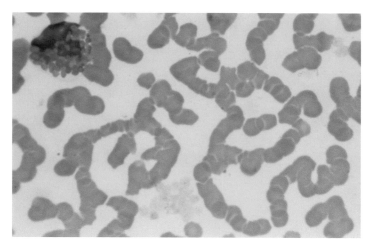

Figure 6.45 Accentuated rouleaux formation in the blood of a healthy Przewalski's wild horse (*Equus ferus przewalskii*). This is a normal finding in equine animals, elephants and some other mammalian species. In these animals the erythrocyte sedimentation rate is always rapid and does not give a useful indication of clinical status. An eosinophil with the strikingly obvious cytoplasmic granules, typical of domestic and wild horses, is present.

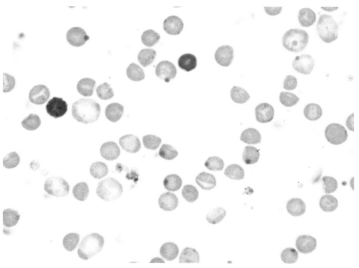

Figure 6.46 Red cells containing Heinz bodies from a healthy domestic cat (*Felis catus*) (new methylene blue stain). It is relatively common to find these inclusions in feline red blood cells and, when present in small numbers, they are generally considered to be without clinical significance. Note the close association between Heinz bodies and cell membrane.

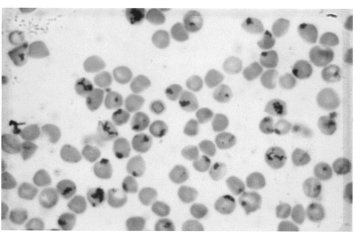

Figure 6.47 Heinz bodies and a few reticulocytes in the blood of a healthy common marmoset (*Callithrix jacchus*) (new methylene blue and rhodanile blue stain).

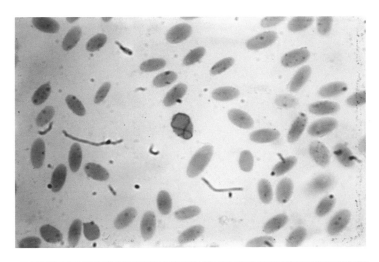

Figure 6.48 A red blood cell containing a Cabot's ring in blood from a healthy llama (*Lama glama*) (new methylene blue stain). These thread-like rings and convolutions are thought to be artefacts representing denatured membrane protein and are frequently found in the blood of normal Camelidae.

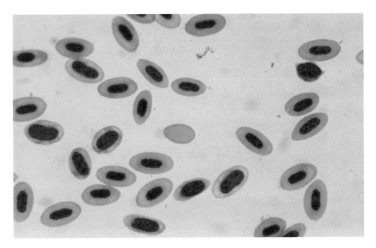

Figure 6.49 An erythroplastid from a healthy budgerigar (*Melopsittacus undulatus*). The presence of a small number of anucleated red blood cells in the blood of birds and reptiles has no pathological significance. Three polychromatic erythroblasts are also present.

6.1.3 Examples of Active Erythropoiesis in Small Mammals, Birds and Reptiles

The red blood cell survival is short in small animals with high basal metabolic rate, and evidence of active erythropoiesis, such as the presence of polychromasia, circulating erythroblasts and, in mammals, a high reticulocyte count, is considered normal in these animals.

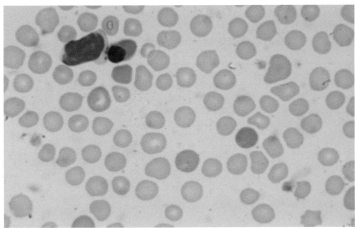

Figure 6.50 Polychromasia and a polychromatic erythroblast in a healthy adult mouse lemur (*Microcebus murinus*). A normal lymphocyte with cytoplasm containing granules is also shown.

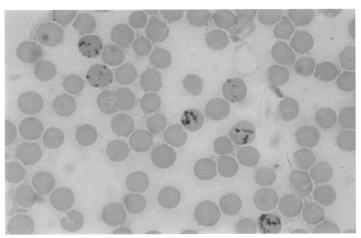

Figure 6.51 Numerous reticulocytes in the blood of a healthy adult red-bellied tamarin (*Saguinus labiatus*) (new methylene blue stain).

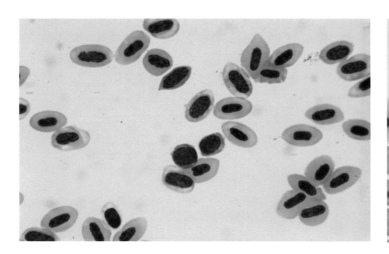

Figure 6.52 Mature red blood cells, basophilic, polychromatic and orthochromatic erythroblasts in the blood of a normal adult budgerigar (*Melopsittacus undulatus*). The nuclear size decreases and the amount of cytoplasm increases as the cells mature. Two of the basophilic erythroblasts have not yet become elliptical.

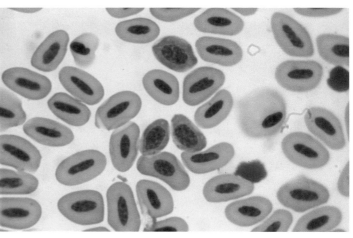

Figure 6.54 Three basophilic erythroblasts in the blood of an immature cattle egret (*Bubulcus ibis*). Regenerative anaemia is often found in young birds and usually resolves without treatment.

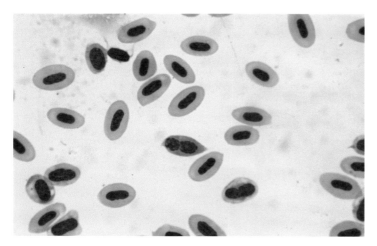

Figure 6.53 A binucleated polychromatic erythroblast from the budgerigar in the previous image.

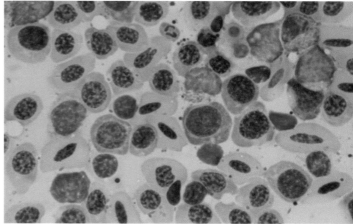

Figure 6.55 Red blood cell precursors in the bone marrow sample from a healthy grey heron (*Ardea cinerea*). All developmental stages from pro-erythroblast to mature red blood cells are present.

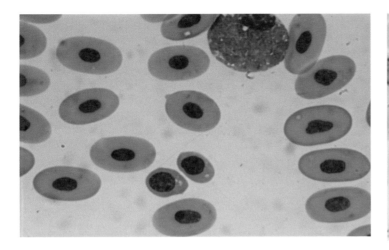

Figure 6.56 Active erythropoiesis in a healthy brown python (*Liasis fuscus*). Two polychromatic erythroblasts are illustrated, which appear microcytic compared with the mature cells. Cytoplasmic vacuolation is often seen in the erythroblasts of normal snakes.

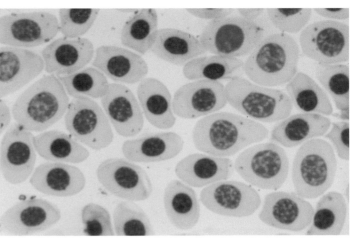

Figure 6.58 Orthochromatic erythroblasts and mature red blood cells from a healthy greater plated lizard (*Gerrhosaurus major*).

6.1.4 Common Artefacts Affecting Red Blood Cells

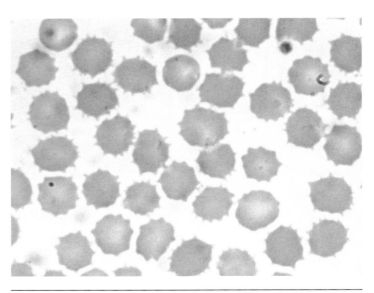

Figure 6.57 A polychromatic erythroblast with vacuolated cytoplasm from a healthy Indian python (*Python molurus*).

Figure 6.59 Crenated red blood cells in an EDTA sample from a normal guinea pig (*Cavia porcellus*). Crenation is frequently seen in mammalian blood samples collected into commercially available EDTA tubes. Crenation of the elliptical cells of Camelidae, birds, reptiles, amphibian and fish does not occur.

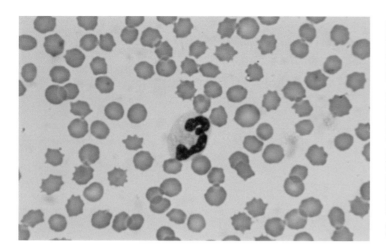

Figure 6.60 Crenation and anisocytosis in an EDTA sample from a normal leopard (*Panthera pardus*).

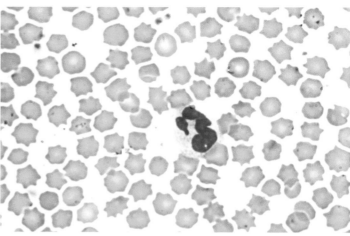

Figure 6.62 Marked crenation in an EDTA sample from a normal black and white ruffed lemur (*Varecia variegata*).

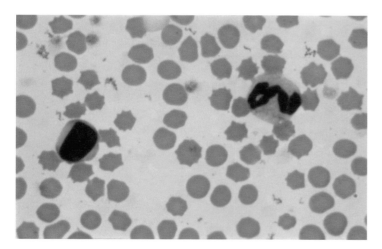

Figure 6.61 Crenated red blood cells in an EDTA blood sample from a domestic pig (*Sus scrofa domesticus*).

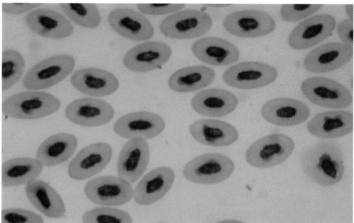

Figure 6.63 Poorly fixed sulphur-crested cockatoo (*Cacatua sulphurea*) red blood cells showing nuclear bleeding.

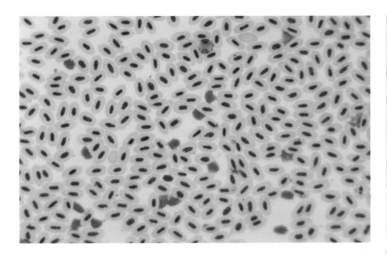

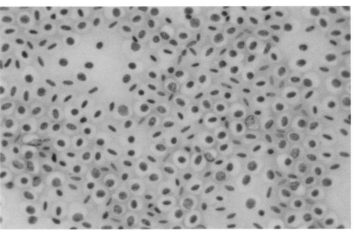

Figure 6.64 Normal red blood cells and amorphous purple staining material originating from the nuclei of disrupted red blood cells from a healthy wood pigeon (*Columba palumbus*) (×40). Breakdown of a small number of red blood cells commonly occurs during the preparation of films from avian and, less often, from reptilian and amphibian blood samples. Unless a large number of cells are affected, this artefact can be disregarded. Care must be taken to avoid the misidentification of red cell nuclei as lymphocytes or thrombocytes.

Figure 6.66 Many haemolysed red blood cells in an EDTA sample from a grey crowned crane (*Balearica regulorum*) (×40). Blood samples from this species and from several other groups of birds and reptiles including ostriches, kookaburras, corvids and tortoises, undergo haemolysis when mixed with EDTA, although not with heparin or citrate. The reason is not known.

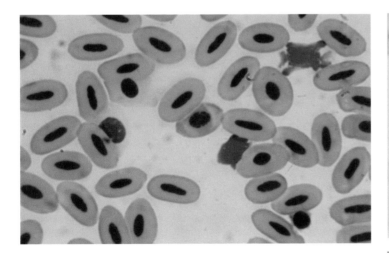

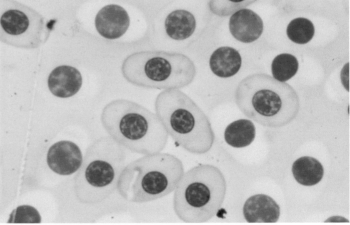

Figure 6.65 Stages in red blood cell breakdown leaning to the presence of amorphous nuclear material in a blood from a normal roseate spoonbill (*Platalea leucorodia*).

Figure 6.67 Red blood cell lysis in an EDTA sample from an Aldabra giant tortoise (*Aldabrachelys gigantea*).

6.1.5 Abnormal Variations in Red Blood Cell Morphology

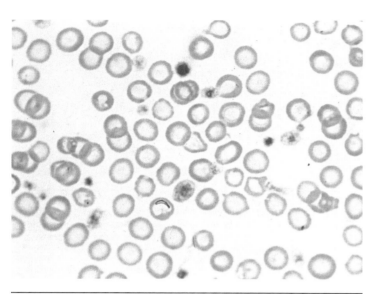

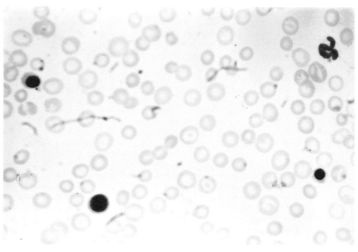

Figure 6.70 Regenerative hypochromic anaemia in a domestic dog (*Canis* lupus *familiaris*). Target cells, two orthochromatic erythroblasts and a normal neutrophil are present.

Figure 6.68 Iron deficiency anaemia in a domestic dog (*Canis lupus familiaris*). The red cells are hypochromic and slightly microcytic. Active erythropoiesis is indicated by the presence of polychromatic cells. The platelets show anisocytosis.

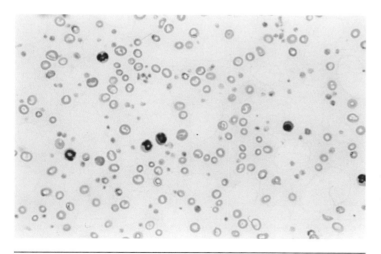

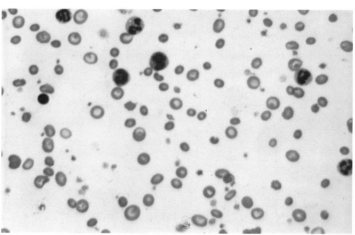

Figure 6.71 Marked anisocytosis, neutrophilia and thrombocytosis in a domestic dog (*Canis* lupus *familiaris*) with severe regenerative hypochromic anaemia associated with hormonal imbalance (×40, Hb 3.3 g/dl).

Figure 6.69 Severe hypochromic anaemia in a domestic cat (*Felis catus*) with *Haemobartonella felis* parasites (×40). The red blood cells also show anisocytosis and one nucleated red blood cell (normoblast, orthochromic, erythroblast) is present.

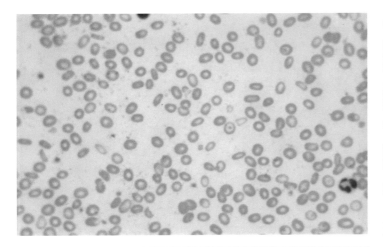

Figure 6.72 Blood film from the dog in the previous image, showing the presence of elliptocytes during the recovery period (×40, Hb 8.1 g/dl).

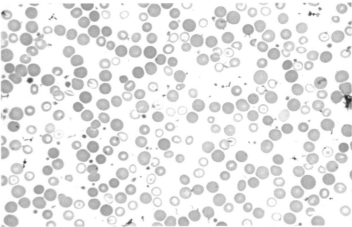

Figure 6.74 The same macaque from the previous image, 10 days after iron treatment (×40, Hb 7.2 g/dl). There is now a population of normochromic normocytic red blood cells.

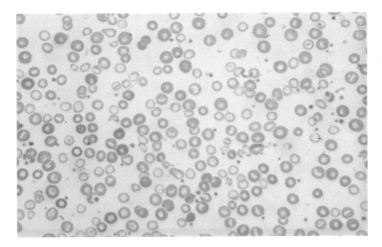

Figure 6.73 Iron deficiency anaemia resulting from chronic blood loss in a lion-tailed macaque (*Macaca silenus*) with menorrhagia (×40, Hb 4.0 g/dl). The red blood cells are hypochromic and microcytic.

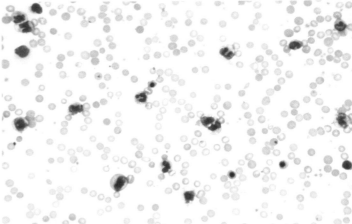

Figure 6.75 Erythroblasts and leucocytosis in a silvery marmoset (*Mico argentatus*) with regenerative haemorrhagic anaemia (×25, Hb 7.2 g/dl).

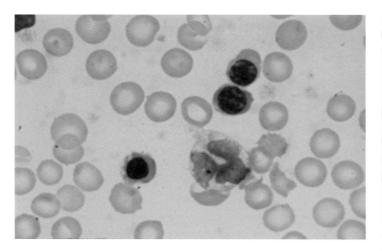

Figure 6.76 A higher magnification of two polychromatic and one ortho-chromic erythroblast from the marmoset in the previous image. The red blood cells show hypochromia and polychromasia. An effete leucocyte is present.

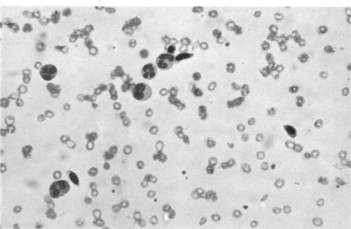

Figure 6.78 Abnormal erythropoiesis in a three-striped night monkey (*Aotus trivirgatus*) with severe regenerative hypochromic anaemia (×40). There are two misshapen and one normal erythroblast in the field.

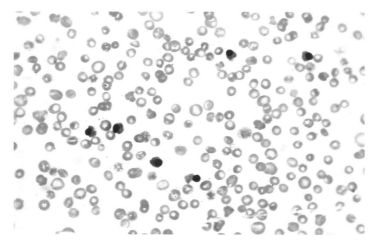

Figure 6.77 Red blood cells showing anisocytosis, hypochromia, stomatocytosis and polychromasia from a three-striped night monkey (*Aotus trivirgatus*) with regenerative anaemia (×40, Hb 5.3 g/dl). Many erythroblasts are present.

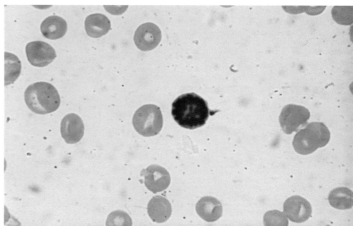

Figure 6.79 A mitotic erythroblast from the owl monkey from the previous image.

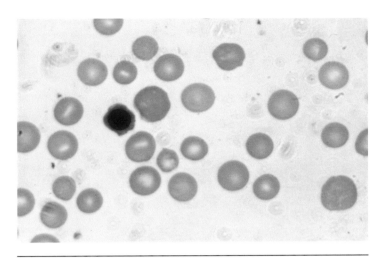

Figure 6.80 Anisocytosis, polychromasia and a polychromatic erythroblast in a red-bellied tamarin (*Saguinus labiatus*) with regenerative haemorrhagic anaemia (Hb 5.3 g/dl).

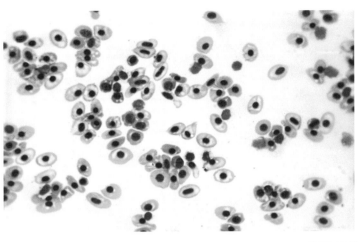

Figure 6.82 Severe hypochromic anaemia in a domestic duck (*Anas platyrhynchos domesticus*) (×40, Hb 3.4 g/dl, MCHC <20 g/dl). The cells show vacuolation, anisocytosis, and shape variation and many abnormally round, condensed nuclei. The presence of polychromatic erythroblasts indicates active erythropoiesis.

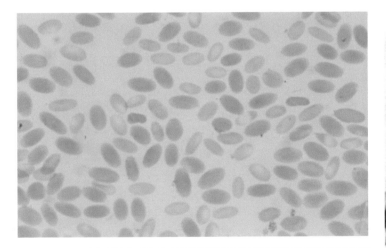

Figure 6.81 Hypochromia and anisocytosis associated with copper deficiency in a Bactrian camel (*Camelus bactrianus*).

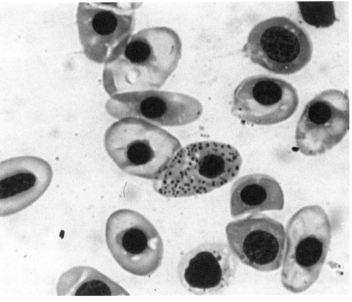

Figure 6.83 A higher magnification of red blood cells from the duck in the previous image. Hypochromia and active erythropoiesis are clearly evident and one cell shows basophilic stippling (punctuate basophilia).

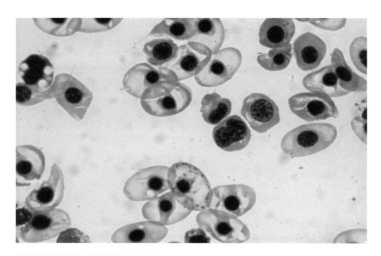

Figure 6.84 Basophilic erythroblasts with abnormal nuclear division from the duck in the previous image.

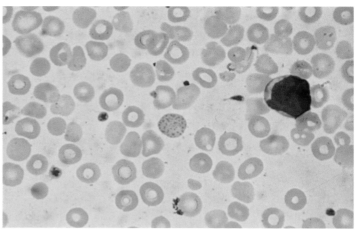

Figure 6.86 A red blood cell showing basophilic stippling from a moose (*Alces alces*) calf with slightly hypochromic anaemia (Hb 10 g/dl). This is not an unusual finding in immature Artiodactyla and appears to be of no pathological significance.

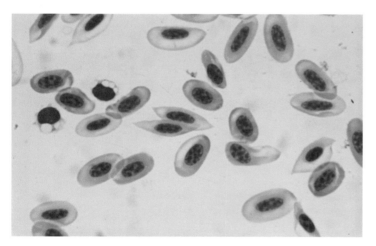

Figure 6.85 Hypochromia, anisocytosis, microcytosis and poikilocytosis in a whooper swan (*Cygnus cygnus*) with lead toxicosis (Hb 5.9 g/dl). Two normal thrombocytes with vacuolated cytoplasm are present in the field.

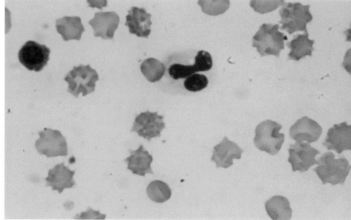

Figure 6.87 Echinocytes and a polychromatic erythroblast from a giant anteater (*Myrmecophaga tridactyla*) suffering from severe, regenerative haemorrhagic anaemia (Hb 2.1 g/dl).

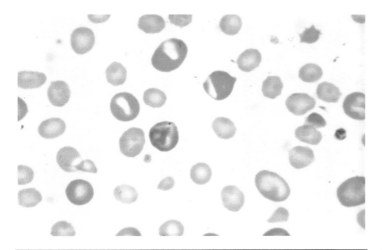

Figure 6.88 Red blood cells showing polychromasia, anisocytosis, poikilocytosis and stomatocytosis from a European hamster (*Cricetus cricetus*) with moderately severe regenerative anaemia (Hb 6.7 g/dl).

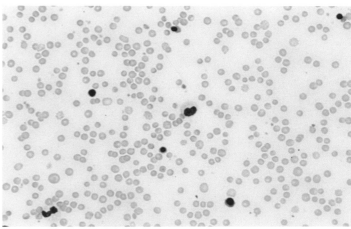

Figure 6.90 Anisocytosis and erythroblasts in a cow (*Bos taurus*) with pyelonephritis and pyuria/haematuria (×40). This animal has neutrophilia with left shift.

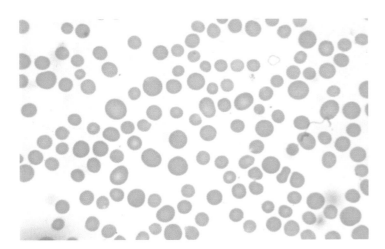

Figure 6.89 Marked anisocytosis in a domestic goat (*Capra aegagrus hircus*) during recovery from severe haemorrhagic anaemia. The macrocytes are the newly formed cells. Polychromasia is not evident and the reticulocyte count was only slightly increased. Like in many other ungulates, reticulocytes of goats do not enter the bloodstream in significant numbers, even after massive blood loss. In these animals, the presence of macrocytes is a more reliable indicator than the reticulocyte count of active erythropoiesis.

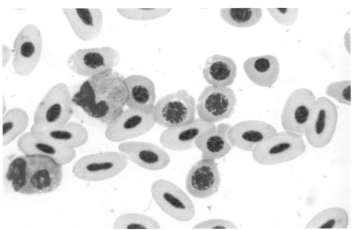

Figure 6.91 Regenerative anaemic in a sarus crane (*Grus antigone*). Four polychromatic erythroblasts and two normal heterophils are present.

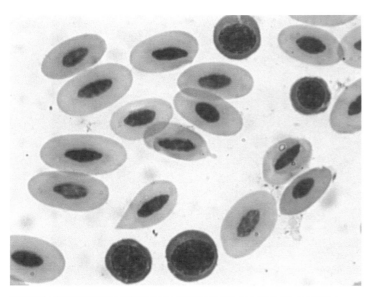

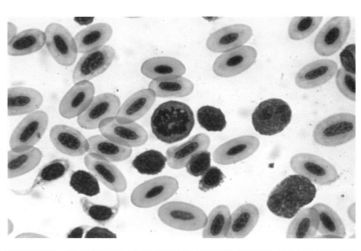

Figure 6.94 Microcytic erythroblasts and an erythroblast in mitosis from the eagle in the previous image. Two thrombocytes are present.

Figure 6.92 Four basophilic erythroblasts from a sacred ibis (*Threskiornis aethiopicus*) with moderately severe, regenerative anaemia associated with chronic infection (Hb 7.7 g/dl). The red blood cells show anisocytosis and some poikilocytosis.

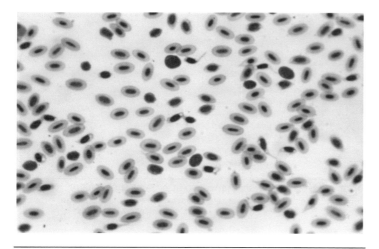

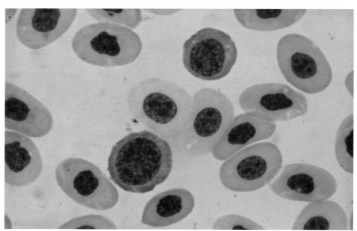

Figure 6.95 Two basophilic erythroblasts from a Gila monster (*Heloderma suspectum*) (Hb 8.1 g/dl). One of these cells appears to be megaloblastic.

Figure 6.93 Abnormal erythropoiesis in a golden eagle (*Aquila chrysaetos*) with severe regenerative anaemia, cause unknown (×40, Hb 5.1 g/dl).

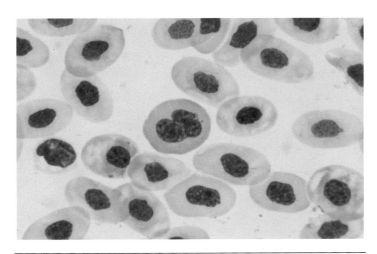

Figure 6.96 A polychromatic megaloblast with abnormal nuclear division from the Gila monster in the previous image.

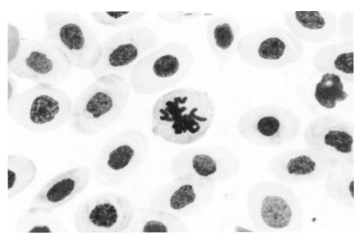

Figure 6.98 A cell in mitosis, probably a polychromatic erythroblast, from a Gila monster in the previous image.

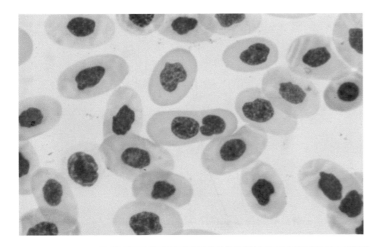

Figure 6.97 A binucleated macrocytes and mature red cells with abnormal nuclei from the Gila monster in the previous case.

Figures 6.99 to 6.102 show erythropoiesis in a Mediterranean spur-thighed tortoise (*Testudo graeca*) which had recently awakened from hibernation. This animal was slightly anaemic (Hb 6.2. g/l).

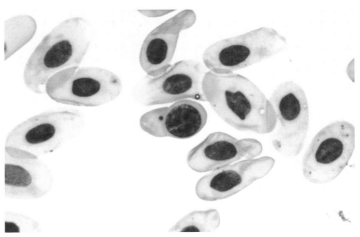

Figure 6.99 A polychromatic erythroblast showing evidence of abnormal nuclear division. Several of the mature cells have folded cytoplasm.

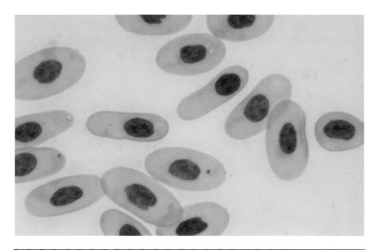

Figure 6.100 Two red blood cells with asymmetrically placed nuclei and a cell with small nuclear fragments in the cytoplasm.

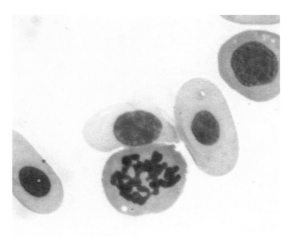

Figure 6.102 A normal erythroblast and an erythroblast in mitosis.

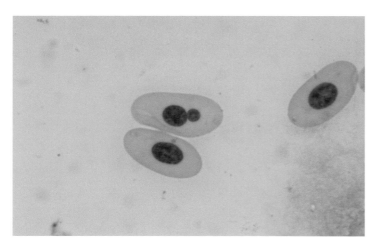

Figure 6.101 A cell containing a large nuclear remnant (Howell-Jolly body?).

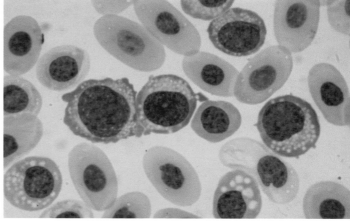

Figure 6.103 Abnormal (megaloblastic?) erythropoiesis in a Mediterranean spur-thighed tortoise (*Testudo graeca*) suffering from anaemia and post-hibernation anorexia (Hb 3.2 g/dl).

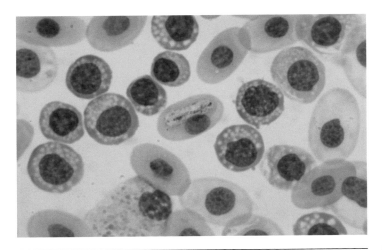

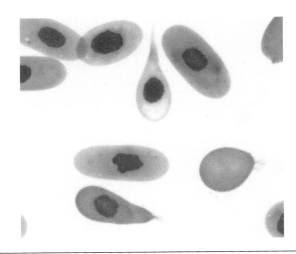

Figure 6.104 Basophilic and polychromatic erythroblasts with vacuolated cytoplasm from the tortoise in the previous image. The red blood cell in the centre of the field contains a haemogregarine parasite which is thought to be non-pathogenic. There is a toxic heterophil at the bottom of the field.

Figure 6.106 Two poikilocytes and an erythroplastid from the snake in the previous image.

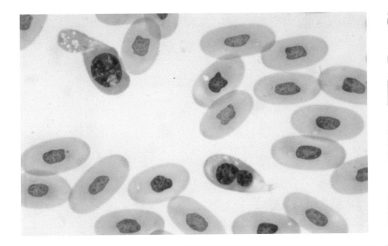

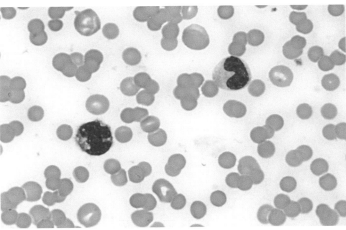

Figure 6.105 Abnormal polychromatic erythroblasts from a boa constrictor (*Boa constrictor*) suffering from chronic weight loss (Hb 6.7 g/dl).

Figure 6.107 Microspherocytes, anisocytosis, agglutination, a monocyte with vacuolated cytoplasm and a band neutrophil.

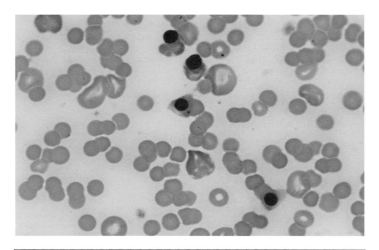

Figure 6.108 Microspherocytes, anisocytosis, agglutination, polychromasia and two polychromatic and two orthochromatic erythroblasts. The erythroblasts appear normal.

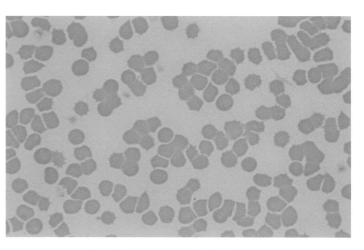

Figure 6.110 Autoagglutination and marked polychromasia in a domestic cat (*Felis catus*) with warm agglutinin disease. (×40).

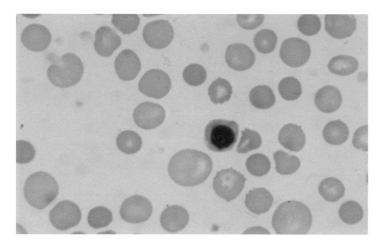

Figure 6.109 Autoimmune haemolytic anaemia in a two-month-old wolf (*Canis lupus*) cub (Hb 9.8 g/dl). Microspherocytes, anisocytosis and an orthochromic erythroblast can be seen.

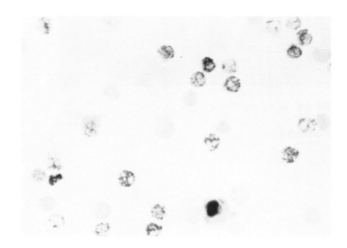

Figure 6.111 Profound reticulocytosis in a domestic cat (*Felis catus*) with Coombs' positive regenerative anaemia (×40, new methylene blue stain). The cat was also Feline leukaemia virus (FeLV)-positive.

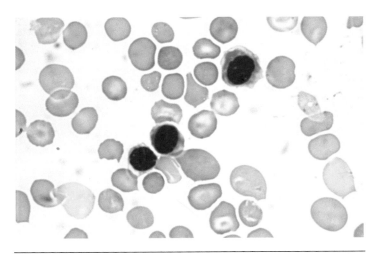

Figure 6.112 Severe, regenerative haemolytic anaemia in a female ring-tailed coati (*Nasua nasua*) (Hb 4.5 g/dl). The red blood cells show spherocytosis and anisocytosis. Three erythroblasts and several targets cells are present.

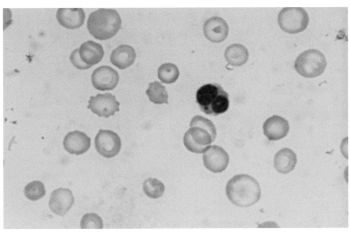

Figure 6.114 A binucleated erythroblast from the coati in the previous image.

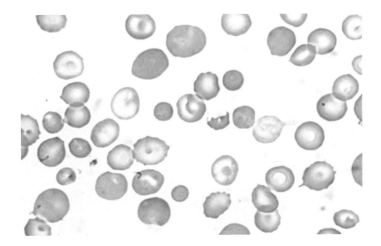

Figure 6.113 Marked anisocytosis. Hypochromia, polychromasia, microspherocytes, schistocytes and target cells ion the coati in the previous image.

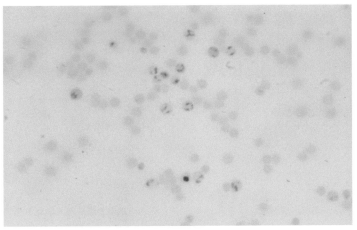

Figure 6.115 Marked reticulocytosis in the coati in the previous image (×40 new methylene blue stain).

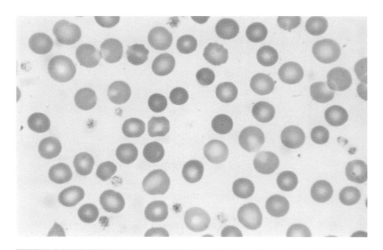

Figure 6.116 Microspherocytes from a black spider monkey (*Ateles paniscus*) with peritonitis associated with chronic amoebiasis (Hb 9.8 g/dl).

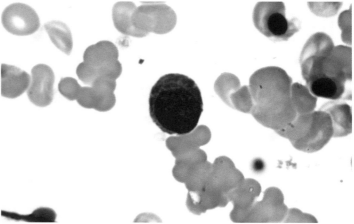

Fig 6.118 A basophilic erythroblast and marked agglutination in the wallaby of the previous image.

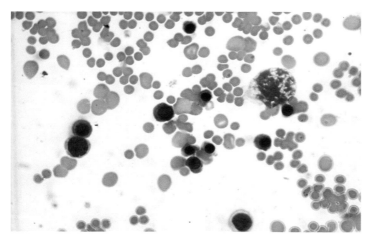

Figure 6.117 Severe, regenerative haemolytic anaemia associated with necrobacillosis in a red-necked wallaby (*Macropus rufogriseus*) (×40, Hb 6.4 g/dl). The red blood cells show anisocytosis and auto agglutination and a large number of erythroblasts are present.

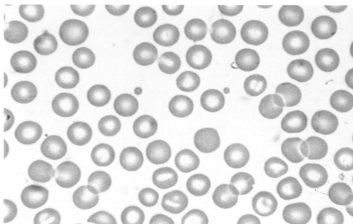

Figure 6.119 Two extremely small microspherocytes from a tree shrew suffering from hypervitaminosis A.

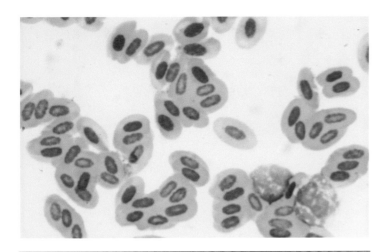

Figure 6.120 Red blood cell agglutination in an anaemic African grey parrot (*Psittacus erithacus*) (Hb 8.4 g/dl). In this case the agglutination was caused by an antibody active at low temperature and the bird was thought to have autoimmune haemolytic anaemia.

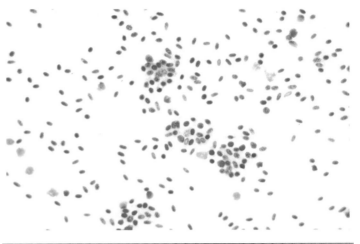

Figure 6.122 Autoagglutination, similar to that in the previous image, in an anaemic lesser sulphur-crested cockatoo (*Cacatua sulphurea*) (×40, Hb 9.0 g/dl).

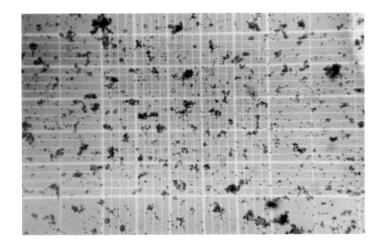

Figure 6.121 Agglutinated red blood cells from the parrot in the previous image, visible in a haemocytometer (×40, phase contrast microscopy).

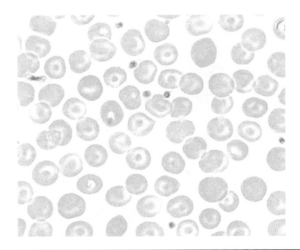

Figure 6.123 Target cells from a domestic dog (*Canis lupus familiaris*).

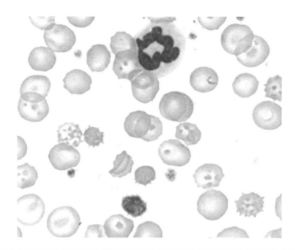

Figure 6.124 Echinocytes, target cells and hypochromia in a hog badger (*Arctonyx collaris*) with anaemia (Hb 6.9 g/dl). The animal was suffering from liver failure and toxaemia secondary to infected burns.

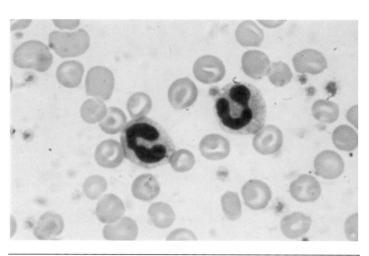

Figure 6.126 Target cells and stomatocytes in a cotton-headed tamarin (*Saguinus oedipus*) with peritonitis (Hb 7.3 g/dl). Two neutrophils with reduced nuclear lobulation are present.

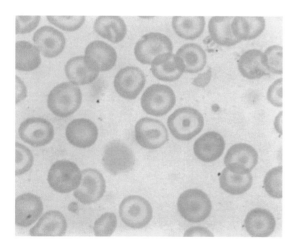

Figure 6.125 Target cells and hypochromia in a prairie marmot (*Cynomys ludovicianus*) with hypochromic anaemia (Hb 10.0 g/dl). The presence of target cells suggested that liver failure was present.

Figure 6.127 Echinocytes from an Indian brown mongoose (*Herpestes fuscus*) with slight non-regenerative normochromic anaemia secondary to kidney failure. The cause of the anaemia was probably depressed erythropoietin production.

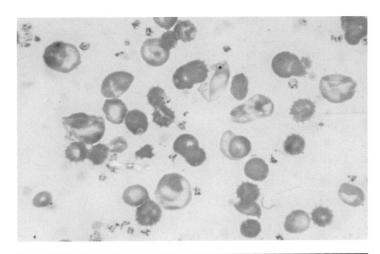

Figure 6.128 Anisocytosis, poikilocytosis, hypochromasia and stomatocytosis in a ring-tailed coati (*Nasua nasua*) with moderate non-regenerative anaemia associated with chronic infection (Hb 7.6 g/dl).

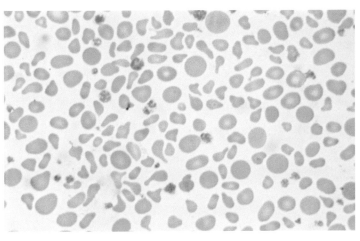

Figure 6.130 Misshapen red blood cells from a newborn domestic goat (*Capra aegagrus hircus*). The larger, discoid cells are presumed to contain adult haemoglobin. The relative number of these increased progressively with time.

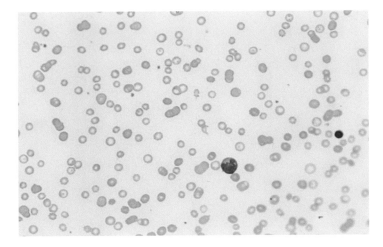

Figure 6.129 Poikilocytes and schistocytes from a domestic dog (*Canis lupus familiaris*) with endocarditis (×25).

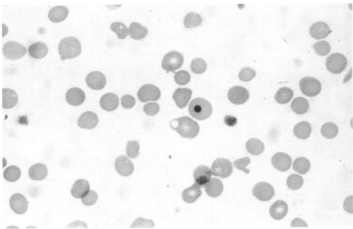

Figure 6.131 Severe regenerative anaemic in a five-day-old roan antelope (*Hippotragus equinus*) calf (Hb 6.2 g/dl). The red cells show anisocytosis. One large Howell-Jolly body is present and one cell appears to have a hole at the periphery.

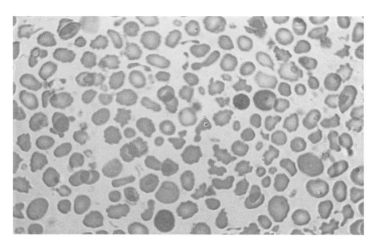

Figure 6.132 Extreme red cell deformation in a one-day-old roan antelope (*Hippotragus equinus*) (Hb 10.1 g/dl). Cells of this type frequently occur in the blood of immature ungulates and appear to be associated with a high neonatal mortality rate. In animals which survive, the abnormal cells are progressively replaced by normal biconcave discocytes containing adult haemoglobin. Unlike sickled red blood cells, this type of red blood cells shape abnormality cannot be reversed by altering the conditions of the blood sample.

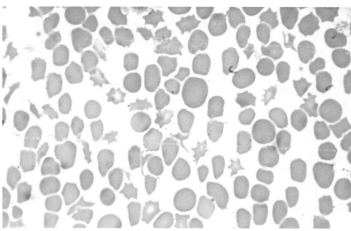

Figure 6.134 Two morphologically distinct populations of red blood cells in a 10-week-old gaur calf (*Bos gaurus*). More than 50% of the cells appear normal and are presumed to be those produced in the post-natal period. The presence of macrocytes indicates active erythropoiesis. This animal was not anaemic (18.0 g/dl).

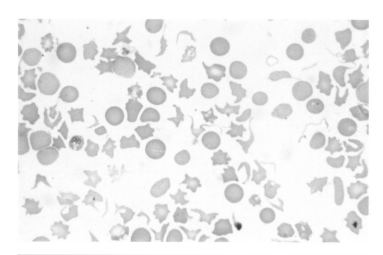

Figure 6.133 Extreme red blood cell deformation and some abnormal cells in a severely ill, one-day-old Arabian oryx (*Oryx leucoryx*) calf.

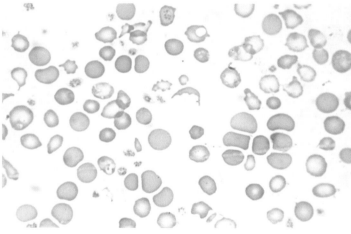

Figure 6.135 Blood film from a 25-day-old scimitar horned oryx (*Oryx dammah*) showing similar red blood cells deformation (Hb 7.1 g/dl) as in the previous image. Some apparently normal cells are present.

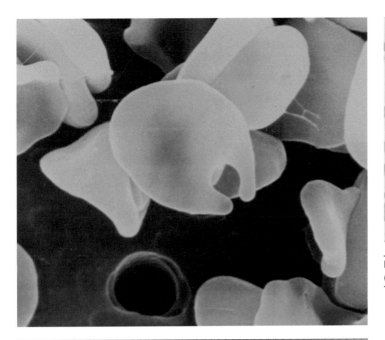

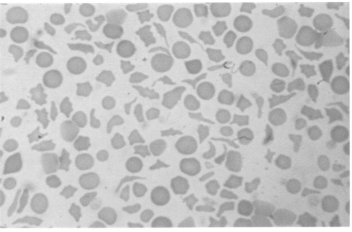

Figure 6.138 Misshapen red blood cells in an eight-week-old fallow deer (*Dama dama*) fawn (Hb 13.1 g/dl). A few sickle cells are also present. These revert to normal biconcave discs when deoxygenated.

Figure 6.136 SEM of the sample of the previous case showing evidence that an early stage in the development of the misshapen red blood cells is the appearance of holes in some of the cells. Progressive increase in the size of the holes and eventual rupture of the cells containing them could result in the bizarre cell shapes observed.

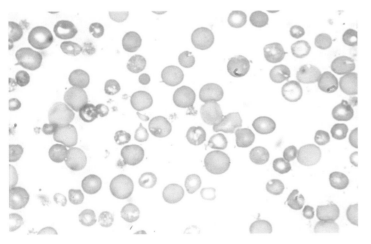

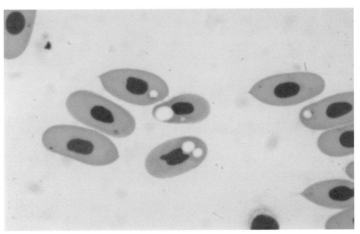

Figure 6.139 Vacuolated red blood cells from a royal python (*Python regius*) with chronic weight loss (Hb 6.0 g/dl). Red blood cells vacuolation is not unusual in normal snakes of this group. One of the red blood cells has a misshapen nucleus. This finding appears to be associated with malnutrition in reptiles.

Figure 6.137 Blood film from the previous case of the scimitar horned oryx (*Oryx dammah*) at the age of seven weeks (Hb 10.1 g/dl). A larger proportion of normal red blood cells is present.

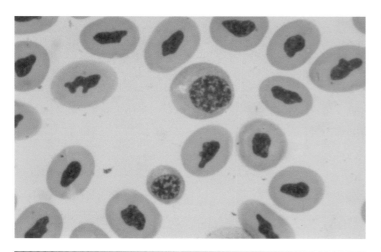

Figure 6.140 A macrocytic and a microcytic polychromatic erythroblast from a boomslang snake (*Dispholidus typus*) with anorexia and necrotic stomatitis. The snake was not anaemic (Hb 13.3 g/dl) but the red blood cell nuclei were deformed.

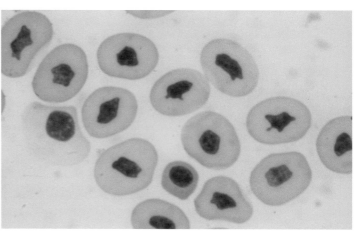

Figure 6.142 Red blood cells with misshapen nuclei from a hawksbill turtle (*Eretmochelys imbricata*) suffering from malnutrition (Hb 10.6 g/dl).

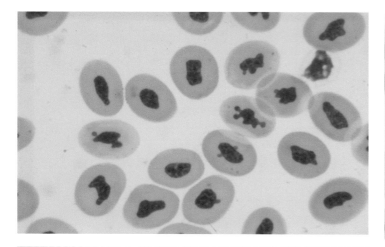

Figure 6.141 Red blood cells from the snake in the previous image with marked nuclear deformation and the budding off of small Howell-Jolly body-like fragments.

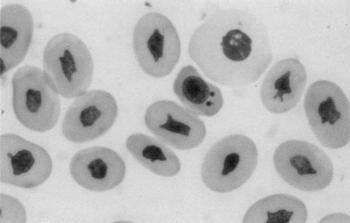

Figure 6.143 A polychromatic erythroblast showing evidence of abnormal nuclear division from the turtle in the previous image.

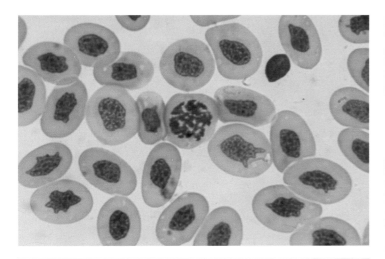

Figure 6.144 Red blood cells with misshapen nuclei from a blue-tongued skink (*Tiliqua scincoides scincoides*) with chronic infection and anorexia (Hb 6.7 g/dl). The red blood cells nuclei appear megaloblastic. A polychromatic erythroblast in mitosis is present.

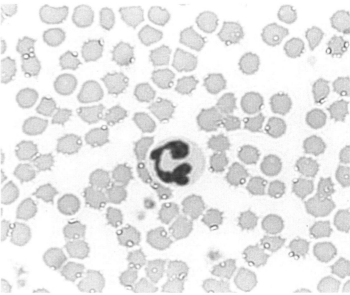

Figure 6.146 Erythrocyte refractile (ER) bodies in an unstained blood film from a domestic cat (*Felis catus*) suffering from paracetamol toxicosis. ER bodies are analogous with Heinz bodies.

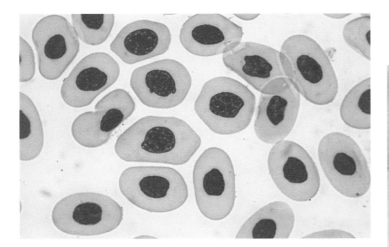

Figure 6.145 Anisocytosis in an Aldabra giant tortoise (*Aldabrachelys gigantea*) suffering from malnutrition and anaemia (Hb 5.2 g/dl). Some of the cells are less elliptical than normal and the nuclei show signs of immaturity.

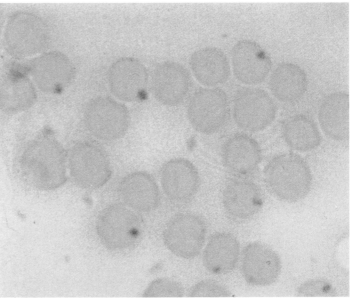

Figure 6.147 Red blood cells from the cat in the previous image stained supravitally with new methylene blue to show the Heinz bodies.

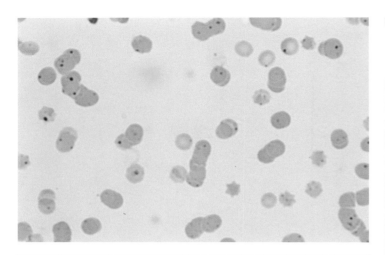

Figure 6.148 *Haemobartonella felis* and Howell-Jolly bodies in red blood cells from a domestic cat (*Felis catus*) (×40).

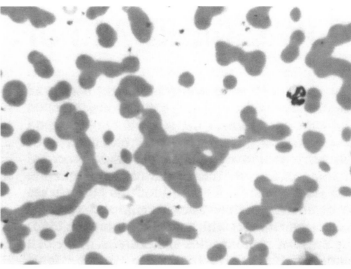

Figure 6.150 Increased rouleaux formation in a domestic dog (*Canis lupus familiaris*) with myeloma.

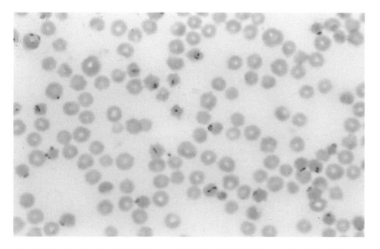

Figure 6.149 Multiple *Babesia divergens* in red blood cells from a Friesian cow (*Bos taurus*).

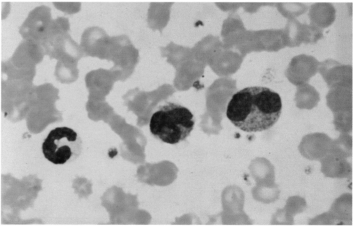

Figure 6.151 Increased rouleaux formation in a domestic dog (*Canis lupus familiaris*) with neoplasia. A normal neutrophil and two neutrophils with decreased nuclear lobulation (left shift) and abnormal cytoplasmic granules are present.

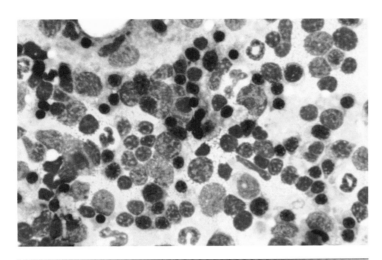

Figure 6.152 Bone marrow preparation from a domestic dog (*Canis lupus familiaris*) with polycythaemia vera rubra.

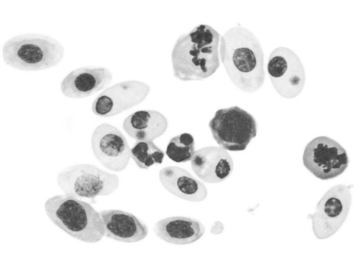

Figure 6.154 This is a higher magnification of the centre of the previous image showing two mitotic RBCs, one recently divided RBC, polychromatophilic RBCs of varying stages of maturity and several RBCs with inclusion bodies (×1000) (courtesy of Helen McCracken).

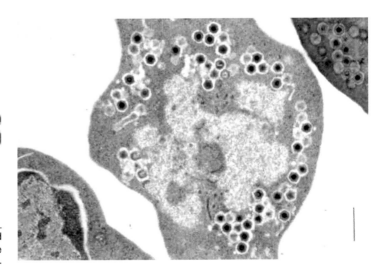

Figure 6.155 Transmission electron microscopy of part of a single RBC. The structure on the lower left is the cell nucleus. The structures in the centre and upper right are separate intracytoplasmic inclusions, each composed of an aggregation of virus particles. Although most infected RBCs had only one inclusion, a low number, including this cell, had two (courtesy of Helen McCracken).

Figure 6.153 Moderate to severe haemolytic anaemia in a diamond python (*Morelia spilota*). Highly pleiomorphic populations of both mature and immature red blood cells (RBC). High polychromasia. A high proportion of the RBCs show circular, acidophilic, intracytoplasmic inclusion bodies (0.5–3 μm in diameter). Transmission electron microscopy demonstrated the inclusion to be aggregations of icosahedral viruses (141–161 nm in diameter). The morphology of the virions was consistent with erythrocytic iridovirus (×400) (courtesy of Helen McCracken).

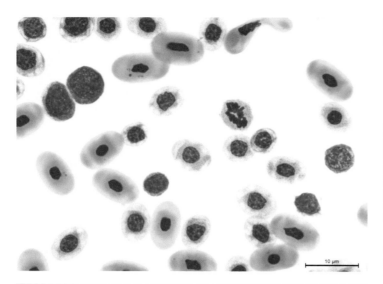

Figure 6.156 Blood smear from a green turtle (*Chelonia mydas*) showing red blood cells in various stages of immaturity all the way back to rubriblast, consistent with adequate erythroid regenerative response (courtesy of Nicole Stacy).

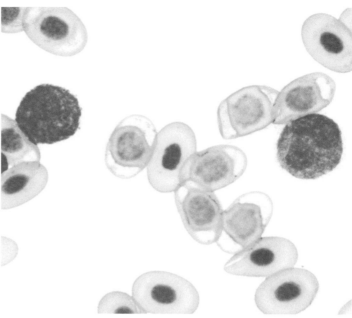

Figure 6.158 Two monocytes in a blood smear of a spotted marsh frog (*Limnodynastes tasmaniensis*) with severe monocytosis and abnormal red blood cells (courtesy of Helen McCracken).

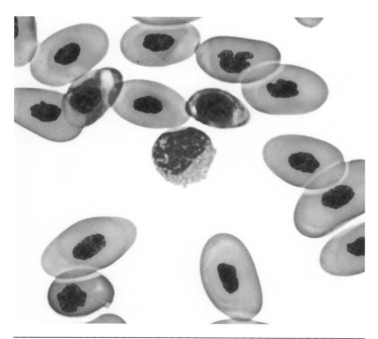

Figure 6.157 A monocyte in a blood smear of a baw baw frog (*Philoria frosti*). The red blood cells show polychromasia (courtesy of Helen McCracken).

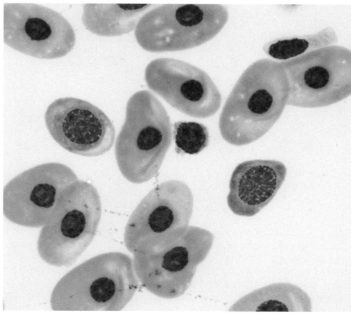

Figure 6.159 Blood smear from an Eastern fiddler ray (*Trygonorrhina guaneria*) showing a typical lymphocyte and two polychromatophilic red blood cells (courtesy of Helen McCracken).

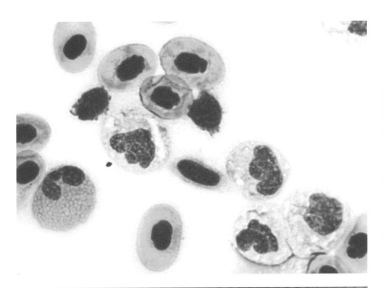

Figure 6.160 Blood smear from a Siberian sturgeon (*Acipenser baerii*). The image shows an eosinophil, 4 toxic neutrophils, 2 lymphocytes, 1 thrombocyte and normal red blood cells (courtesy of Nicole Stacy).

6.1.6 Species Variation in Normal Granulocytes

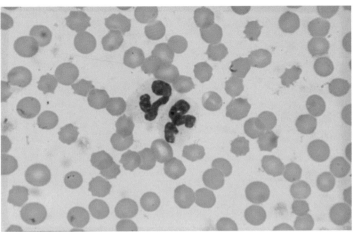

Figure 6.162 Two neutrophils with agranular cytoplasm and indistinct nuclear lobulation from a jaguar (*Panthera onca*). These cells are typical of the neutrophils found in feline species.

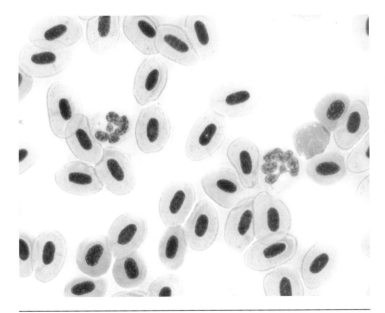

Figure 6.161 Blood smear from a healthy wild sockeye salmon (*Oncorhynchus nerka*). The photograph shows two neutrophils and erythrocytes (mature and with polychromasia). A small amount of irregular drying/staining artefacts are present of the red blood cells (courtesy of Nicole Stacy).

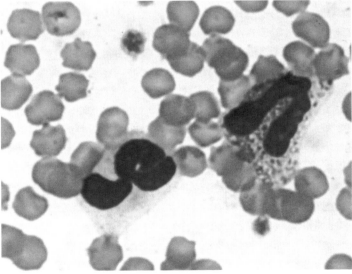

Figure 6.163 An agranular neutrophil and an eosinophil with small red cytoplasmic granules from a domestic cat (*Felis catus*). The nuclear lobes are poorly defined in both cells.

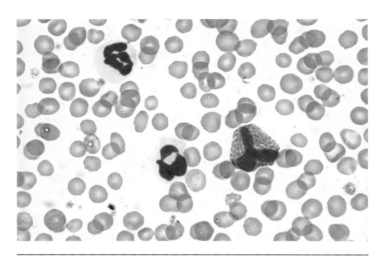

Figure 6.164 Two agranular neutrophils and an eosinophil from a cheetah (*Acinonyx jubatus*) (×40).

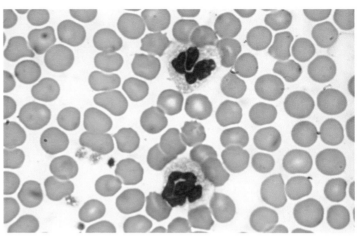

Figure 6.166 Two neutrophils from a polar bear (*Ursus maritimus*). The cytoplasm contains aggregates of pink-staining material and the nuclear lobes are well defined.

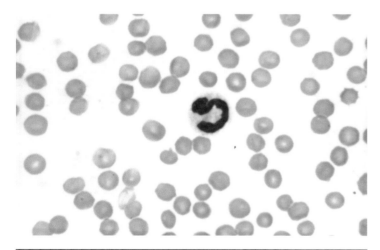

Figure 6.165 A neutrophil from a healthy ferret (*Mustela putorius furo*).

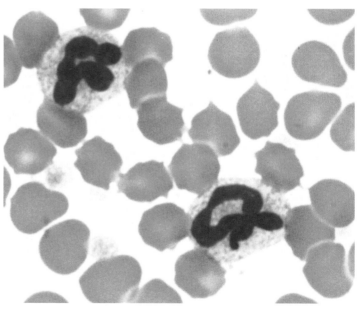

Figure 6.167 Two neutrophils from a healthy domestic dog (*Canis lupus familiaris*). The cells have faint, pink, cytoplasmic granules.

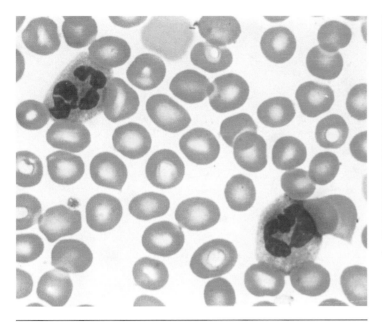

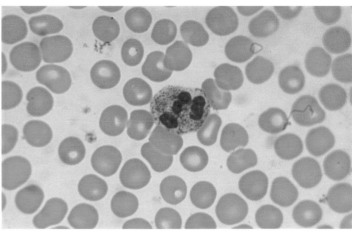

Figure 6.170 A neutrophil from a chimpanzee (*Pan troglodytes*) with some basophilic granules.

Figure 6.168 Two neutrophils from a healthy human (*Homo sapiens*) individual.

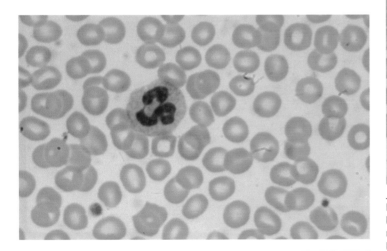

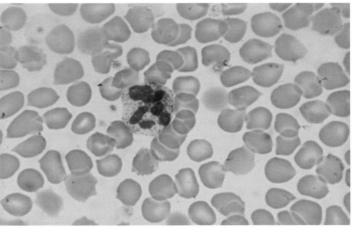

Figure 6.171 A neutrophil with orange-staining granules and a relatively large number of well-defined nuclear lobes from a white-headed saki monkey (*Pithecia pithecia*). The neutrophils of Old World monkeys tend to have a greater number of lobes than those of humans.

Figure 6.169 A neutrophil with pink granules and well-defined nuclear lobes from a Bornean orangutan (*Pongo pygmaeus*).

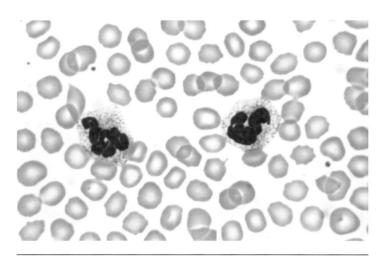

Figure 6.172 Two neutrophils from a healthy rhesus macaque (*Macaca mulatta*).

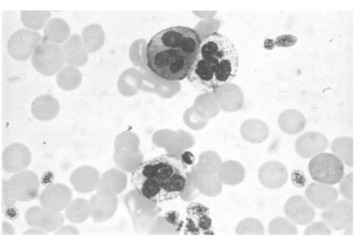

Figure 6.174 Two neutrophils with many orange and a few purple granules and an eosinophil with indistinct red granules from a healthy three-striped night monkey (*Aotus trivirgatus*).

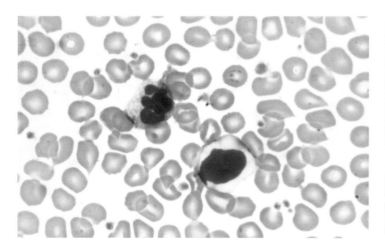

Figure 6.173 A neutrophil and a lymphocyte from a healthy crab-eating macaque (*Macaca fascicularis*).

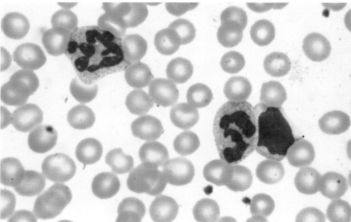

Figure 6.175 Two neutrophils and a lymphocyte from a healthy common marmoset (*Callithrix jacchus*).

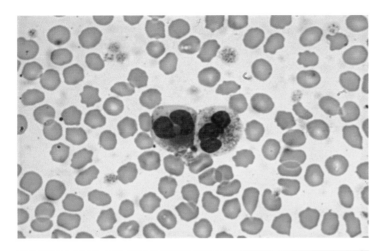

Figure 6.176 A neutrophil and an eosinophil from a brown lemur (*Eulemur fulvus*). These cells have staining characteristics similar to those of the three-striped night monkey (*Aotus trivirgatus*).

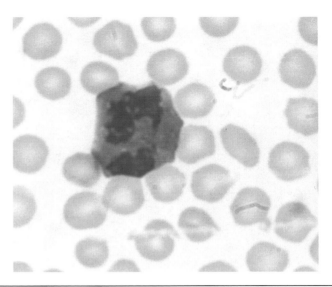

Figure 6.178 Comparison with true eosinophil from the rabbit in the previous image. The esosinophils have larger, more numerous and somewhat poorly defined, eosinophilic granules, some of which may overlay the nucleus. The nucleus usually has fewer lobes than that of the 'pseudoeosinophil'.

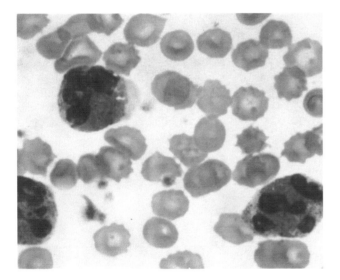

Figure 6.177 Three neutrophils with strongly eosinophilic cytoplasmic granules from a rabbit (*Oryctolagus cuniculus*). These cells are sometimes known as 'pseudoeosinophils' and can be confused with true eosinophils.

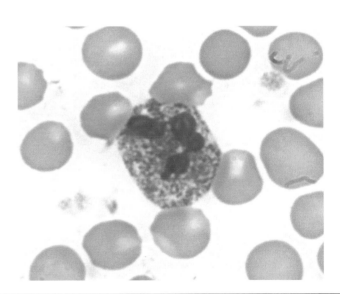

Figure 6.179 A 'pseudoeosinophilic' neutrophil from a capybara (*Hydrochoerus hydrochaeris*). The neutrophil granules of many hystrico-morph rodents show these staining characteristics.

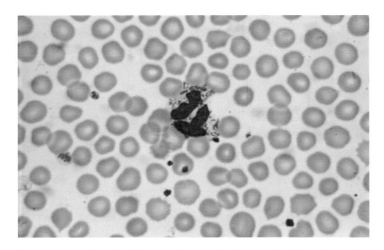

Figure 6.180 A 'pseudoeosinophilic' neutrophil from a healthy guinea pig (*Cavia porcellus*).

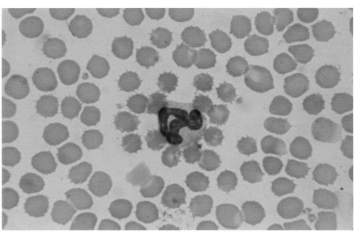

Figure 6.182 A neutrophil from a healthy brown rat (*Rattus norvegicus*). The nuclear lobes are poorly defined and there are small, pale pink cytoplasmic granules. The red cells are crenated.

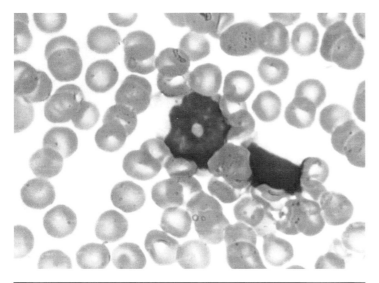

Figure 6.181 A neutrophil with agranular cytoplasm and a ring-form nucleus from a mouse (*Mus musculus*). Compared with hystriocomorph and sciuromorph rodents, the nuclei of myomorph rodents' neutrophils have poorly defined lobes.

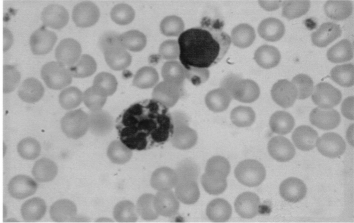

Figure 6.183 A neutrophil with a multilobed nucleus (right shift) and a normal lymphocyte from a healthy grey squirrel (*Sciurus carolinensis*).

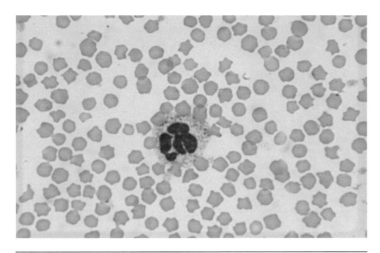

Figure 6.184 A neutrophil with small basophilic granules and distinct nuclear lobes from a roan antelope (*Hippotragus equinus*). The red cells are crenated.

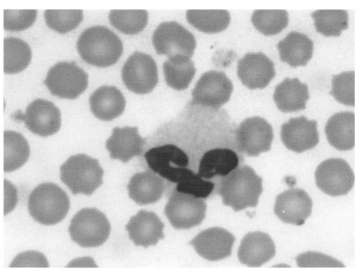

Figure 6.186 An agranular neutrophil from a cow (*Bos taurus*).

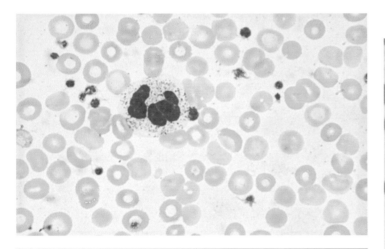

Figure 6.185 A neutrophil from a moose (*Alces alces*) with similar characteristics to the antelope in the previous image.

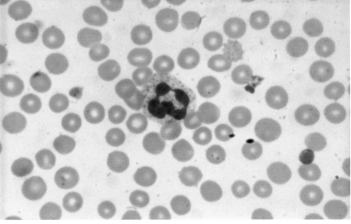

Figure 6.187 A neutrophil from a healthy eland (*Taurotragus oryx*). The cell is indistinguishable from those of domestic cattle.

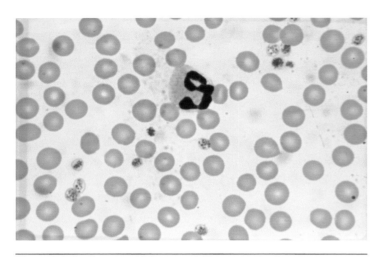

Figure 6.188 A neutrophil from a yak (*Bos grunniens*).

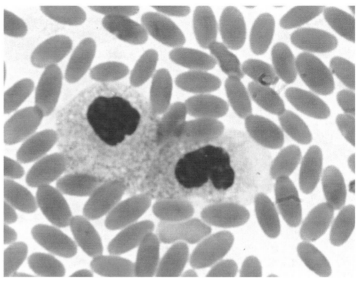

Figure 6.190 Neutrophils with condensed, bilobed or unlobed nuclei from Bactrian camels. The appearance of the nuclei is similar to that seen in humans, dogs, cats and rabbits showing the Pelger-Huët phenomenon. Neutrophils of this type occur in some but not all Bactrian camels (*Camelus bactrianus*). Possible patterns of inheritance have not yet been studied. There is no evidence of associated impairment of cells function.

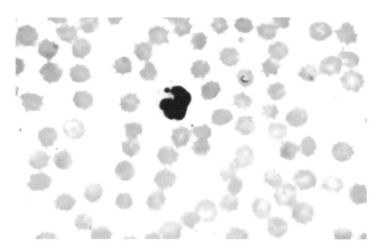

Figure 6.189 A neutrophil with a condensed nucleus with a drumstick appendage from a female common hippopotamus (*Hippopotamus amphibious*). The red cells show marked crenation.

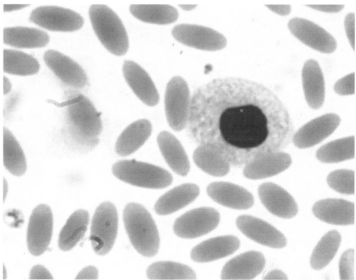

Figure 6.191 Neutrophils with condensed, bilobed or unlobed nuclei from Bactrian camels. The appearance of the nuclei is similar to that seen in humans, dogs, cats and rabbits showing the Pelger-Huët phenomenon. Neutrophils of this type occur in some but not all Bactrian camels (*Camelus bactrianus*). Possible patterns of inheritance have not yet been studied. There is no evidence of associated impairment of cells function.

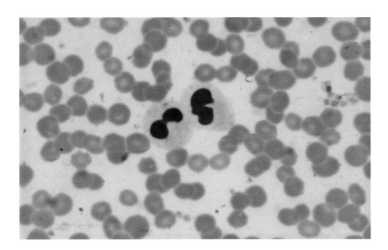

Figure 6.192 Two neutrophils showing a Pelger-Huët-like phenomenon from a white rhinoceros (*Ceratotherium simum*). As in Bactrian camels, these cells are found in some but not all individuals of the species and appear to function normally.

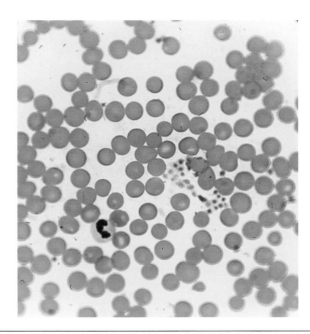

Figure 6.194 A blood film from a captive clinically normal South American sea lion (*Otaria flavescens*) showing a neutrophil and platelet clumping (courtesy of Guillermo Sanchez Contreras).

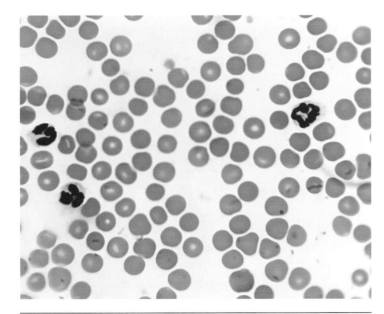

Figure 6.193 Three neutrophils from a captive clinically normal bottlenose dolphin (*Tursiops truncatus*) (courtesy of Guillermo Sanchez Contreras).

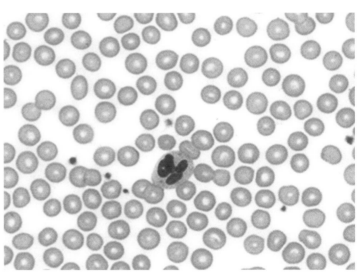

Figure 6.195 A neutrophil in a blood smear of a ferret (*Mustela putorius furo*) (courtesy of Jaume Martorell).

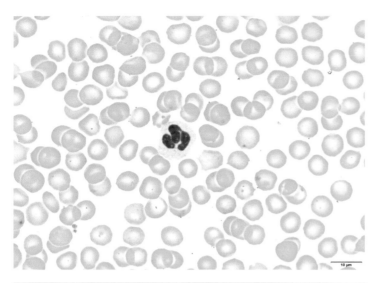

Figure 6.196 Photograph of a neutrophil in a blood film from an Eastern grey kangaroo (*Macropus giganteus*) (courtesy of Karen Jackson).

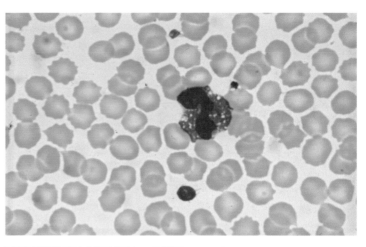

Figure 6.198 An eosinophil with cytoplasmic vacuoles from a terrier dog (*Canis lupus familiaris*). Cells of this type are normal in some canine breeds.

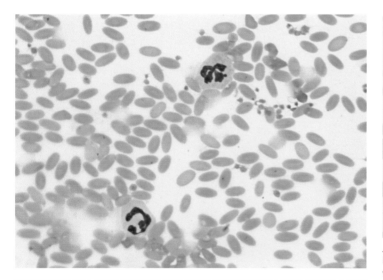

Figure 6.197 Normal neutrophils in a blood smear from a clinically normal captive alpaca (*Vicugna pacos*) (courtesy of Margarita Colburn).

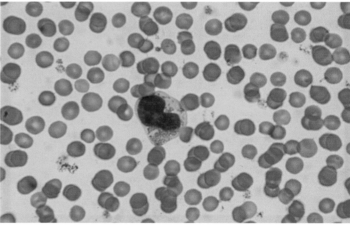

Figure 6.199 An eosinophil with distinct, small eosinophilic granules which do not completely fill the blue-grey cytoplasm from a blesbok (*Damaliscus pygargus phillipsi*).

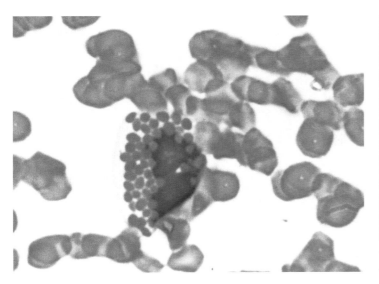

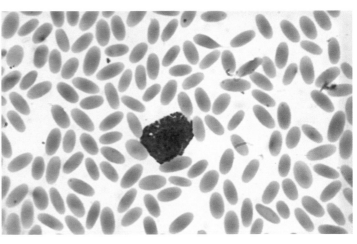

Figure 6.202   A basophil from a Bactrian camel (*Camelus bactrianus*). The granules show variation in staining intensity and do not mask the nucleus.

Figure 6.200   An eosinophil from a horse (*Equus caballus*). The cytoplasm is completely filled with large, spherical, strongly eosinophilic granules. A high degree of red cell rouleaux formation, as shown, is normal in horses.

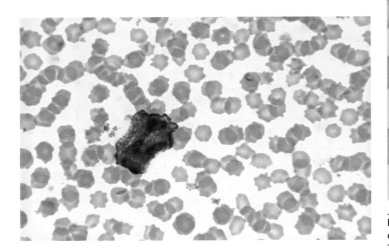

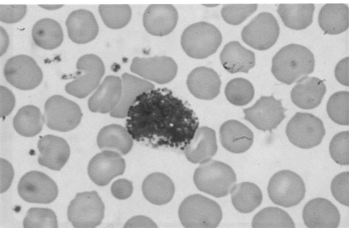

Figure 6.203   A basophil with distinct spherical granules showing differences in staining density from a two-toed sloth (*Choloepus didactylus*). This species has relatively large red blood cells.

Figure 6.201   A basophil from a gaur (*Bos gaurus*). The granules do not completely mask the lobed nucleus.

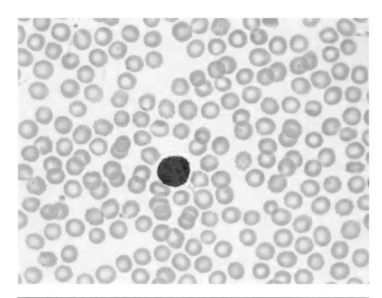

Figure 6.204 A typical basophil from a clinically normal guinea pig (*Cavia porcellus*) (courtesy of Nico Schoemaker and Yvonne RA van Zeeland).

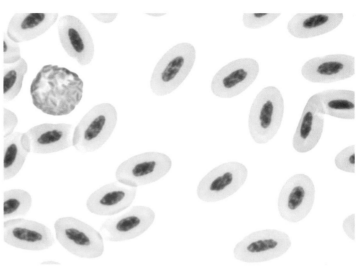

Figure 6.206 A typical avian heterophil from a captive clinically normal Chilean flamingo (*Phoenicopterus chilensis*). The intracytoplasmic granules are typically elongated and brick red in colour (courtesy of Jean-Michel Hatt).

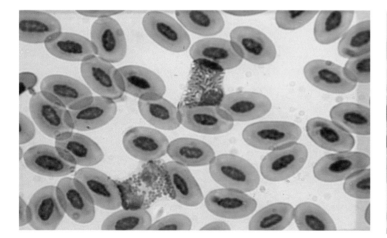

Figure 6.205 In birds, reptiles, amphibians and fish, the cells analogous with mammalian neutrophils have strongly eosinophilic cytoplasmic granules and are known as heterophils. In some species, it can be difficult to distinguish between heterophils and eosinophils. In most birds, heterophils have brick red, oval or spiculate granules, whereas the eosinophilic granules are usually round and more brightly eosinophilic. A typical avian heterophil and eosinophil from a Javan fish owl (*Ketupa ketupu*) are shown. The heterophil has brick red, spiculate granules. The eosinophil has bright red, round granules. Both cells have bilobed nuclei.

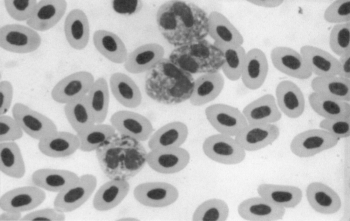

Figure 6.207 Four heterophils from an apparently healthy domestic fowl (*Gallus gallus domesticus*). The nucleus of one of these cells shows decreased lobulation, suggesting sub-clinical inflammatory disease.

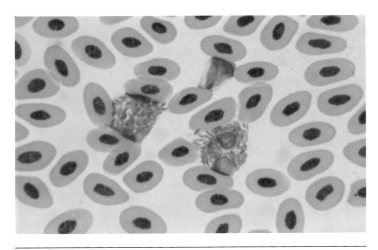

Figure 6.208 Two heterophils and a lymphocyte from a domestic turkey (*Meleagris gallopavo*).

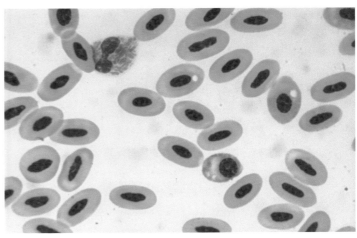

Figure 6.210 A heterophil from a domestic duck (*Anas platyrhynchos domesticus*). A polychromatic erythroblast is also present.

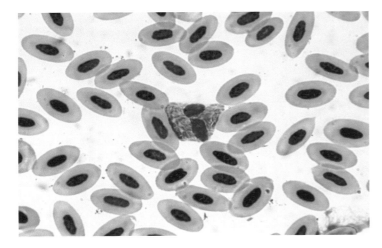

Figure 6.209 A heterophil from a domestic goose (*Anser anser*).

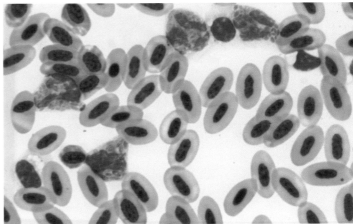

Figure 6.211 Heterophils and two thrombocytes from a healthy pigeon (*Columba livia*), The red blood cells show polychromasia.

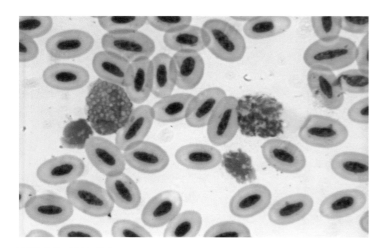

Figure 6.212 A heterophil with distinct brick-red granules and an eosinophil with distinctive round, brighter red granules from a kori bustard (*Ardeotis kori*). Both cells have bilobed nuclei. Both cells have bilobed nuclei. Two disrupted red blood cells nuclei are also present.

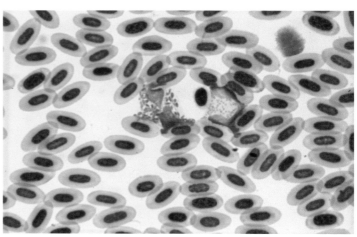

Figure 6.214 A heterophil and an eosinophil from a healthy canary (*Serinus canaria*). The eosinophil has sparse, round, orange-staining cytoplasmic granules in contrast to those of the heterophil which are brick red and spiculate.

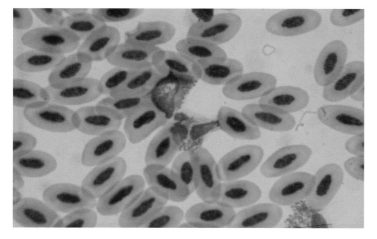

Figure 6.213 A heterophil with indistinct, brick red, spiculate granules and an eosinophil with brighter red, spherical granules from a common buzzard (*Buteo buteo*). Both cells have bilobed nuclei.

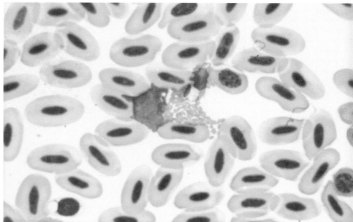

Figure 6.215 An intact heterophil with brick-red granules and a broken eosinophil with oval orange granules from a white stork (*Ciconia ciconia*).

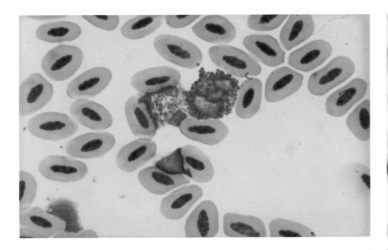

Figure 6.216 A heterophil with typical brick red, spiculate granules and a trilobed nucleus, a slightly disrupted eosinophil with bluish, spherical granules and a bilobed nucleus from a lesser sulphur-crested cockatoo (*Cacatua sulphurea*). The basophilic nature of the eosinophil granules is characteristic of most psittacine birds.

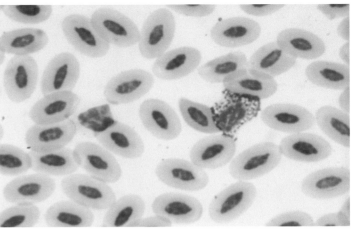

Figure 6.218 A basophil with sparse purple granules from a Javan fish owl (*Ketupa ketupu*).

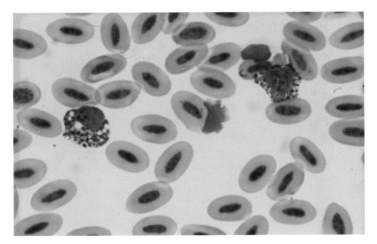

Figure 6.217 Two normal basophils, for comparison, from the cockatoo in the previous image.

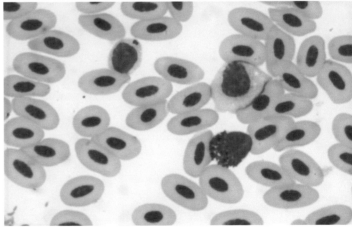

Figure 6.219 A basophil with dark purple granules, a lymphocyte with a cytoplasmic granule and a monocyte from a hooded crane (*Grus monacha*).

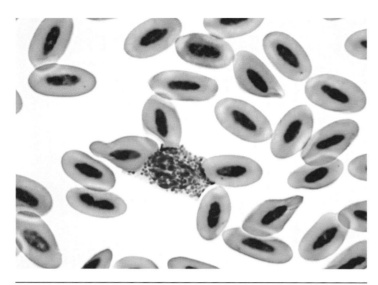

Figure 6.220 Blood film from a great pied hornbill (*Buceros bicornis*) showing an eosinophil with small double-coloured granules typical of the species (courtesy of Helene Pendl).

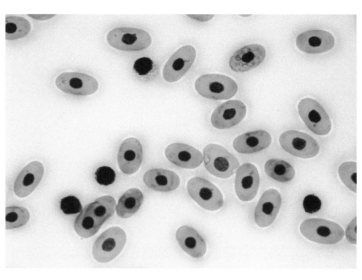

Figure 6.222 Four basophils in a blood film from a yellow-bellied slider (*Trachemys scripta scripta*) fixed and stained using Diff-Quick rapid staining method (courtesy of David Perpiñan).

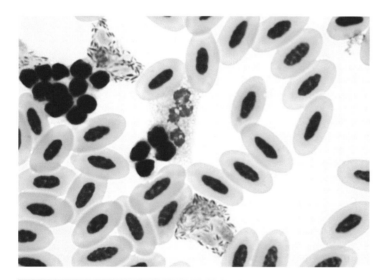

Figure 6.221 Photomicrograph of the blood smear of a Eurasian oyster-catcher (*Haematopus ostralegus*) showing two thrombocytes aggregates, two normal heterophils and an eosinophil with basophilic granules typical of shore and sea birds (courtesy of Helene Pendl).

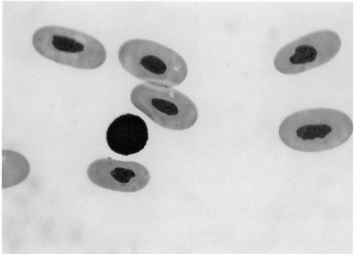

Figure 6.223 Blood film from a green iguana (*Iguana iguana*) showing an eosinophil (courtesy of David Perpiñan).

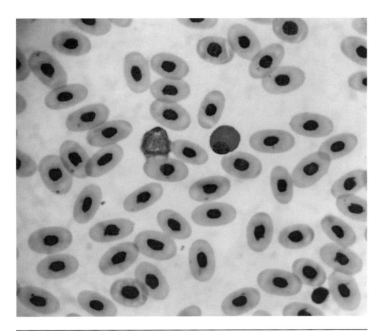

Figure 6.224 A heterophil (left) and an eosinophil (right) in a blood smear of a South American rattlesnake (*Crotalus durissus*) (courtesy of Carlos Troiano).

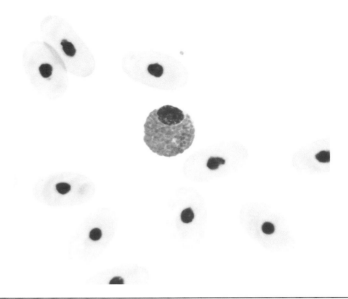

Figure 6.226 Blood smear from an American alligator (*Alligator mississippiensis*) showing an eosinophil with excellent granule shape detail (courtesy of Helen McCracken).

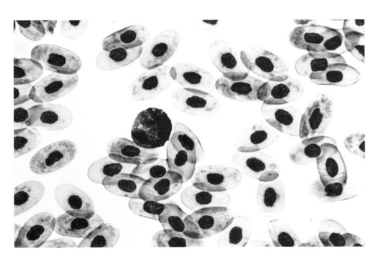

Figure 6.225 Basophil from a boa constrictor (*Boa constrictor*) in a blood smear stained with May-Grünwald-Giemsa method (courtesy of Carlos Troiano).

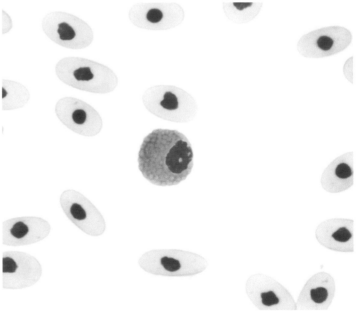

Figure 6.227 An eosinophil showing good granular detail in a blood film from a Philippine crocodile (*Crocodylus mindorensis*) (courtesy of Helen McCracken).

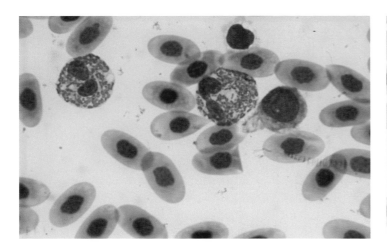

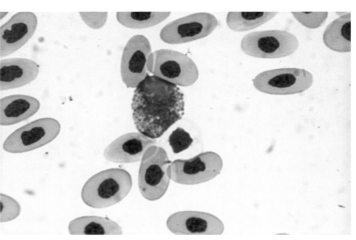

Figure 6.228 Two heterophils from a healthy green iguana (*Iguana iguana*). The cells have bilobed nuclei and bacilliform, eosinophilic cytoplasm granules and are somewhat similar to the heterophils found in birds. Heterophils with lobed nuclei occur in many but not all species of Sauria (lizards), whereas those of Testudines (tortoises, terrapins and turtles), Crocodylia (crocodiles and alligators) and Serpentes have unlobed nuclei.

Figure 6.230 An eosinophil from a healthy green iguana (*Iguana iguana*). Unlike the eosinophils of most other reptiles, these have round, bluish granules.

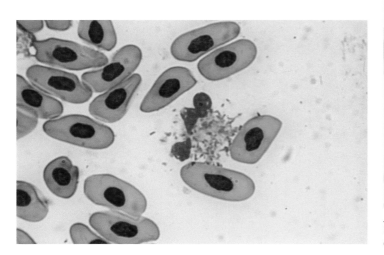

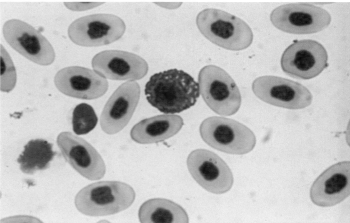

Figure 6.231 A basophil, for comparison, from the iguana in the previous image.

Figure 6.229 A ruptured heterophil from the iguana in the previous image showing the bacilliform, eosinophilic granules and lobed nucleus.

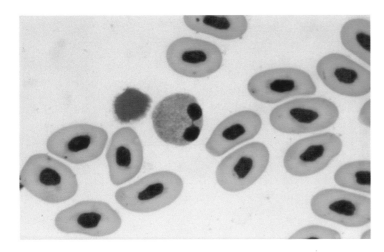

Figure 6.232 Cells with bilobed nuclei, resembling mammalian neutrophils, found occasionally in the blood of some species of Sauria, in this case a green iguana (*Iguana iguana*). The origin of these cells is not known; possibly they are necrotic granulocytes.

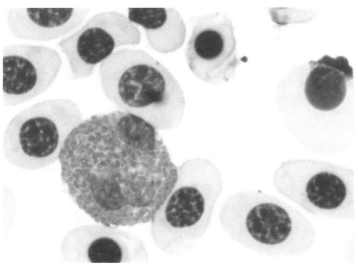

Figure 6.234 A heterophil from a Sudan plated lizard (*Gerrhosaurus major*). The nucleus is bilobed.

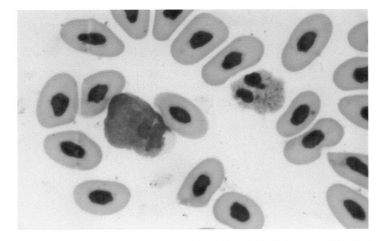

Figure 6.233 Cells with bilobed nuclei, resembling mammalian neutrophils, found occasionally in the blood of some species of Sauria, in this case a green iguana (*Iguana iguana*). The origin of these cells is not known; possibly they are necrotic granulocytes. A normal heterophil with coalescent granules is present.

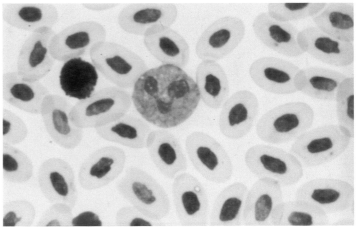

Figure 6.235 A heterophil with a trilobed nucleus and a basophil completely filled with granules from a black-pointed tegu (*Tupinambis teguizin*).

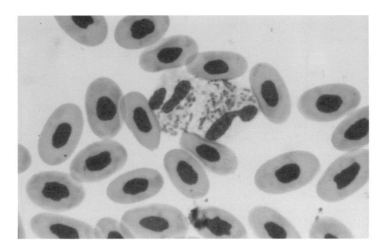

Figure 6.236 Two heterophils with bilobed nuclei and sparse eosinophilic granules from a healthy shingleback skink (*Tiliqua rugosa*). Cells of this type are typical of skinks.

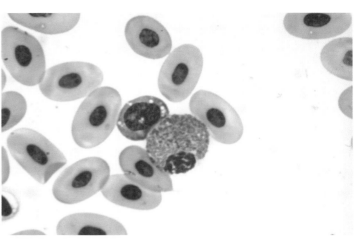

Figure 6.238 A heterophil with an unlobed, excentric nucleus from a healthy Indian python (*Python molurus*). A polychromatic erythroblast with cytoplasmic vacuoles is also present.

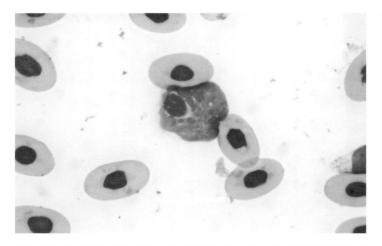

Figure 6.237 A heterophil from a healthy Nile monitor lizard (*Varanus niloticus*). Monitors and water dragons are examples of Sauria which have heterophils with unlobed nuclei, similar to those of Chelonia, Serpentes and Crocodylia.

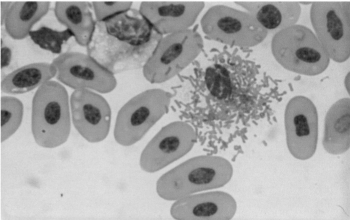

Figure 6.239 A ruptured heterophil from a brown python (*Liasis fuscus*) showing the round nucleus and bacilliform eosinophilic granules. The cell on the left is an azurophil.

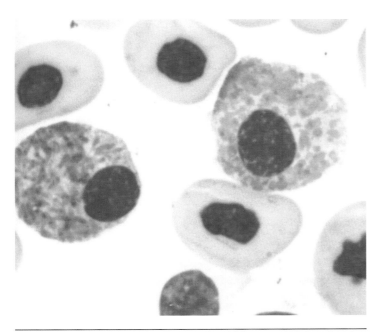

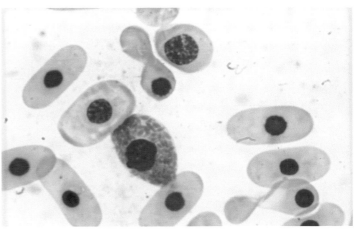

Figure 6.242 A heterophil from a Mississippi alligator (*Alligator mississippiensis*).

Figure 6.240 A heterophil and an eosinophil from an Aldabra giant tortoise (*Aldabrachelys gigantea*). The heterophil has irregular, bacilliform, magenta-coloured granules, while those of the eosinophil are round and pale red. Both cells have unlobed nuclei.

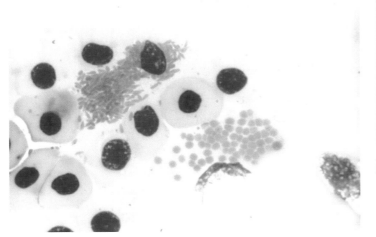

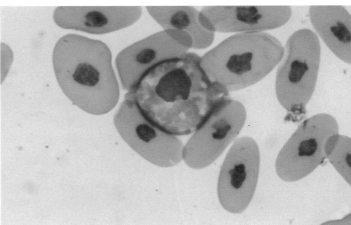

Figure 6.243 An eosinophil from a Chinese alligator (*Alligator sinensis*).

Figure 6.241 A ruptured heterophil and a ruptured eosinophil from an Indian peacock soft-shelled turtle (*Nilssonia hurum*) showing differences in granule shape and staining characteristics.

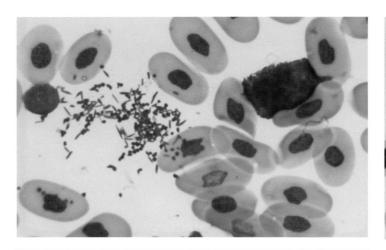

Figure 6.244 An intact and a ruptured basophil from a boa constrictor (*Boa constrictor*). Both spiculate and round granules are present.

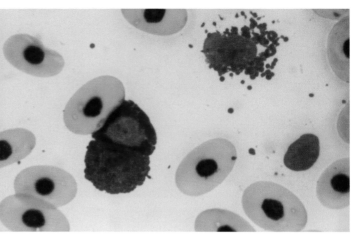

Figure 6.246 Three basophils from a healthy red-eared terrapin (*Trachemys scripta elegans*). The morphological differences in the two intact cells may reflect different stages in maturation. Alternatively, it is possible that mast cells occur normally in the peripheral blood of some terrapins. The basophil is a relatively abundant cell in this species.

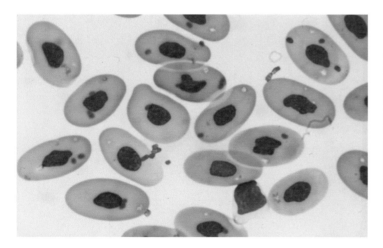

Figure 6.245 Blood smear from the boa constrictor from the previous image, showing granules from a ruptured basophil which, under some circumstances, might be mistaken for parasites, bacteria or red blood cell nuclear remnants.

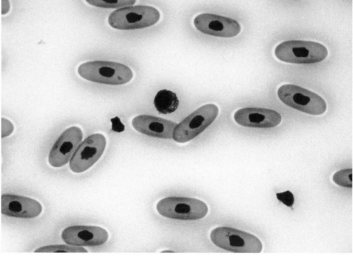

Figure 6.247 A basophil in a blood smear from a loggerhead sea turtle (*Caretta caretta*). In this species, the basophil appears to be the smallest granulocytes in common with other similar species (courtesy of David Perpiñan).

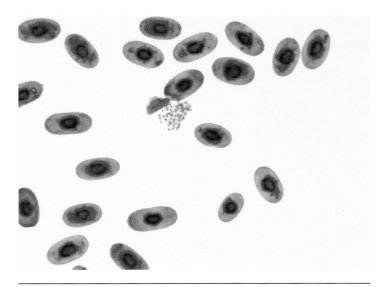

Figure 6.248 A normal eosinophil in a blood film from a common box turtle (*Terrapene carolina*) prepared using Diff Quick™. The granules are rounded and uniform in size. Small vacuoles are present in the cytoplasm (courtesy of David Perpiñan).

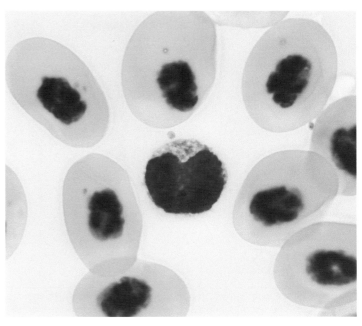

Figure 6.250 An eosinophil in a blood film from a clinically normal axolotl (*Ambystoma mexicanum*) (courtesy of Helen McCracken).

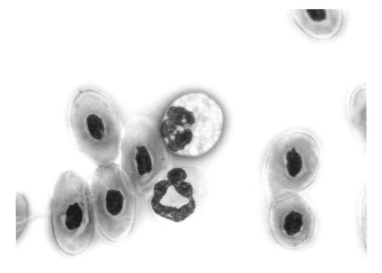

Figure 6.249 Photomicrograph of an eosinophil and a neutrophil in a blood smear from a gopher frog (*Lithobates capito*) (courtesy of Nicole Stacy).

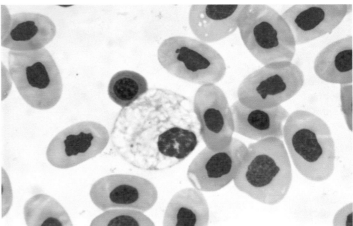

Figure 6.251 Blood film from a motorbike frog (*Litoria moorei*) showing an eosinophil with normal highly packed intracytoplasmic round eosinophilic granules (courtesy of Helen McCracken).

6.1.7 Common Artefacts Affecting Granulocytes

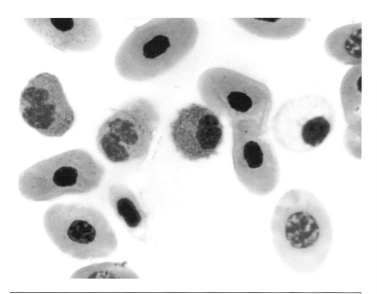

Figure 6.252 Blood smear from an Atlantic guitarfish (*Rhinobatos lentiginosus*). The microphotograph shows a basophil, fine eosinophilic granulocytes (heterophil equivalent), coarse eosinophilic granulocytes (eosinophil equivalent) and neutrophil (from left to right). One thrombocyte is also present (courtesy of Nicole Stacy).

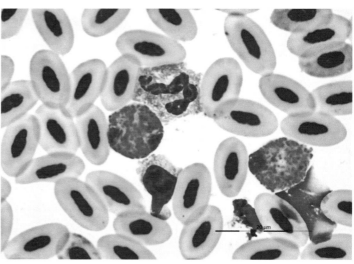

Figure 6.254 Two heterophils, one eosinophil, one lymphocyte in a blood film of a greylag goose (*Anser anser*). The eosinophils in Anseriformes have delicate rod-shaped granules, in contrast to the round granules in other species (courtesy of Helene Pendl).

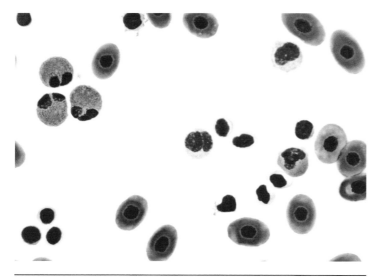

Figure 6.253 Blood film from a spotted eagle ray (*Aetobatus narinari*). The image shows a coarse eosinophilic granulocyte (upper left), a fine eosinophilic granulocyte (mid-right), thrombocytes, lymphocytes and two monocytes (courtesy of Nicole Stacy).

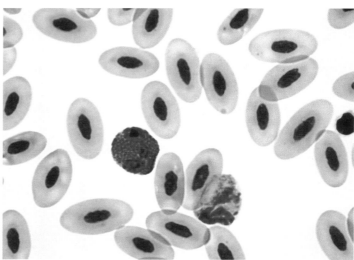

Figure 6.255 Blood smear from a common buzzard (*Buteo buteo*) showing an eosinophil to the left and a heterophil to the right. For unknown reasons members of the family Buteoidae may physiologically have large numbers of eosinophils, up to 30%. A correlation to clinical pathology could not be determined so far (courtesy of Helene Pendl).

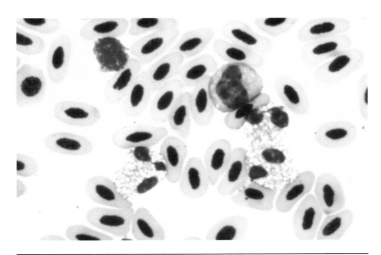

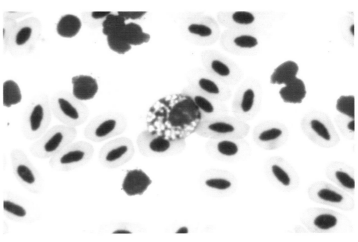

Figure 6.256 Three degranulated heterophils in a blood film from a sarus crane (*Grus antigone*). The cytoplasm contains irregular spaces surrounded by aggregations of eosinophilic material. This artefact is usually associated with inadequate fixation. The lobed nature of the heterophil nuclei is clearly evident. A disrupted red blood cell nucleus and a normal monocyte are also present.

Figure 6.258 A partially degranulated basophil from a sarus crane (*Grus antigone*). Several disrupted red blood cell nuclei are present.

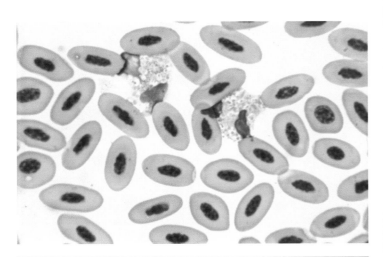

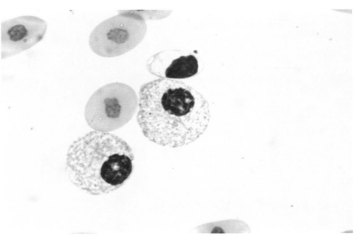

Figure 6.257 Two degranulated heterophils in a poorly fixed blood film from an emu (*Dromaius novaehollandiae*). The cytoplasm, of one heterophil contains some bacilliform granules, a number of round vacuoles containing small round eosinophilic granules and areas of amorphous eosinophilic material. The second heterophil shows a more advanced degree of deterioration in that bacilliform granules are absent and the nucleus is degenerated.

Figure 6.259 Two degranulated heterophils and a normal thrombocyte from a Nile crocodile (*Crocodylus niloticus*). The unlobed nature of the heterophil nuclei is clearly evident. The nuclei of the red blood cells are poorly stained.

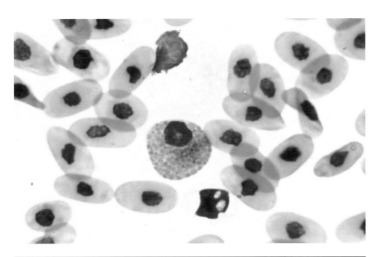

Figure 6.260 A disintegrated degranulated heterophil, a normal eosinophil and a normal thrombocyte from the crocodile in the previous image.

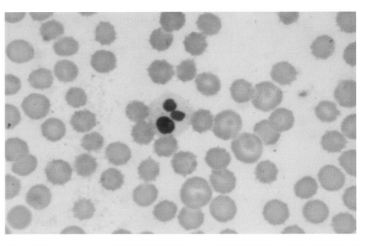

Figure 6.262 Necrotic white blood cells, probably neutrophils, in a domestic dog (*Canis lupus familiaris*) blood. The sample had been stored in EDTA for 24 hours.

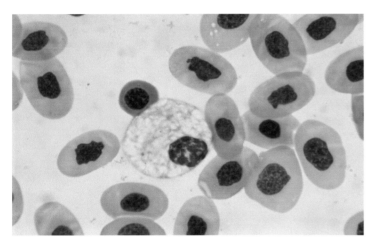

Figure 6.261 A degranulated heterophil from a Gila monster (*Heloderma suspectum*). The adjacent small cell is a microcytic polychromatic normoblast.

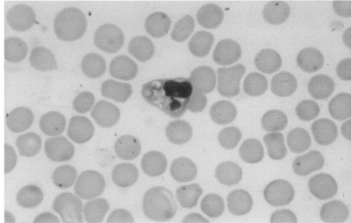

Figure 6.263 Necrotic white blood cells, probably neutrophils, in a domestic dog (*Canis lupus familiaris*) blood. The sample had been stored in EDTA for 24 hours.

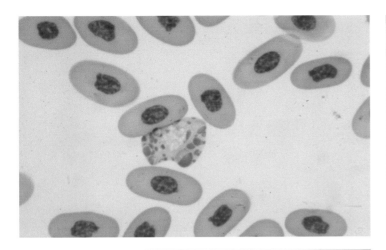

Figure 6.264 A necrotic white blood cell in a stored blood sample from a Chinese alligator (*Alligator sinensis*).

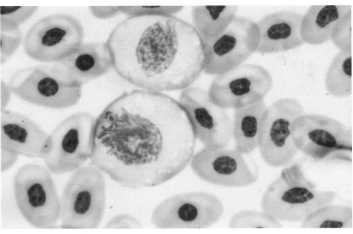

Figure 6.266 Two cells of unknown origin contaminating a 'blood sample' collected at *post-mortem* examination from a Mozambique spitting cobra (*Naja mossambica*). These are not blood cells.

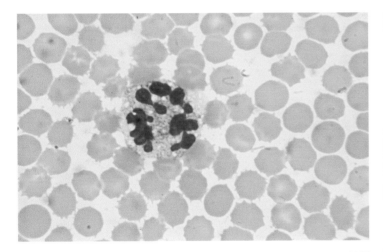

Figure 6.265 A structure resembling a giant neutrophil from a chimpanzee (*Pan troglodytes*) blood sample which had been in transit for several days.

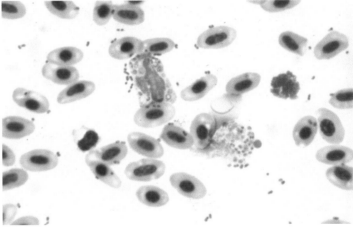

Figure 6.267 Postal sample from a sulphur-crested cockatoo (*Cacatua galerita*), contaminated with *Escherichia coli*, visible as extracellular basophilic bodies.

6.1.8 Pathological Responses Involving the Granulocytes

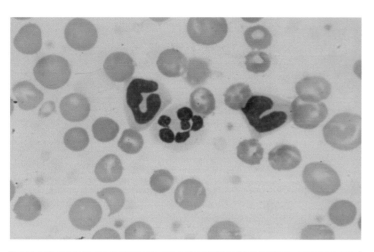

Figure 6.268 A normal neutrophil and two band neutrophils from a grey wolf (*Canis lupus*) with infected skin lesions and marked neutrophilia (neutrophils 62.8 × 10⁹/l). The finding of neutrophils with reduced nuclear lobulation can be described as left shift. Except in mammals which show Pelger-Huët phenomenon and in reptilian species which normally have heterophils with round nuclei, a left shift is a reliable indication of increase cell production, usually in response to infections and other inflammatory conditions. This wolf was also severely anaemic and the red blood cells show marked anisocytosis, polychromasia, stomatocytosis and microspherocytosis.

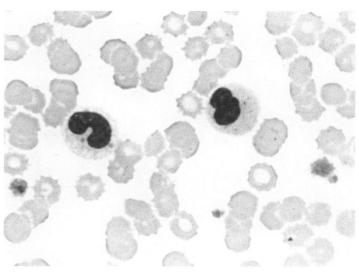

Figure 6.270 Two band neutrophils from a domestic dog (*Canis lupus familiaris*) with acute pancreatitis. The cytoplasm of the cell on the right contains toxic granules and that of the cell on the left contains two Döhle bodies. The toxic granules indicate a disturbance in granule formation, resulting in decreased intracellular lysosomal activity. Döhle (Amato) bodies are small blue areas in the cytoplasmic reticulum containing RNA. They are thought to represent focal failure of cytoplasmic maturation.

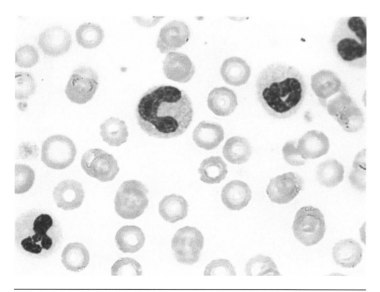

Figure 6.269 Neutrophils in a domestic dog (*Canis lupus familiaris*) with peritonitis. A normal neutrophil and three slightly larger neutrophils showing a left shift are present. The red blood cells show anisocytosis and hypochromia.

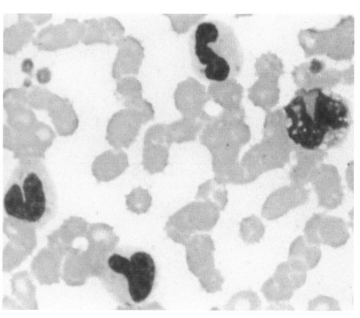

Figure 6.271 Band neutrophils and a vacuolated monocyte from a domestic dog (*Canis lupus familiaris*) with bacterial endocarditis. The red blood cells show increased rouleaux formation.

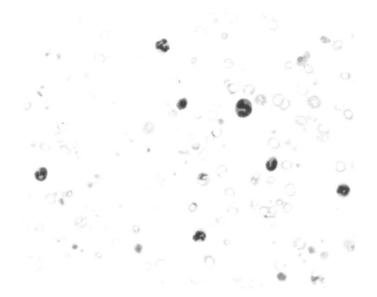

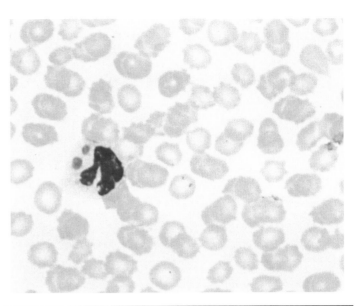

Figure 6.272 Neutrophils with a left shift in a domestic dog (*Canis lupus familiaris*) with a Coomb's positive anaemia (×40). The red blood cells show marked anisocytosis and hypochromia. One normoblast (erythroblast) is present and the platelet count appears to be increased. Granulocytosis and thrombocytosis are not unusual in regenerative anaemias, caused by generalised marrow stimulation.

Figure 6.274 Two eosinophilic inclusion bodies in the cytoplasm of a neutrophil from a domestic dog (*Canis lupus familiaris*) with distemper. These inclusions, which consist of viral nucleocapsid tubules, can occur also in red blood cells, lymphocytes and monocytes in case of canine distemper.

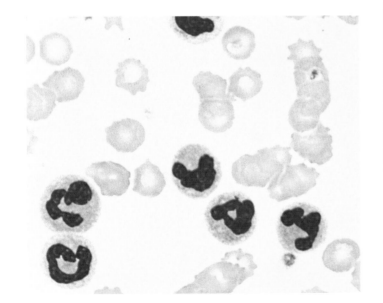

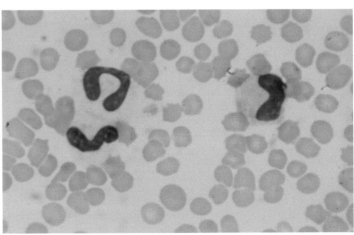

Figure 6.275 Three band neutrophils from a lion (*Panthera leo*) with slight neutrophilia associated with hepatitis (neutrophils 14.4 × 10⁹/l).

Figure 6.273 Neutrophilia in a domestic dog (*Canis lupus familiaris*) with inflammatory disease. The cells do not show a left shift (mature neutrophilia). There is increased red cell rouleaux formation and the platelet count appears to be reduced.

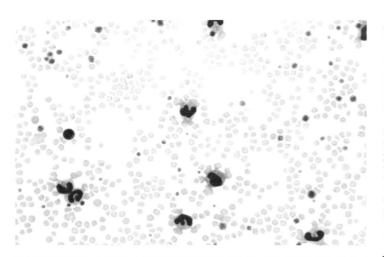

Figure 6.276 Neutrophils showing a left shift from a domestic cat (*Felis catus*) with pyothorax (×40).

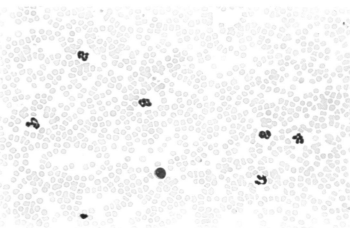

Figure 6.278 Mature neutrophilia in a domestic cat (*Felis catus*) with a bite abscess (×40).

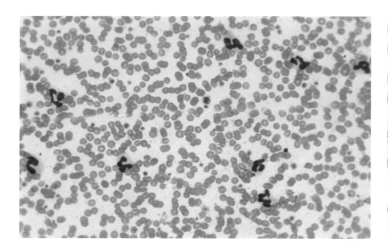

Figure 6.277 Mature neutrophilia in a domestic cat (*Felis catus*) with gastritis (×40, neutrophils 23.7 × 10^9/l).

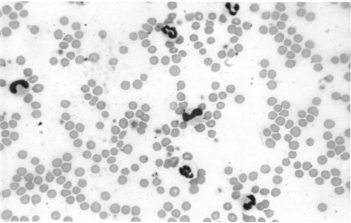

Figure 6.279 Neutrophilia in a cheetah (*Acinonyx jubatus*) with acute peritonitis (×40, neutrophils 27.3 × 10^9/l). Two band neutrophils and four normal neutrophils are shown.

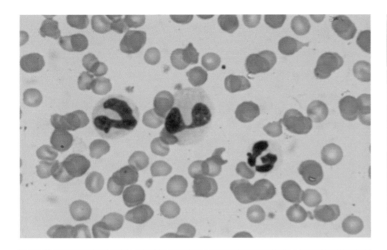

Figure 6.280   A higher magnification of a normal neutrophil and two neutrophils showing a left shift from the cheetah in the previous image.

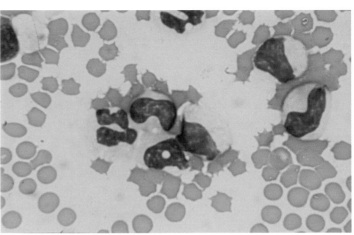

Figure 6.282   Neutrophilia in a cheetah (*Acinonyx jubatus*) with acute *Vaccinia variola* (cowpox) infection (neutrophils 44.3 × 10⁹/l). The cells have vacuolated cytoplasm and show a marked left shift. One cell with a ring-form nucleus is present.

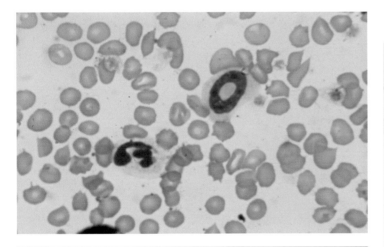

Figure 6.281   A normal neutrophil and a neutrophil with a ring-form nucleus from the cheetah in the previous image. Neutrophils with ring-form nuclei are characteristic of inflammatory diseases in carnivores but occur normally in some rodents.

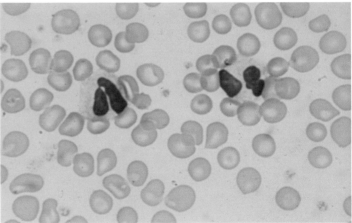

Figure 6.283   Two neutrophils showing a left shift from a ring-tailed coati (*Nasua nasua*) with a chronic bacterial infection (neutrophils 17.1 × 10⁹/l).

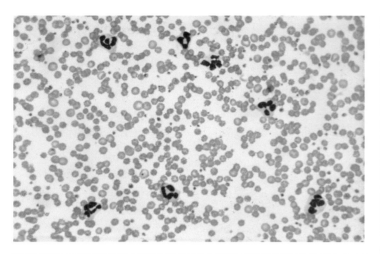

Figure 6.284 Mature neutrophils in a crab-eating mongoose (*Herpestes urva*) with a tooth root abscess (×40, neutrophils 19.2 × 10⁹/l).

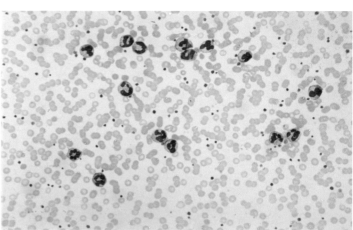

Figure 6.286 Mature neutrophilia in a Barbary macaque (*Macaca sylvanus*), with tracheitis (×40), neutrophils 33.2 × 10⁹/l). This animal also had an abnormally high platelet count.

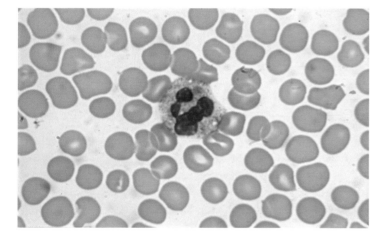

Figure 6.285 A neutrophil from a chimpanzee (*Pan troglodytes*) with an infected digit. The neutrophil count in this animal was normal (5.3 × 10⁹/l) but the presence of infection was suggested by the observation of Döhle bodies and vacuoles on the cells.

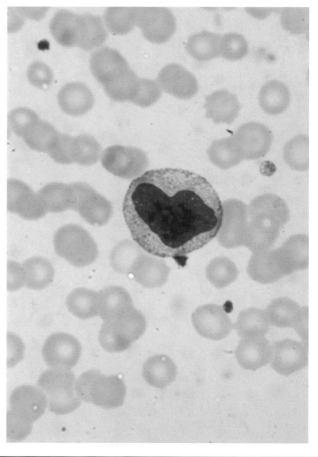

Figure 6.287 A neutrophil metamyelocyte from the macaque in the previous image.

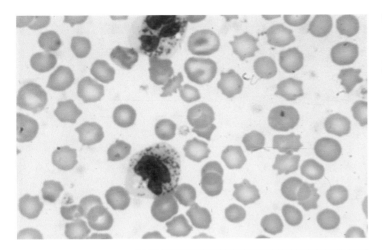

Figure 6.288 A neutrophil showing toxic granulation and vacuolation from a red slender loris (*Loris tardigradus*) with a chronic facial abscess (neutrophils 9.4 × 10⁹/l).

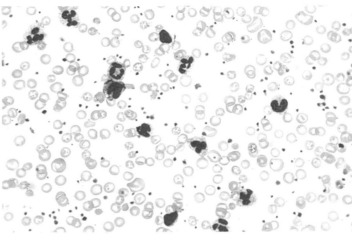

Figure 6.290 Neutrophilia in a three-striped night monkey (*Aotus trivirgatus*) with infected, haemorrhagic skin lesions (×40, neutrophils 23.6 × 10⁹/l). Hypochromic anaemia and thrombocytosis are also shown.

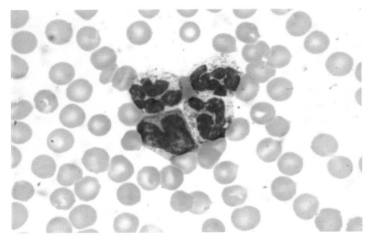

Figure 6.289 Neutrophilia in a common squirrel monkey (*Saimiri sciureus*) with *Yersinia pseudotuberculosis* infection (neutrophils 37.1 × 10⁹/l). Three neutrophils and a monocyte are shown.

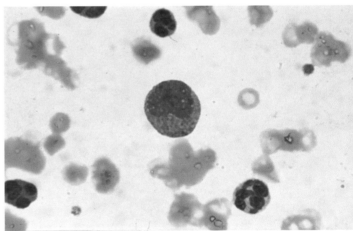

Figure 6.291 A neutrophil myelocyte from the monkey in the previous image.

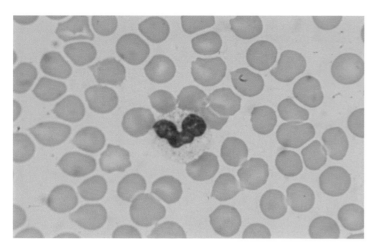

Figure 6.292 A neutrophil showing a left shift from a Californian sea lion (*Zalophus californianus*) with meningitis (neutrophils 16.9 × 10⁹/l). There are several small (Döhle) bodies in the cytoplasm.

Figure 6.294 Band neutrophils from a cow (*Bos taurus*) with pyelonephritis, pyuria and haematuria (×40).

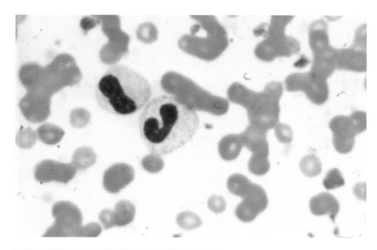

Figure 6.293 Two band neutrophils from a collared peccary (*Pecari tajacu*) with infected fight wounds (neutrophils 14.5 × 10⁹/l). The red blood cells show increased rouleaux formation.

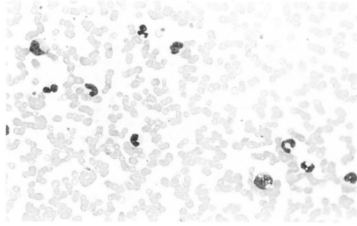

Figure 6.295 Neutrophilia and monocytosis in a cow (*Bos taurus*) with acute mastitis (×40).

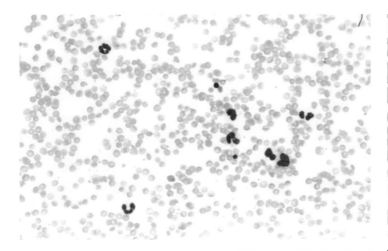

Figure 6.296 Neutrophilia with a left shift secondary to traumatic reticulitis in a cow (*Bos taurus*), which was anaemic.

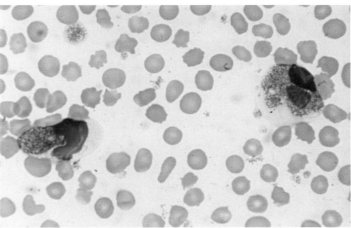

Figure 6.298 Two neutrophils showing toxic granulation and cytoplasmic streaming from a nilgai (*Boselaphus tragocamelus*) with colitis (neutrophils 20.6 × 10⁹/l).

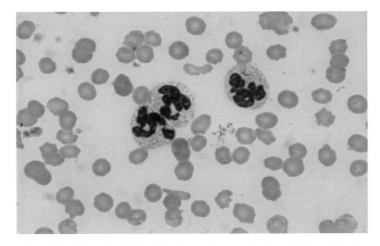

Figure 6.297 Mature neutrophilia in a bighorn sheep (*Ovis canadensis*) with chronic infection (neutrophils 31.9 × 10⁹/l) The cells show an increased number of nuclear lobes (right shift).

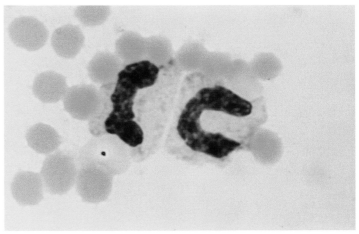

Figure 6.299 Two band neutrophils with basophilic cytoplasm from a horse (*Equus caballus*) with inflammatory disease.

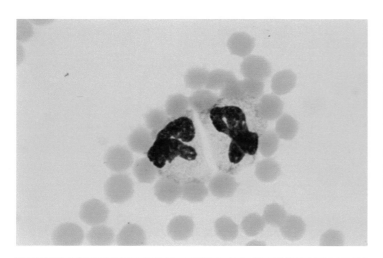

Figure 6.300 Two overlapping band cells with basophilic cytoplasm from a horse (*Equus caballus*) with inflammatory disease.

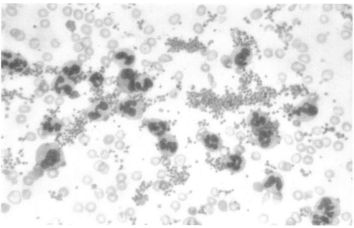

Figure 6.302 Neutrophilia in a Guaira spiny rat (*Proechimys guairae*) with infected fight wounds (×40, neutrophils 19.8 × 10⁹/l). This animal also had regenerative hypochromic anaemia and thrombocytosis.

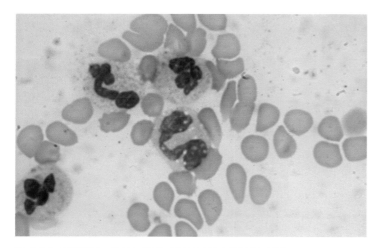

Figure 6.301 Neutrophilia in a Jamaican coney (*Geocapromys brownii*) with chronic corneal ulceration (neutrophils 19.6 × 10⁹/l). There is no apparent left shift but the cell in the bottom left hand corner of the field contains several Döhle bodies. A monocyte with cytoplasmic vacuoles is also present.

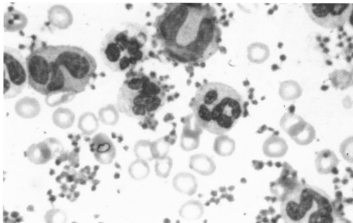

Figure 6.303 A higher magnification of three mature neutrophils and two neutrophils with a left shift from the Guaira spiny rat (*Proechimys guairae*) from the previous image.

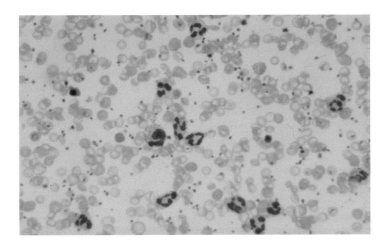

Figure 6.304 Neutrophilia in a plains vizcacha (*Lagostomus maximus*) with inflammatory disease (×40, neutrophils 36.9 × 10^9/l). This animal had severe hypochromic anaemia and thrombocytosis.

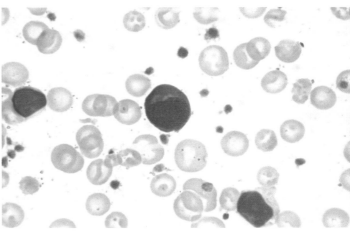

Figure 6.306 A myeloblast from the vizcacha in the previous image.

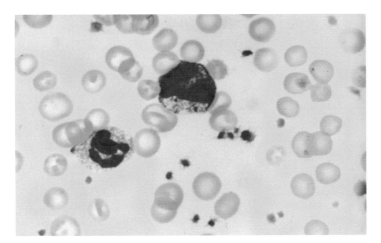

Figure 6.305 A higher magnification of a mature neutrophil and a band neutrophil with toxic granulation from the vizcacha in the previous image. The red blood cells show anisocytosis and hypochromia.

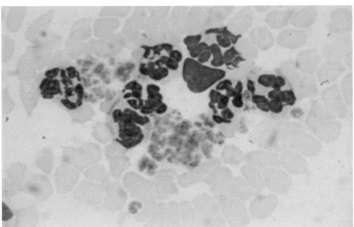

Figure 6.307 Neutrophils showing a right shift from a large tree shrew (*Tupaia tana*) with hypervitaminosis A.

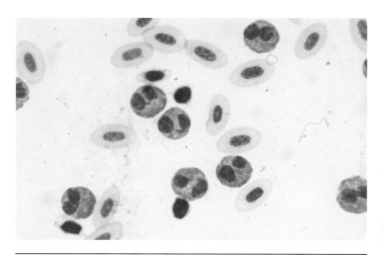

Figure 6.308 Heterophilia in blue and gold macaw (*Ara ararauna*) with a chronic respiratory infection (heterophils 41.3 × 10⁹/l). This bird was severely anaemic.

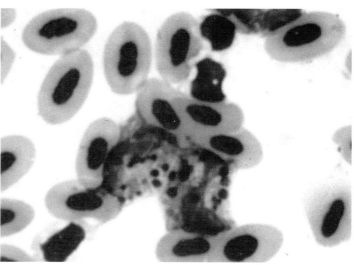

Figure 6.310 Two abnormal heterophils from a yellow fronted Amazon parrot (*Amazona ochrocephala*) with an upper respiratory tract infection (heterophils 44.1 × 10⁹/l). The cells contain irregular eosinophilic granules and an extra complement of round, strongly basophilic granules of varying sizes. These are considered to indicate immature or toxicity.

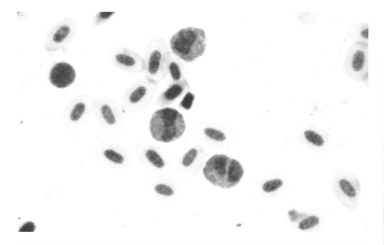

Figure 6.309 Two heterophils showing a left shift (unlobed nuclei) and a normal heterophil from the macaw in the previous image.

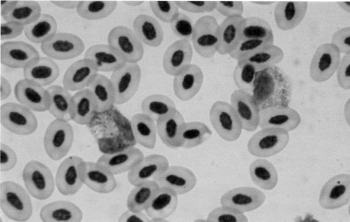

Figure 6.311 Toxic granules in the blood film from a Buffon's macaw (*Ara ambiguus*) with chronic viral hepatitis and peritonitis (heterophils 8.5 × 10⁹/l). Both heterophils have irregular cytoplasmic granules and one has an unlobed nucleus.

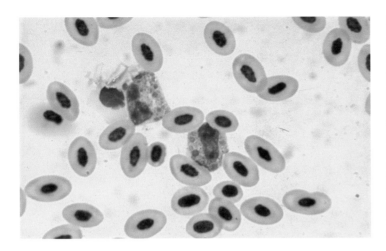

Figure 6.312 Two abnormal heterophils from the macaw in the previous image. Some of the cytoplasmic granules are round and comparatively large and have basophilic tinge.

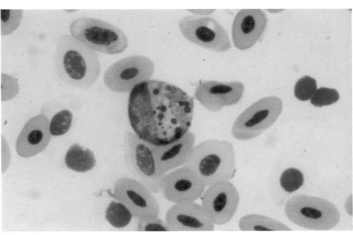

Figure 6.314 A heterophil myelocyte from the ostrich in the previous image.

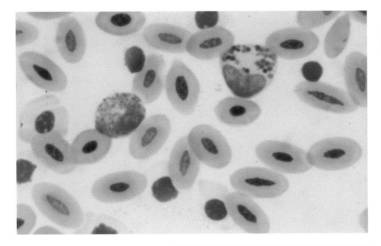

Figure 6.313 Abnormal heterophils from an ostrich (*Struthio camelus*) suffering from anorexia and weight loss (heterophils 3.7 × 10^9/l). The cause of the illness was not known but the appearance of the heterophils in the absence of an increased count suggested a degenerative white blood cell response to an infectious condition.

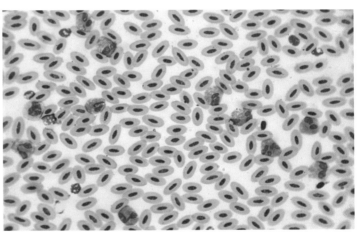

Figure 6.315 Heterophilia in a rosy flamingo (*Phoenicopterus roseus*) with cellulitis (×40, heterophilia 17.6 × 10^9/l).

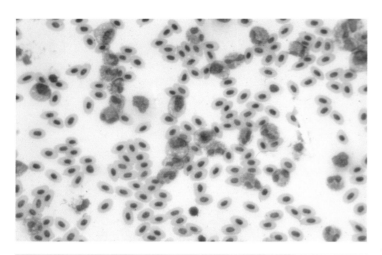

Figure 6.316 Heterophilia and monocytosis in a domestic fowl (*Gallus gallus domesticus*) with *Mycobacterium avium* infection (×40, heterophils 134.1 × 10^9/l).

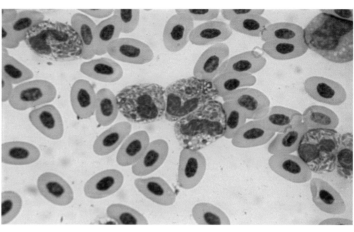

Figure 6.318 Heterophils from the bird in the previous image showing varying degrees of nuclear and granule abnormalities.

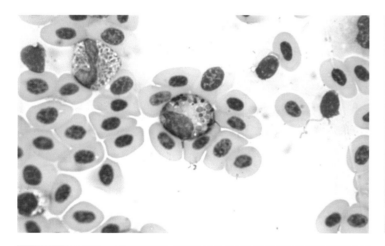

Figure 6.317 An immature heterophil with an unlobed nucleus, a reduced number of eosinophilic granules and a number of round basophilic granules from the bird in the previous image. A normal heterophil is also present.

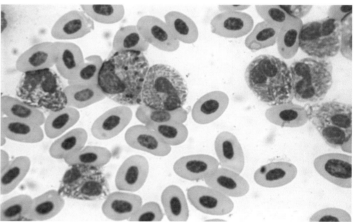

Figure 6.319 Heterophils from the bird in the previous image showing varying degrees of nuclear and granule abnormalities.

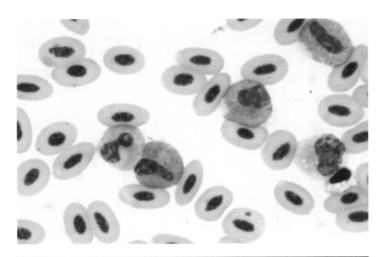

Figure 6.320 Heterophils showing a left shift and a normal eosinophil (left) from a hooded crane (*Grus monacha*) with *Mycobacterium avium* infection (heterophils 30.1 × 10⁹/l).

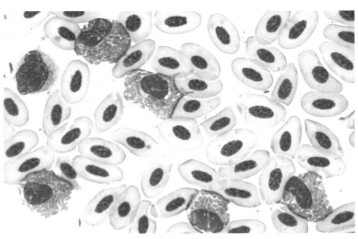

Figure 6.322 Three heterophils with unlobed nuclei from a sarus crane (*Grus antigone*) with *Mycobacterium avium* infection (heterophils 33.6 × 10⁹/l). An unidentified disintegrating mononuclear cell is also present.

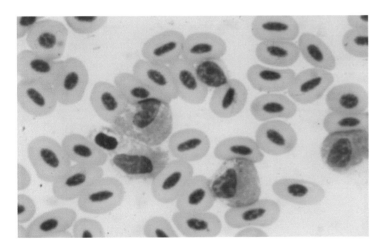

Figure 6.321 Two more band heterophils, a band eosinophil, a normal lymphocyte and a normal thrombocyte from the crane in the previous image.

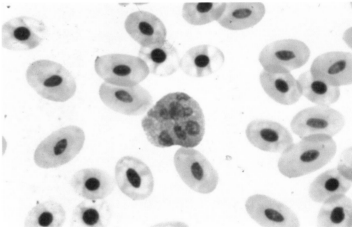

Figure 6.323 A giant heterophil with a hypersegmented nucleus from an African penguin (*Spheniscus demersus*) with aspergillosis. The bird had heterophilia (20.6 × 10⁹/l), monocytosis and normochromic anaemia.

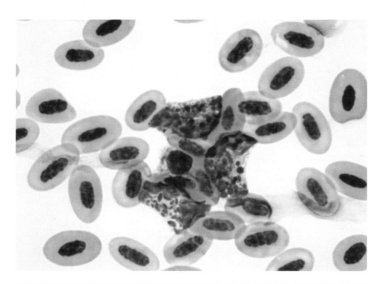

Figure 6.324 Blood film from a blue fronted amazon (*Amazona aestiva*) showing toxic left shift of heterophils. The two heterophils on top and right have a basophilic cytoplasm and besides the mature elliptic eosinophilic tertiary granules, basophilic large round secondary and dust-like dark primary granules. The number of mature granules exceeds the number of immature granules which classifies the two cells as late metamyelocytes. The erythrocytes have an irregular fading colouration of the cytoplasm, several cells are shorter and smaller than normal with a higher N/C ratio – microcytic hypochromic anaemia (courtesy of Helene Pendl).

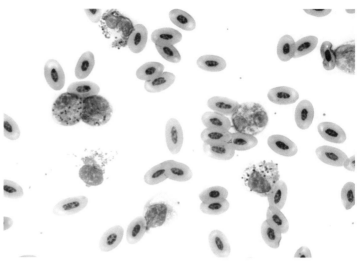

Figure 6.326 Photomicrograph of immature, mildly toxic heterophils and a reactive lymphocyte in a blood film from a West African crowned crane (*Balearica pavonina*) (courtesy of Nicole Stacy).

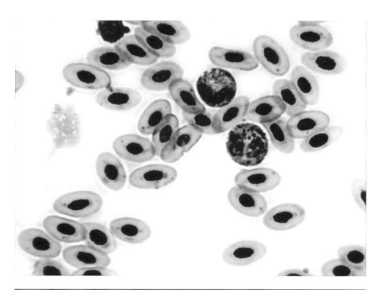

Figure 6.325 Toxic heterophils in blood smear from a common pheasant (*Phasianus colchicus*). The nucleus of the heterophil above shows marked loss of lobulation. The heterophil below show mature and immatures granules of different sizes (courtesy of Mary Pinborough).

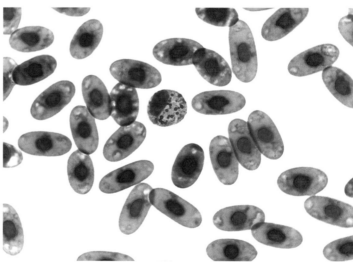

Figure 6.327 Metamyelocyte from a Greek or spur-thighed tortoise (*Testudo graeca*). Irregular colouration of the cytoplasm in basically all erythrocytes with dull cloudy to vacuolated pattern. 'Vacuoles' of different size, paleness and irregular forms, unclear boundaries (transitional changes from pale to non-pale). Multifocally small deeply basophilic spots, some of them in clusters of 3–4 associated with the vacuole like structures – dust artefact overlaying the cells rather unlikely as the background of the slide is completely devoid of dust particles or other non-vital structures. These all could be artefacts of the fixation and staining or drying procedure (courtesy of David Perpiñan).

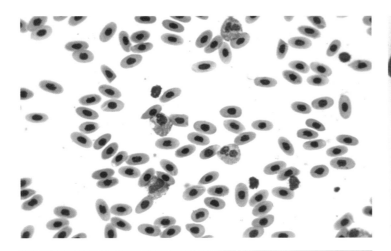

Figure 6.328 Heterophilia with left shift in a green iguana (*Iguana iguana*) with a fractured hind limb (×40) (heterophils 12.3 × 10⁹/l). The heterophils of this species normally have lobed nuclei.

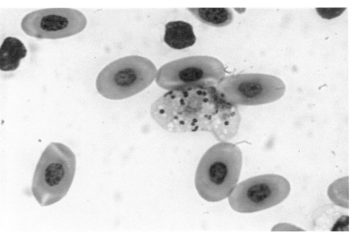

Figure 6.330 Abnormal heterophils from an Australian water dragon (*Itellagama lesueurii*) with osteomyelitis (heterophils 1.1 × 10⁹/l). In water dragons the heterophil nucleus is normally round.

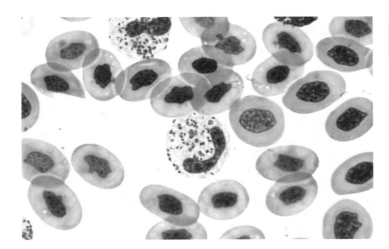

Figure 6.329 Heterophils with band-form nuclei and abnormal (toxic?) cytoplasmic granules from a blue-tongued skink (*Tiliqua scincoides*) with multiple abscesses (heterophils 3.1 × 10⁹/l). The heterophils of skinks normally have distinctively bilobed nuclei. Heterophilia is relatively uncommon in reptiles with inflammatory diseases and the presence of heterophils with morphological abnormalities is often the most useful indicator of infection in these animals.

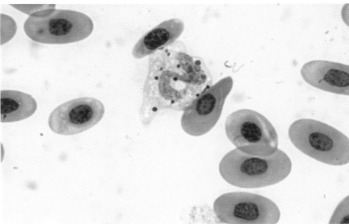

Figure 6.331 Abnormal heterophils from an Australian water dragon (*Itellagama lesueurii*) with osteomyelitis (heterophils 1.1 × 10⁹/l). In water dragons the heterophil nucleus is normally round.

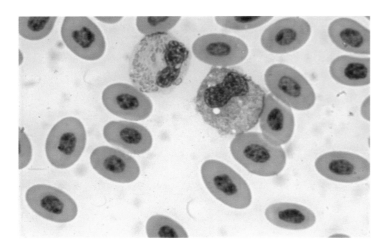

Figure 6.332 Abnormal heterophils from a Chinese water dragon (*Physignathus cocincinus*) with blepharitis (heterophils 9.2 × 10⁹/l). The nuclei are slightly lobed with clumping of the chromatin. The cytoplasm is grey and the granules are irregular in shape and reduced in number. One of the cells shows cytoplasmic vacuolations.

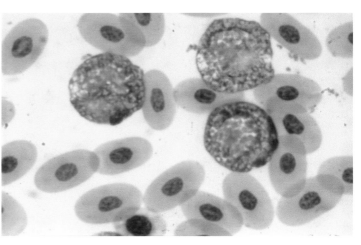

Figure 6.334 A higher magnification of three abnormal heterophils from the snake in the previous image. The cells are macrocytic and the unlobed nuclei have not taken up the stain.

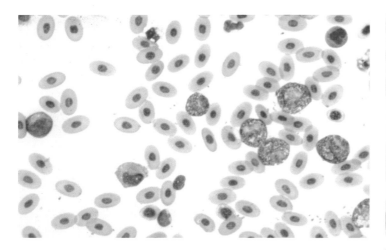

Figure 6.333 Heterophilia in an Eastern fox snake (*Pantherophis gloydi*) with severe spreading necrosis of the tail (heterophils 23.6 × 10⁹/l). This snake also had marked azurophilia.

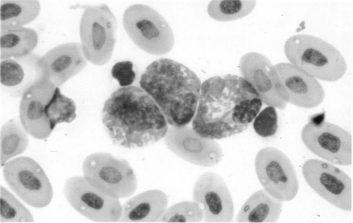

Figure 6.335 Three heterophils from a carpet python (*Morelia spilota*) with cellulitis, multiple abscesses and muscle necrosis (heterophils 3.4 × 10⁹/l). The cytoplasmic granules are irregular in shape and there is variation in stain uptake. This snake also had heterophilia.

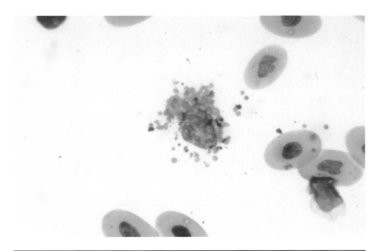

Figure 6.336 A ruptured heterophil from the snake in the previous image. Revealing the presence of abnormal granules.

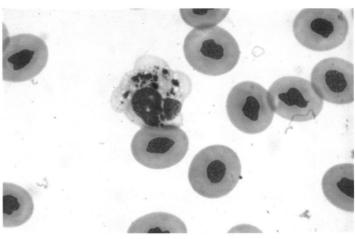

Figure 6.338 Abnormal heterophils from a Montpellier snake (*Malpolon monspessulanus*) with ascites and cloacal haemorrhage. This snake had a low heterophil count (heterophils 0.3 × 10^9/l), anaemia and azurophilia. The heterophils show abnormal nuclear lobulation (the heterophils of snakes normally have round nuclei), grey cytoplasm and toxic granulation.

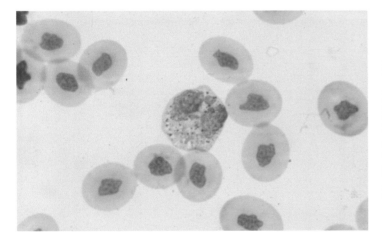

Figure 6.337 Abnormal heterophils from a Montpellier snake (*Malpolon monspessulanus*) with ascites and cloacal haemorrhage. This snake had a low heterophil count (heterophils 0.3 × 10^9/l), anaemia and azurophilia. The heterophils show abnormal nuclear lobulation (the heterophils of snakes normally have round nuclei), grey cytoplasm and toxic granulation.

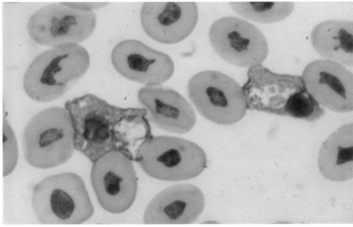

Figure 6.339 Abnormal heterophils from a hawksbill sea turtle (*Eretmochelys imbricata*) with suspected liver failure (heterophils 3.6 × 10^9/l). The nuclei of the red blood cells are distorted.

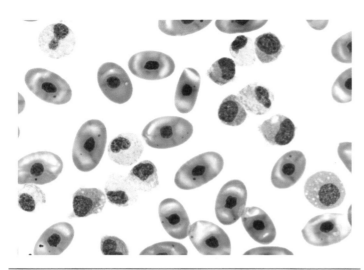

Figure 6.340 Blood smear from a Florida soft shell turtle (*Apalone ferox*) showing immature, mildly toxic heterophils and monocytes (courtesy of Nicole Stacy).

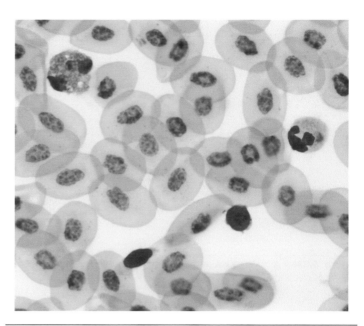

Figure 6.342 Low power microphotograph of the blood smear from a motorbike frog (*Litoria moorei*) showing an eosinophil, a neutrophil, a lymphocyte and a thrombocyte (courtesy of Helen McCracken).

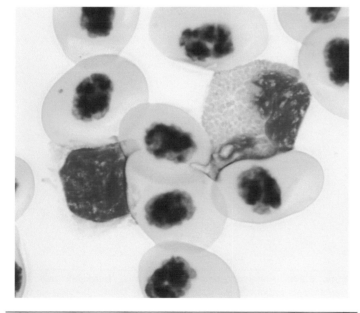

Figure 6.341 A blood smear from an axolotl (*Ambystoma mexicanum*) showing a suspected heterophil and an intact lymphocyte (courtesy of Helen McCracken).

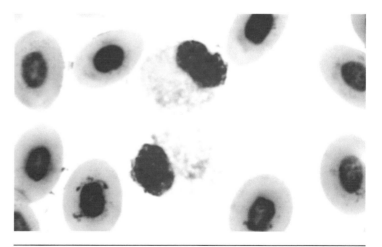

Figure 6.343 Microphotograph of the blood smear from a koi or nishikigoi (*Cyprinus carpio*), a coloured variety of the Amur carp. Two moderately toxic neutrophils can be observed in the preparation. Top neutrophil left-shifted, bottom neutrophil mature and mildly toxic (courtesy of Nicole Stacy).

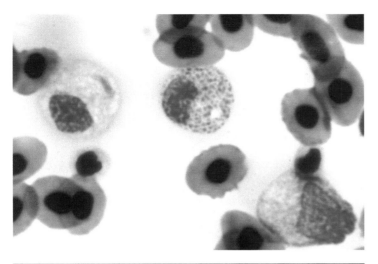

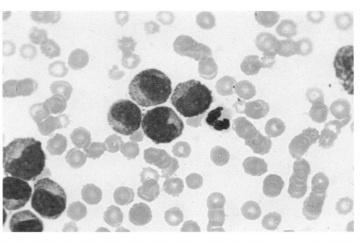

Figure 6.344 Blood smear from a porkfish (*Anisotremus virginicus*) showing, from left to right, a mature and slightly toxic neutrophil, an eosinophil and a monocyte (courtesy of Nicole Stacy).

Figure 6.346 Promyelocytes and a normal neutrophil from a domestic dog (*Canis lupus familiaris*) with chronic myeloid (granulocytic) leukaemia. The red blood cells show increased rouleaux formation and platelets are absent.

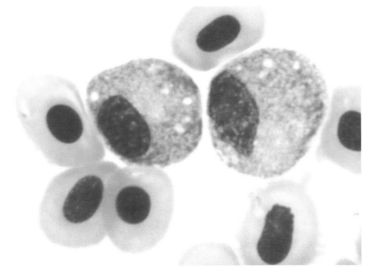

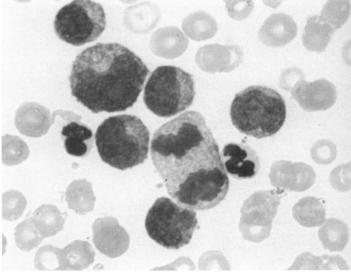

Figure 6.345 Two markedly toxic neutrophils in the blood film from a goldfish (*Carassius auratus*). The neutrophil on the left is mature with a degree of toxicity No 3. The neutrophil on the right is immature with a degree of toxicity No 3 (courtesy of Nicole Stacy).

Figure 6.347 Five myeloblasts, two normal neutrophils and a mitotic figure from the dog in the previous image.

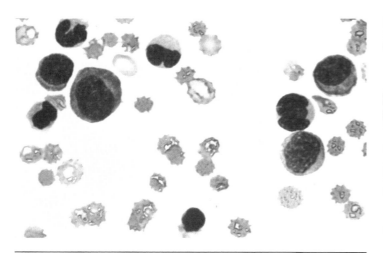

Figure 6.348 Promyelocytes and band neutrophils in a domestic dog (*Canis lupus familiaris*) with chronic myeloid leukaemia. The red blood cells show crenation and anisocytosis and platelets are absent.

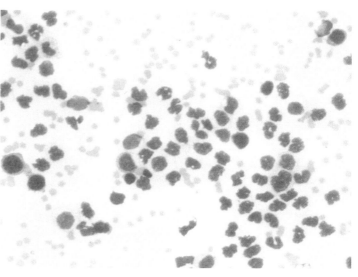

Figure 6.350 Chronic myeloid leukaemia in a domestic cat (*Felis catus*) (×40). This animal was strongly FeLV positive.

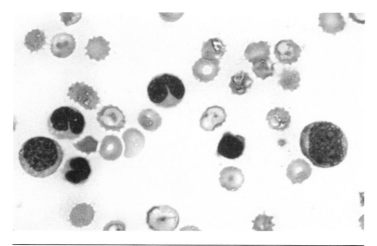

Figure 6.349 Myeloblasts and band neutrophil from the dog in the previous image.

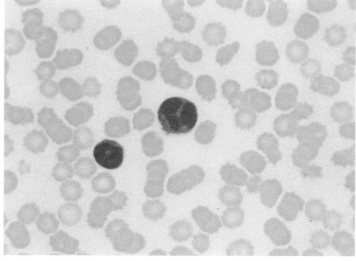

Figure 6.351 Chronic myeloid leukaemia in a domestic cat (*Felis catus*).

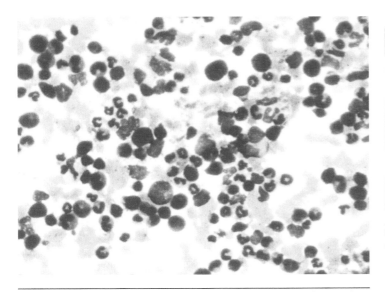

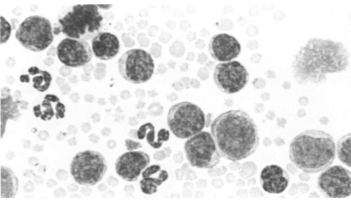

Figure 6.354   Blood cells from the dog from the previous image at higher magnification.

Figure 6.352   Bone marrow preparation from the cat in the previous image with chronic myeloid leukaemia.

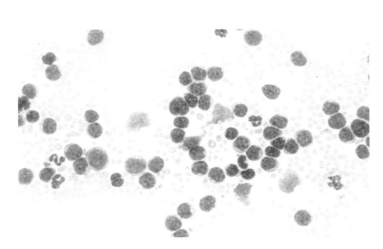

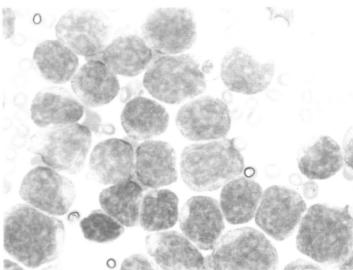

Figure 6.353   Chronic myeloid leukaemia in transition to acute myeloid leukaemia in a domestic dog (*Canis lupus familiaris*) (×40). Many myeloblasts and five normal neutrophils are present. There are also several disintegrated cells.

Figure 6.355   A pleiomorphic population of myeloblasts in the peripheral blood of domestic dog (*Canis lupus familiaris*) with acute myeloid leukaemia.

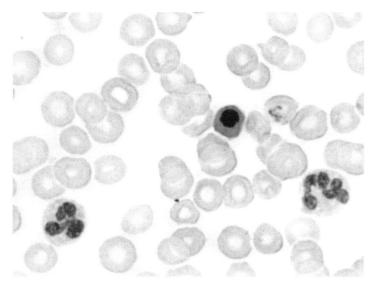

Figure 6.356 A normoblast (erythroblast) and abnormally segmented neutrophils in the blood of a domestic dog (*Canis lupus familiaris*) with myeloid dysplasia.

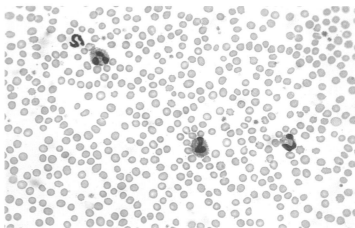

Figure 6.358 Eosinophilia in a cheetah (*Acinonyx jubatus*) (×40, eosinophils 4.1 × 10⁹/l). One normal neutrophil is present.

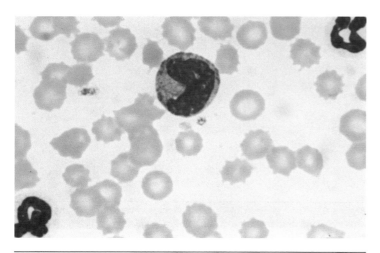

Figure 6.357 Two normal neutrophils and an abnormal metamyelocyte (?) in a domestic dog (*Canis lupus familiaris*) with testicular tumour (neutrophils 41.7 × 10⁹/l).

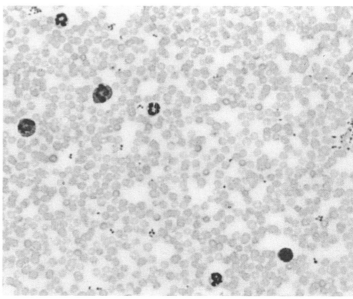

Figure 6.359 Eosinophilia in a domestic cat (*Felis catus*) with bronchial asthma (×40). The eosinophil nuclei are multilobed.

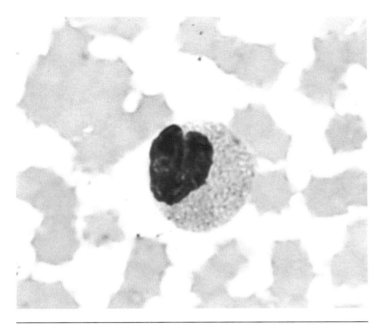

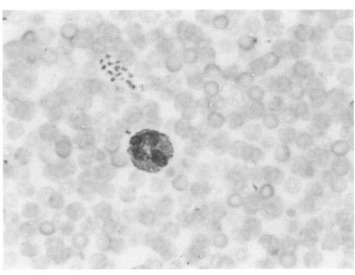

Figure 6.362   An eosinophil from a domestic dog (*Canis lupus familiaris*) with *Filaroides osleri* parasites.

Figure 6.360   A higher magnification of an eosinophil from the cat in the previous image. The cytoplasm appears to contain Döhle bodies.

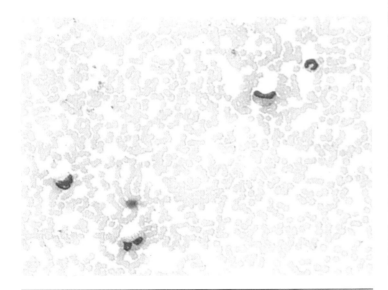

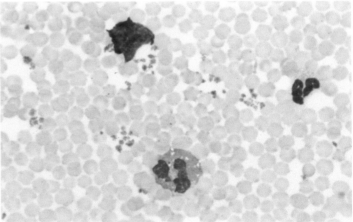

Figure 6.363   Eosinophilia and basophilia in a sheep (*Ovis aries*) with heavy *Haemonchus contortus* burden. There is also a normal neutrophil in the field.

Figure 6.361   Eosinophilia in a domestic cat (*Felis catus*) with clinically significant *Aelurostrongylus abstrusus* infestation.

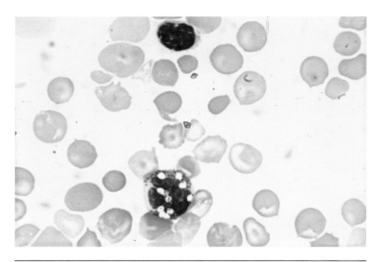

Figure 6.364 A vacuolated eosinophil from a ring-tailed coati (*Nasua nasua*) with haemolytic anaemia of unknown aetiology (eosinophils 0.01 × 10⁹/l).

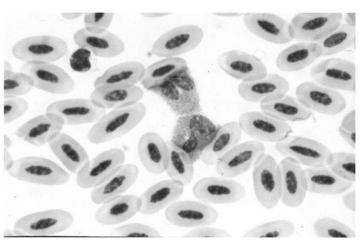

Figure 6.366 Eosinophilia in a lappet-faced vulture (*Torgos tracheliotos*) (eosinophils 1.2 × 10×⁹/l). This bird was clinically normal but was probably suffering from a sub-clinical parasite infestation.

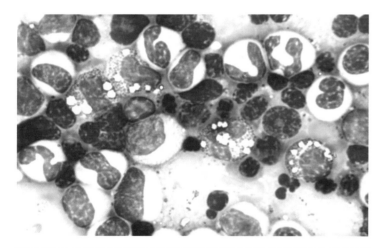

Figure 6.365 Myeloid hyperplasia and vacuolated eosinophil precursors in the bone marrow of the coati in the previous image.

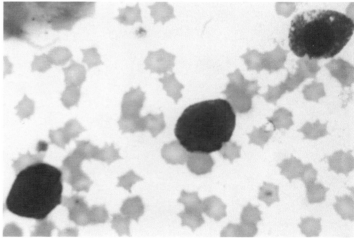

Figure 6.367 Mast cells in the peripheral blood of a domestic cat (*Felis catus*) with a solid mast cell tumour (mast cells 10.1 × 10⁹/l). In cats and dogs, mast cells in the peripheral blood can be distinguished from basophils by their unlobed nuclei and the very large number of basophilic granules. There were very few of these cells in the bone marrow of this cat but many in the spleen and liver.

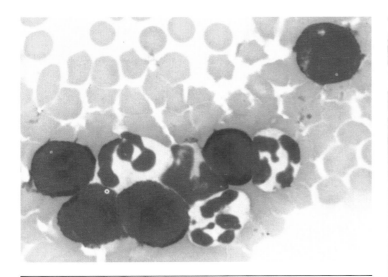

Figure 6.368 Mast cells in the 'tail' of blood film from the cat in the previous image.

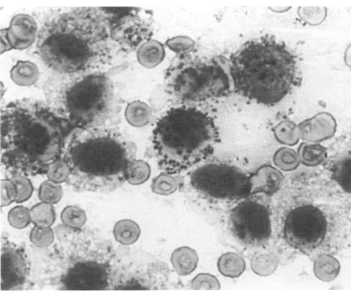

Figure 6.370 Mast cells in a buffy coat preparation from a domestic dog (*Canis lupus familiaris*) with a solid mast cell tumour.

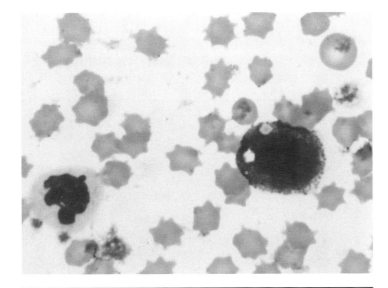

Figure 6.369 A vacuolated mast cell from the cat in the previous image.

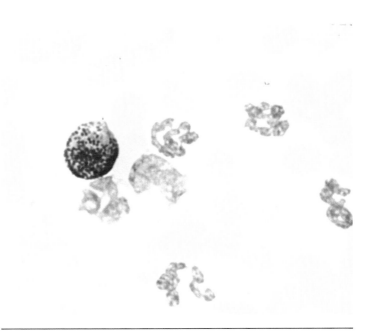

Figure 6.371 A mast cell and several unstained neutrophils in the blood of a domestic dog (*Canis lupus familiaris*) with a solid mast cell tumour (toluidine blue stain).

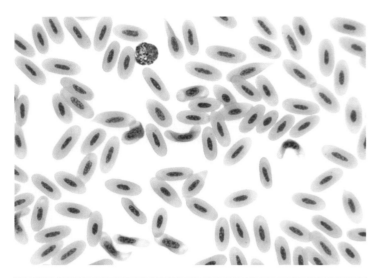

Figure 6.372 Photomicrograph of the blood smear of a bald eagle (*Haliaeetus leucocephalus*) showing a basophil and frequent misshapen erythrocytes, especially in polychromatophils. Some tear-drop-shaped erythrocytes are also present. The morphology of misshapen erythrocytes has been associated with lead toxicosis in trumpeter swans (*Cygnus buccinator*) and Canada geese (*Branta canadensis*) and prompted testing for lead level. The patient showed a lead level of 4.2 mg/l consistent with lead toxicosis (courtesy of Nicole Stacy).

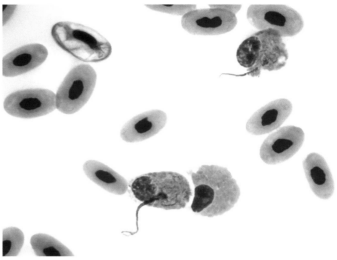

Figure 6.374 Photomicrograph of a blood smear from a gopher tortoise (*Gopherus polyphemus*) showing an erythrocyte with haemogregarine, two heterophils with whip-like projections and one eosinophil. The tortoise was injured with systemic inflammation (courtesy of Nicole Stacy).

6.1.9 Normal Variation in Lymphocytes Morphology

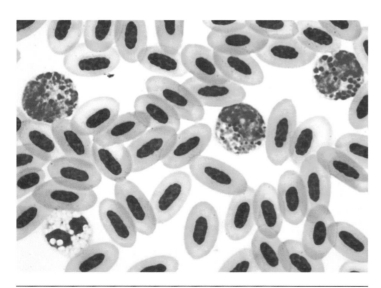

Figure 6.373 Blood smear from a red-fronted or Jardine's parrot (*Poicephalus gulielmi*). In the preparation it is possible to see an eosinophil with unstained granules, probably due to fixation deficiency, three heterophils with toxic left shift in various degrees (+1 cytoplasmic basophilia, +2 swelling of granules, +3 toxic, i.e. immature basophilic granulation and lack of nuclear lobulation, sign of left shift, hypochromasia of erythrocytes) (courtesy of Helene Pendl).

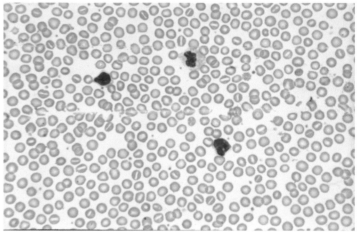

Figure 6.375 Two normal lymphocytes and a monocyte from a juvenile olive baboon (*Papio anubis*) (×40).

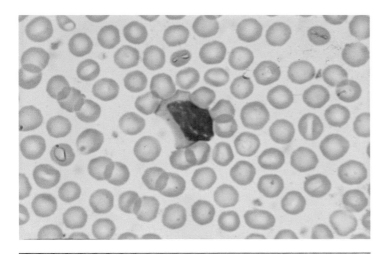

Figure 6.376 A higher magnification of a normal large lymphocyte from the baboon in the previous image. The cell is deformed by the surrounding cells.

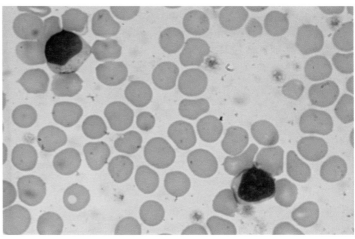

Figure 6.378 Two normal lymphocytes from a healthy juvenile white-headed saki monkey (*Pithecia pithecia*).

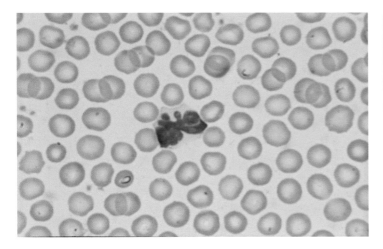

Figure 6.377 An atypical lymphocyte from the baboon in the previous image.

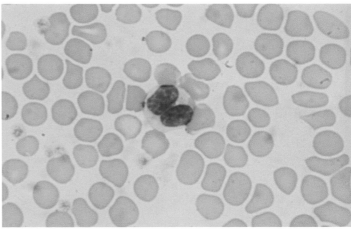

Figure 6.379 A binucleated lymphocyte from the monkey in the previous image. These occur occasionally in the blood of normal primates.

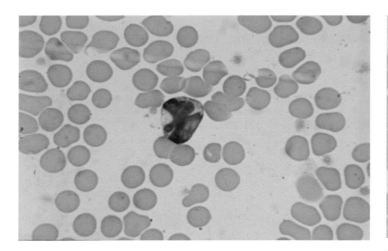

Figure 6.380 An atypical lymphocyte from a healthy adult brown capuchin monkey (*Cebus apella*).

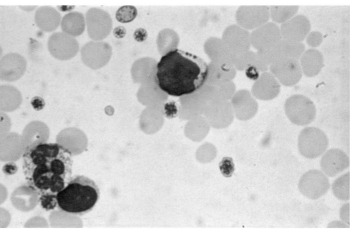

Figure 6.382 Two normal lymphocytes and a normal neutrophil from a three-striped night monkey (*Aotus trivirgatus*). The cytoplasm of one of the lymphocytes contains distinctive basophilic granules at the periphery.

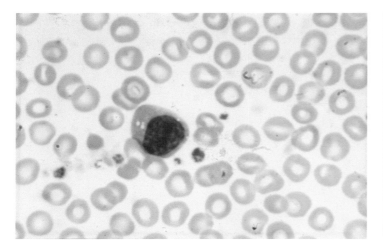

Figure 6.381 A lymphocyte with small cytoplasmic vacuoles from a healthy orangutan (*Pongo pygmaeus*).

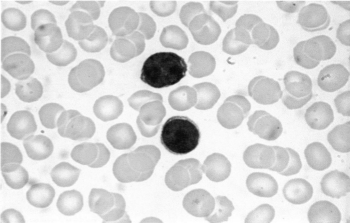

Figure 6.383 Two normal lymphocytes from a red-necked wallaby (*Macropus rufogriseus*). In this species, the nuclear chromatin typically forms dark-staining blocks and the amount of cytoplasm present is often small.

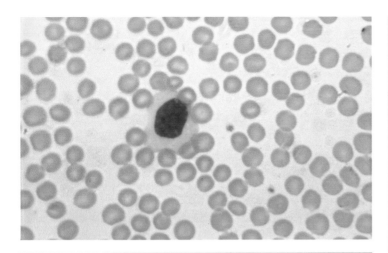

Figure 6.384 A large lymphocyte from a healthy masked palm civet (*Paguma larvata*).

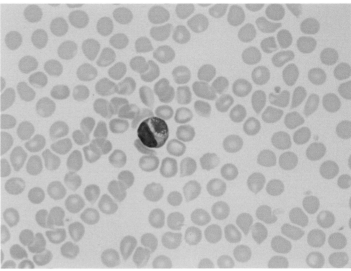

Figure 6.386 A Foà-Kurloff body in the blood smear from a guinea pig (*Cavia porcellus*). Foà-Kurloff bodies are single, large cytoplasmic inclusions which occur under normal conditions in a small proportion of the lymphocytes of this species (courtesy of Nico Schoemaker and Yvonne R A van Zeeland).

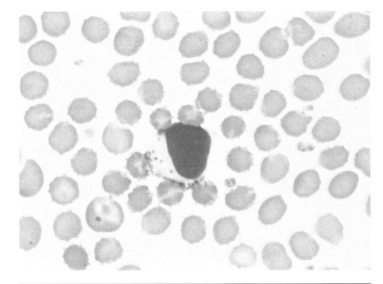

Figure 6.385 A lymphocyte with azurophilic cytoplasmic granules from a healthy black rat (*Rattus rattus*).

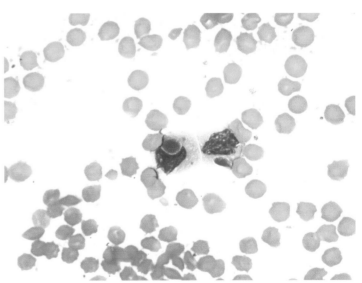

Figure 6.387 Blood smear from a capybara (*Hydrochoerus hydrochaeris*) showing a Foà-Kurloff body. This is the only species, other than guinea pigs, in which these inclusions have been found (courtesy of Jean-Michel Hatt).

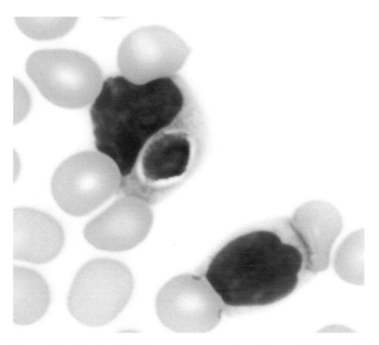

Figure 6.388 Microphotograph of a blood smear of a capybara (*Hydrochoerus hydrochaeris*) showing a magnified view of a Foà-Kurloff body. A normal lymphocyte can also be seen to the right (courtesy of Nicole Stacy).

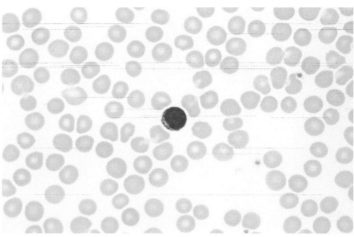

Figure 6.390 Blood smear from a clinically normal ferret (*Mustela putorius furo*) showing a typical lymphocyte for this species. Lymphocytes are the smallest of the mononuclear cells in most animal species (courtesy of Nico Schoemaker and Yvonne R A van Zeeland).

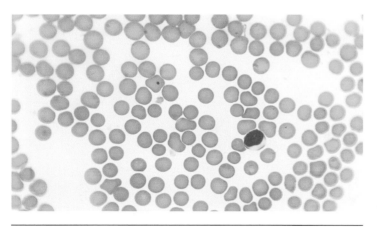

Figure 6.389 A normal lymphocyte in the blood smear of a bottlenose dolphin (*Tursiops truncatus*). A Howell-Jolly body is also present in a red blood cell (courtesy of Guillermo Sanchez).

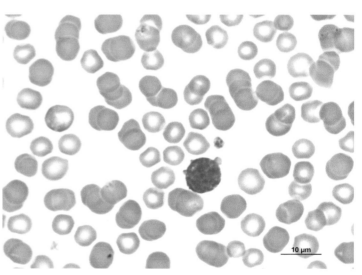

Figure 6.391 A lymphocyte in the blood film of a European hedgehog (*Erinaceus europaeus*). The photograph also shows some polychromatic red blood cells and some platelets (courtesy of Jaume Martorell).

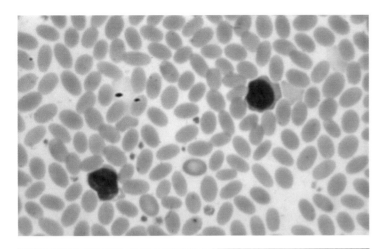

Figure 6.392 Two lymphocytes from a healthy Arabian camel (*Camelus dromedarius*).

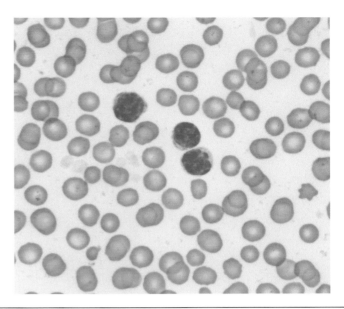

Figure 6.394 Three normal lymphocytes in a blood smear from a short-beaked echidna (*Tachyglossus aculeatus*) (courtesy of Helen McCracken).

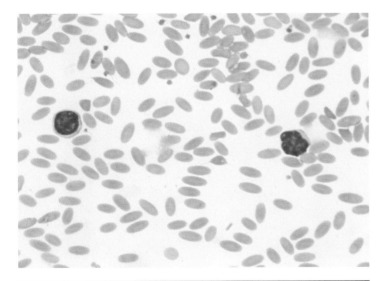

Figure 6.393 Microphotograph of a blood smear from a clinically normal captive alpaca (*Vicugna pacos*) showing two normal lymphocytes and numerous platelets (courtesy of Margarita Colburn).

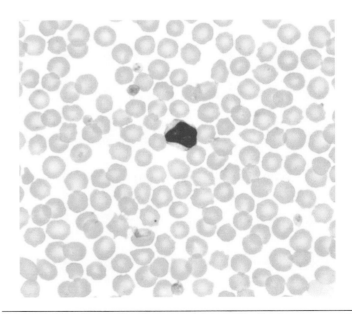

Figure 6.395 Blood film from a platypus (*Ornithorhynchus anaticus*) showing a single medium-sized lymphocyte (courtesy of Helen McCracken).

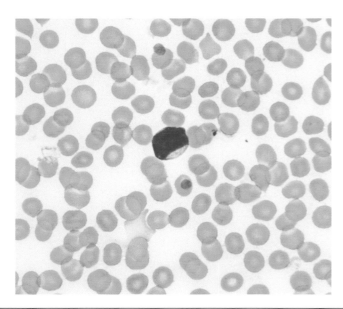

Figure 6.396 Normal lymphocyte in a blood smear from a common brushtail possum (*Trichosurus vulpecula vulpecula*) (courtesy of Helen McCracken).

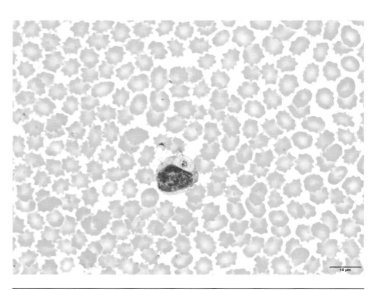

Figure 6.398 Lymphocyte with an inclusion body in a blood film from a short-beaked echidna (*Tachyglossus aculeatus*). All of the red blood cells are crenated, probably due to drying of the film at the time of preparation. Echidnas and the platypus are the only living mammals that lay eggs (courtesy of Karen Jackson).

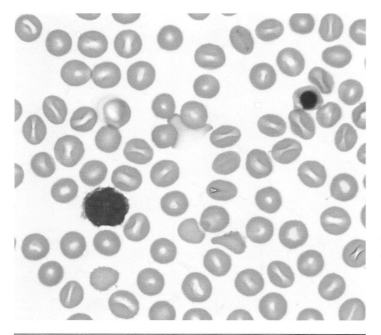

Figure 6.397 Blood smear from a Southern koala (*Phascolarctos cinereus victor*) showing a large lymphocyte. Koala lymphocytes have pleiomorphic appearance with small, medium and large cells observed. Nucleated red blood cells are a common finding in koala blood smears (courtesy of Helen McCracken).

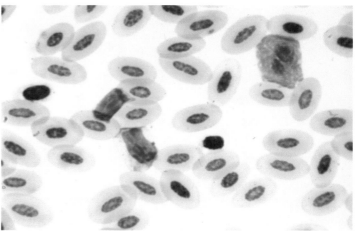

Figure 6.399 Two normal lymphocytes from a common moorhen (*Gallinula chloropus*). These show the typical appearance of avian lymphocytes, which often look deformed in stained blood films. Two thrombocytes with smaller, more condensed nuclei are also present.

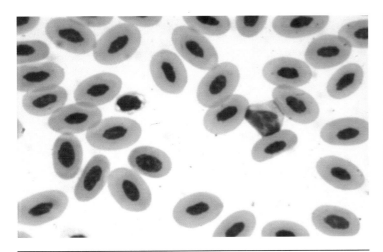

Figure 6.400   A lymphocyte and a normal thrombocyte from a Buffy fish owl (*Ketupa ketupu*).

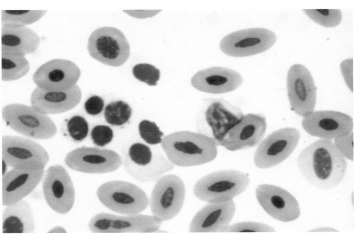

Figure 6.402   A normal lymphocyte, an aggregate of four thrombocytes showing various stages of activation, two nuclei from red blood cells which have lost their cytoplasm, a polychromatic erythroblast and a ruptured red blood cell from a healthy ostrich (*Struthio camelus*).

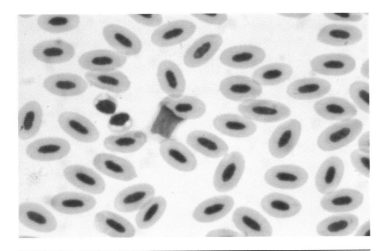

Figure 6.401   A normal lymphocyte and two thrombocytes from a common hill mynah (*Gracula religiosa*).

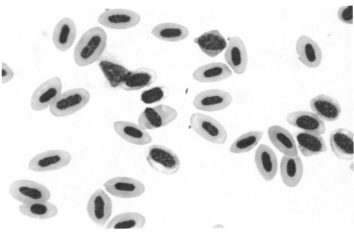

Figure 6.403   A lymphocyte, a thrombocyte (with cytoplasmic vacuoles) and several polychromatic red blood cells from a budgerigar (*Melopsittacus undulatus*).

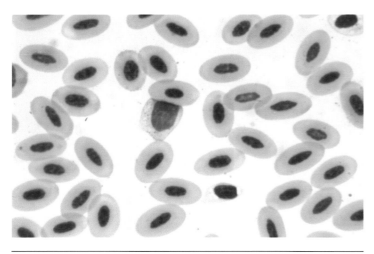

Figure 6.404 A normal lymphocyte from a rosy flamingo (*Phoenicopterus roseus*).

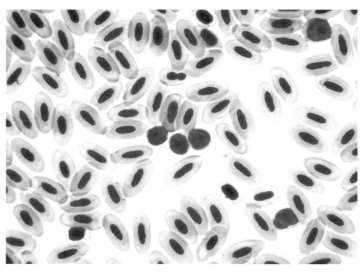

Figure 6.406 Lymphocytes in a blood film of an emperor goose (*Anser canagicus*) (courtesy of Michelle OBrien).

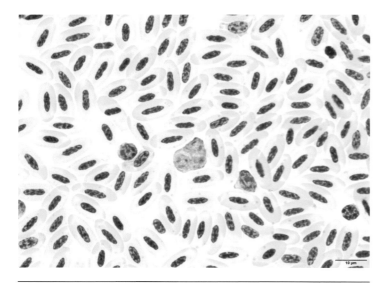

Figure 6.405 Blood smear from a monk parakeet or Quaker parrot (*Myiopsitta monachus*) showing three normal lymphocytes and a normal monocyte in the middle (courtesy of Karen Jackson).

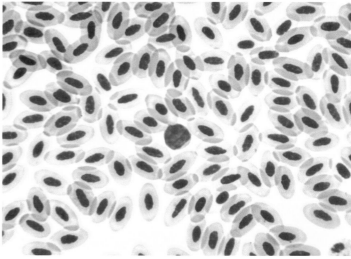

Figure 6.407 A single large lymphocyte in the blood smear of a captive Nene or Hawaiian goose (*Branta sandvicensis*) (courtesy of Michelle OBrien).

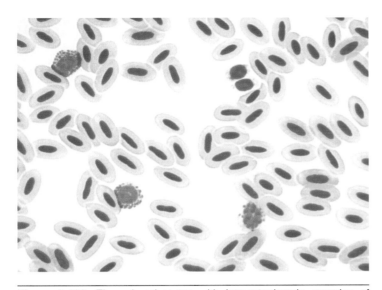

Figure 6.408 Three lymphocytes with intracytoplasmic granules of diverse diameter in a blood smear of a captive Laysan duck or Laysan teal (*Anas laysanensis*). Two large thrombocytes are present on the right upper corner. The presence of intracytoplasmic granules in lymphocytes appears to be a common occurrence in this species (courtesy of Michelle OBrien).

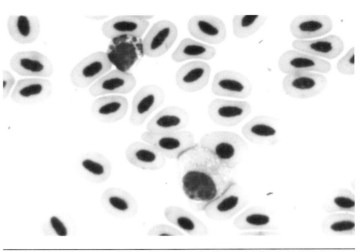

Figure 6.410 A small lymphocyte with cytoplasmic granules and a monocyte from a sarus crane (*Grus antigone*).

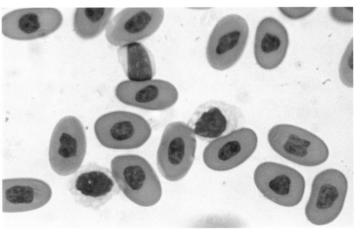

Figure 6.411 A normal lymphocyte (top) and two thrombocytes from a healthy taipan snake (*Oxyuranus microlepidotus*). Reptilian lymphocytes usually appear less deformable than those of birds. The thrombocytes can be distinguished by their irregular, vacuolated cytoplasm.

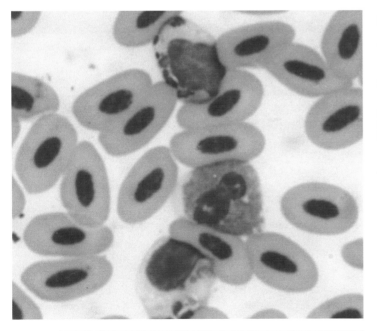

Figure 6.409 Two lymphocytes and a heterophil from a hooded crane (*Grus monacha*) with *Mycobacterium avium* infection. Several large basophilic granules are present in the cytoplasm of the lymphocytes. These also occur in lymphocytes from healthy cranes. The heterophil has a band nucleus.

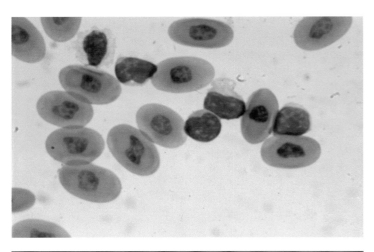

Figure 6.412 Four lymphocytes and a thrombocyte from the snake in the previous image. Two of the lymphocytes have indented nuclei and one shows an irregularity of the cytoplasmic membrane.

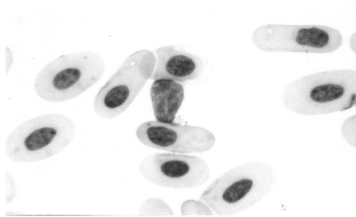

Figure 6.414 A lymphocyte from a healthy Greek tortoise (*Testudo graeca*).

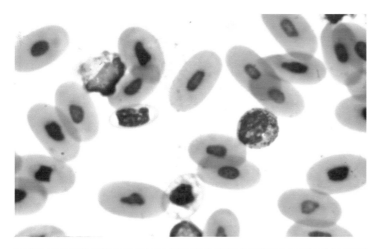

Figure 6.413 A lymphocyte with abundant cytoplasmic granulation and a normal thrombocyte from a healthy reticulated python (*Malayopython reticulatus*).

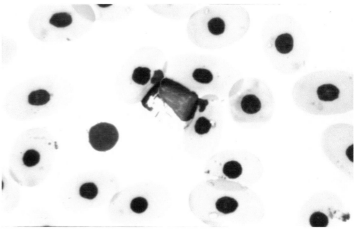

Figure 6.415 A lymphocyte with rod-shaped cytoplasmic granules from a healthy red-eared terrapin (*Trachemys scripta elegans*).

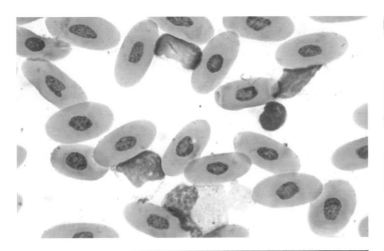

Figure 6.416 Three lymphocytes, a thrombocyte and a monocyte from a healthy rhinoceros iguana (*Cyclura cornuta*). The cytoplasm of two of the lymphocytes contains faint eosinophilic granules. The monocyte shows signs of disintegration.

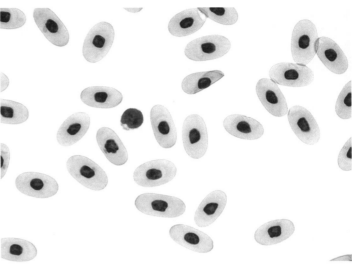

Figure 6.418 A small lymphocyte from a blood film of a gopher tortoise (*Gopherus polyphemus*) The species is native to the southeastern United States (courtesy of David Perpiñan).

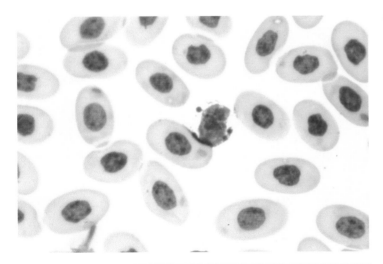

Figure 6.417 A lymphocyte with large basophilic cytoplasmic granules from a healthy Australian water dragon (*Intellagama lesueurii*).

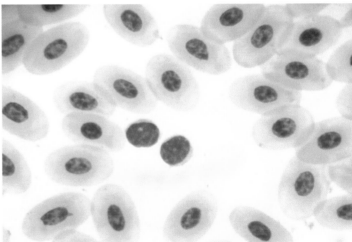

Figure 6.419 A blood smear from an Argentine black and white tegu (*Salvator merianae*), also known as the Argentine giant tegu, showing two medium-sized lymphocytes in the centre of the preparation (courtesy of Juan Carlos Troiano).

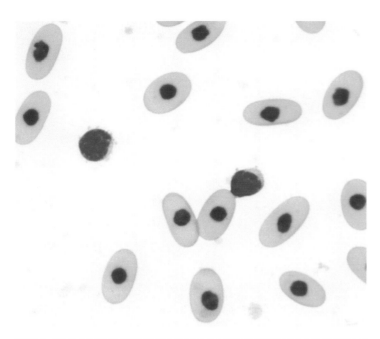

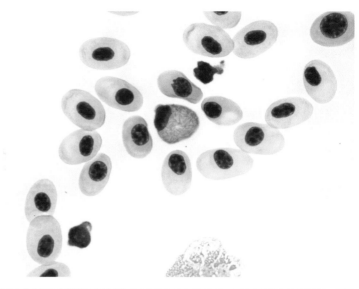

Figure 6.422 Blood film from a horn shark (*Heterodontus francisci*) showing one heterophil in the centre of the image and two lymphocytes with cytoplasmic membrane blebs (courtesy of Jean-Michel Hatt).

Figure 6.420 Two relatively small normal lymphocytes in a blood film from a Philippine crocodile (*Crocodylus mindorensis*). The lymphocytes show the cytoplasmic membrane 'blebs' often seen with this cell type in many reptile species (courtesy of Helen McCracken).

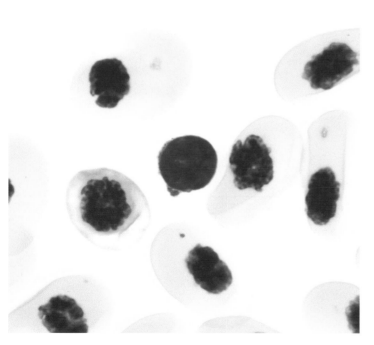

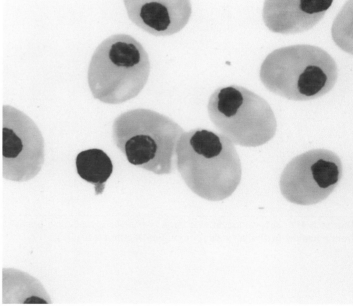

Figure 6.423 A small lymphocyte with cytoplasmic membrane blebs in a blood film of a Port Jackson shark (*Heterodontus portusjacksoni*) (courtesy of Helen McCracken).

Figure 6.421 An almost perfectly round medium-sized lymphocyte with cytoplasmic membrane bleb in the magnification of a blood smear from an axolotl (*Ambystoma mexicanum*) (courtesy of Helen McCracken).

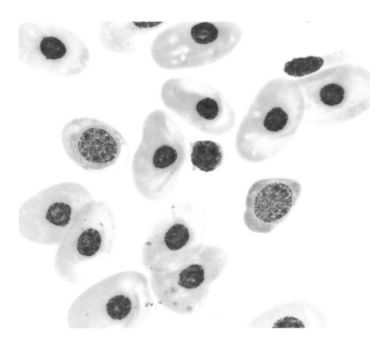

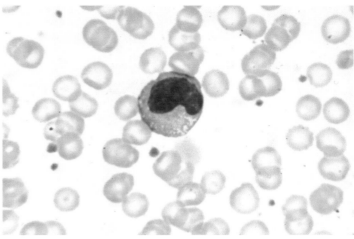

Figure 6.426 An abnormal lymphocyte from the macaque in the previous image.

Figure 6.424 Blood smear from an Eastern fiddler ray (*Trygonorrhina guaneria*) showing a very small lymphocyte with cytoplasmic membrane blebs (courtesy of Helen McCracken).

6.1.10 Abnormalities in the Lymphocytes Associated with Disease

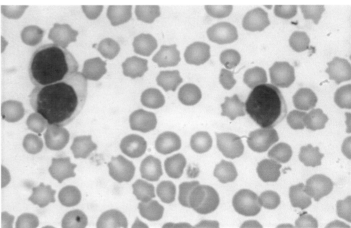

Figure 6.427 Three normal lymphocytes from a red slender loris (*Loris tardigradus*) with lymphocytosis (lymphocytes 13.5 × 10⁹/l). The cells show tendency to aggregate. This animal had a facial abscess but the cause of the lymphocytosis was not known.

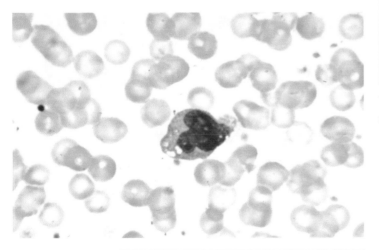

Figure 6.425 An abnormal lymphocyte from a crab-eating macaque (*Macaca fascicularis*), with chronic peritonitis (lymphocytes 4.0 × 10⁹/l). The cell has an irregularly shaped nucleus and the cytoplasm shows vacuolation.

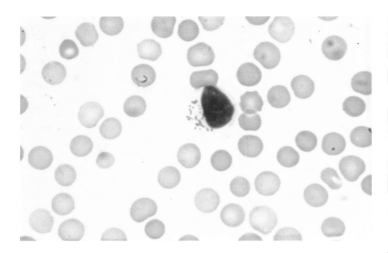

Figure 6.428 A lymphocyte with numerous cytoplasmic granules from the loris in the previous image.

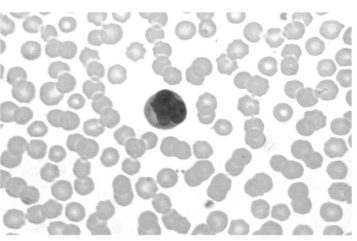

Figure 6.430 An active lymphocyte (virocyte, immunoblast) from a masked palm civet (*Paguma larvata*) with respiratory infection (lymphocytes 5.3×10^9/l). The cell has strongly basophilic cytoplasm.

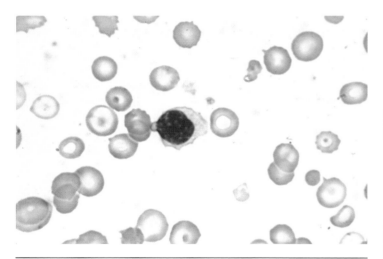

Figure 6.429 A lymphocyte with an irregular cytoplasmic membrane from a ring-tailed coati (*Nasua nasua*) with autoimmune haemolytic anaemia.

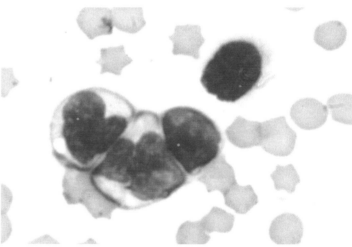

Figure 6.431 Abnormal lymphocytes from a domestic cat (*Felis catus*) with lymphoma. This animal had lymphopenia (lymphocytes 0.4×10^9/l) and the cells show a tendency to aggregate together. The two larger cells with nuclear cleavage are lymphoma cells.

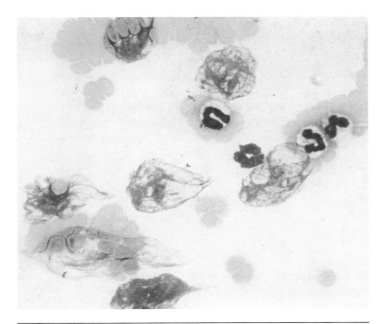

Figure 6.432 Three normal neutrophils and many basket cells, presumed to be disintegrating lymphocytes, in the 'tail' of a blood film from a domestic dog (*Canis lupus familiaris*) with lymphoma (lymphocytes 4.5 × 10⁹/l).

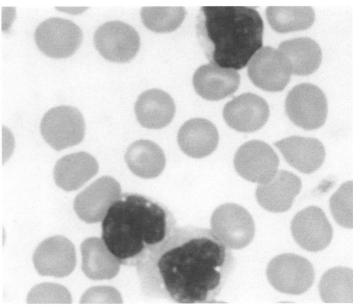

Figure 6.434 A higher magnification of lymphocytes from the dog in the previous image. The cells show nuclear cleavage and cytoplasmic vacuolation typical of lymphoma cells and a tendency to aggregate together.

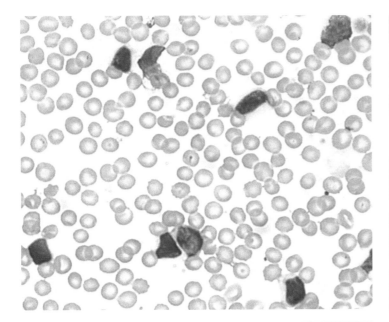

Figure 6.433 Lymphocytosis in a domestic dog (*Canis lupus familiaris*) with untreated lymphocytic lymphoma (×40, lymphocytes 60.8 × 10⁹l). Thrombocytopenia is also evident.

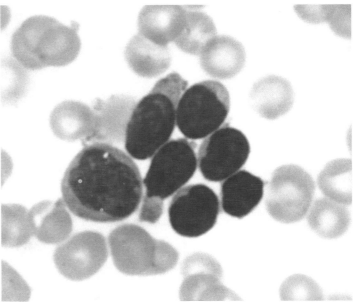

Figure 6.435 A group of six abnormal lymphocytes, probably lymphoblasts, from the dog in the previous image.

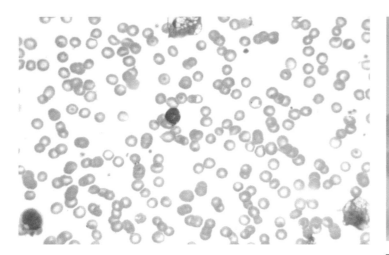

Figure 6.436 A blood film from the dog in the previous image after treatment with steroids (×40, lymphocytes 6.1 × 10⁹/l). Two lymphocytes and two disintegrating white blood cells (basket cells) are shown. Thrombocytopenia is still evident from the lack of platelets on the film and the dog died a few days later from acute haemorrhage.

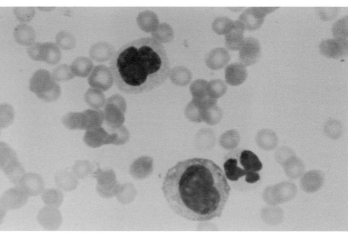

Figure 6.438 Images of a blood film from a horse (*Equus caballus*) with lymphosarcoma.

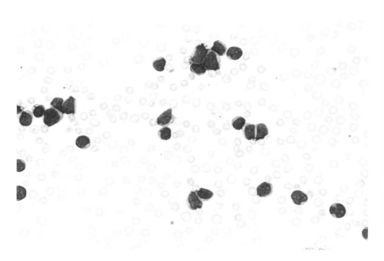

Figure 6.437 Lymphoblastic overflow in the blood of a domestic cat (*Felis catus*) with multicentric lymphosarcoma.

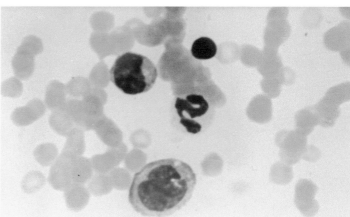

Figure 6.439 Images of a blood film from a horse (*Equus caballus*) with lymphosarcoma.

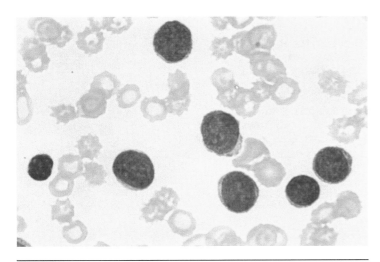

Figure 6.440 One normal lymphocyte and many lymphoblasts from a domestic dog (*Canis lupus familiaris*) with lymphoblastic leukaemia. The red blood cells show hypochromia, crenation and ovalocytosis. The dog also had thrombocytopenia.

Figure 6.442 Chronic lymphocytic leukaemia (×40). The lymphocyte count is increased but the cell morphology is unremarkable.

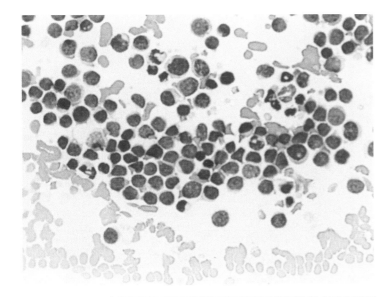

Figure 6.441 General view of a blood film from a case of unclassified lymphoid leukaemia.

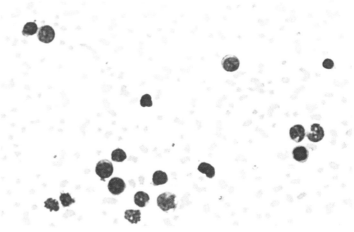

Figure 6.443 Lymphocytosis with some cells showing abnormal morphology in a case of chronic lymphocytic leukaemia (×40).

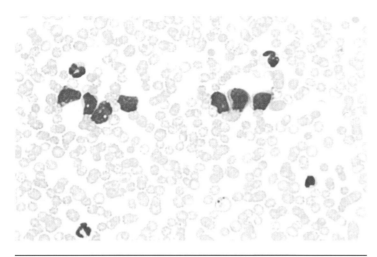

Figure 6.444 Chronic lymphocytic leukaemia (×40).

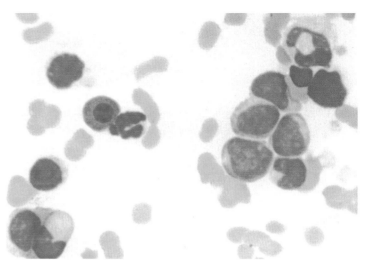

Figure 6.446 A higher magnification of the lymphoblasts in the cat in the previous image.

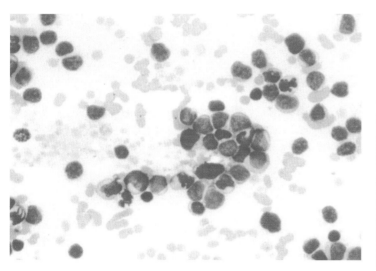

Figure 6.445 Lymphoblasts in a case of acute lymphoblastic leukaemia (×40).

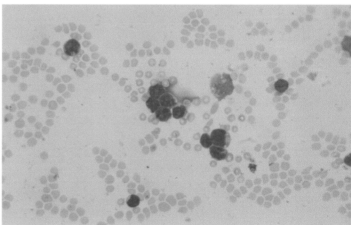

Figure 6.447 Acute lymphoblastic leukaemia in a grey mouse lemur (*Microcebus murinus*), (×40, lymphocytes 138.9 × 10⁹/l). The lymphoblasts show a marked tendency to aggregate. This animal also had severe, regenerative normocytic anaemia and thrombocytopenia.

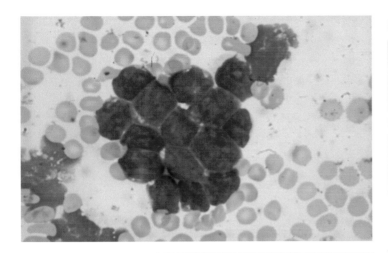

Figure 6.448 A higher magnification of an aggregate of lymphoblasts from the lemur in the previous image.

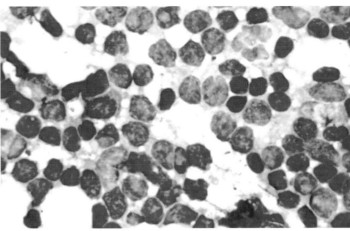

Figure 6.450 A bone marrow sample from the lemur in the previous image obtained immediately after death, showing massive lymphoblast infiltration.

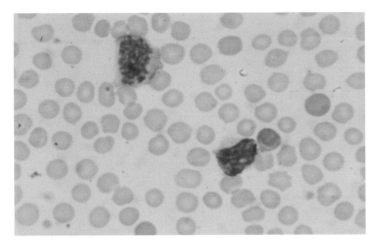

Figure 6.449 Two abnormal lymphocytes from the lemur in the previous image. The red blood cells show polychromasia.

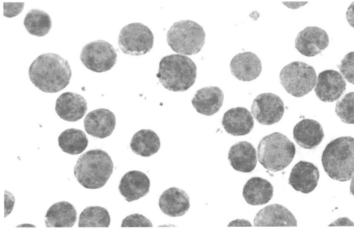

Figure 6.451 Lymphoid leukaemia in a domestic sheep (*Ovis aries*).

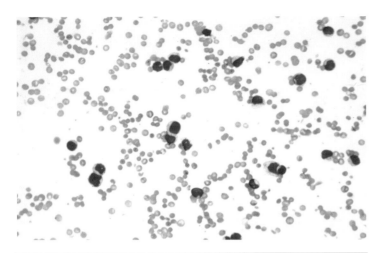

Figure 6.452 Lymphocytic leukaemia in a rat (*Rattus rattus*) (×25, lymphocytes 49.2 × 10⁹/l). The lymphocytes have irregular nuclei with a loose chromatin pattern and the cytoplasm of many is vacuolated and contains prominent basophilic granules. This animal also had neutropenia, thrombocytopenia and severe regenerative macrocytic anaemia.

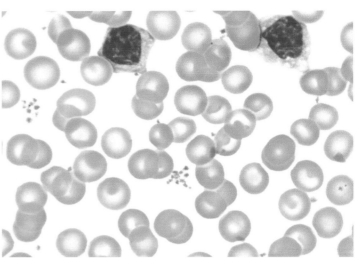

Figure 6.454 Microphotograph of the blood smear from an African elephant (*Loxodonta africana*) showing two reactive lymphocytes, and small platelets free and in clumps. Elephants often show target cells and mild anisocytosis in blood smears (courtesy of Nicole Stacy).

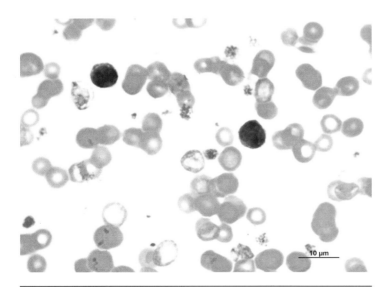

Figure 6.453 Blood film from a European hedgehog (*Erinaceus europaeus*) with two small activated lymphocytes. There are also several medium-sized to large platelets present in the preparation (courtesy of Jaume Martorell).

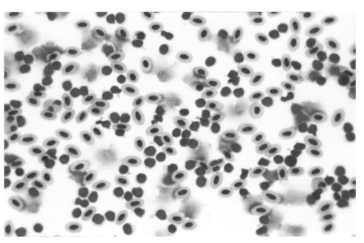

Figure 6.455 Lymphocytosis in a mallard duck (*Anas platyrhynchos*). with lymphoid leucosis (lymphocytes 127.0 × 10⁹/l). The cells show some tendency to aggregate.

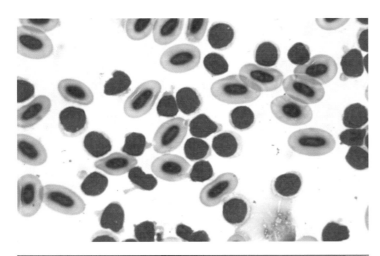

Figure 6.456 Lymphoblasts from the duck in the previous image, some of which show irregularities of the cytoplasmic membrane.

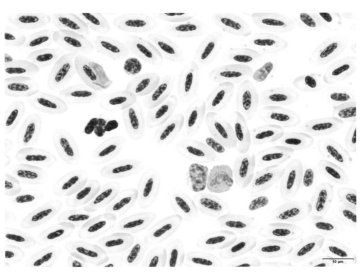

Figure 6.458 Blood smear from a Quaker parrot (*Myiopsitta monachus*) showing a normal heterophil, a normal small lymphocyte and a reactive lymphocyte and a small cluster of thrombocytes (courtesy of Karen Jackson).

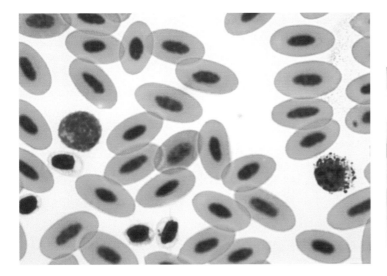

Figure 6.457 Activated lymphocyte with cytoplasmic basophilia and with an evident Golgi apparatus visible as a light blue area adjacent to the nucleus. A basophil is also present. The sample came from a three-year-old male eider duck (*Somateria mollissima*) kept in a zoo, losing weight despite a good appetite. No endoparasites were detected. Further diagnostic work up revealed the bird was undergoing avian malaria (courtesy of Jean-Michel Hatt).

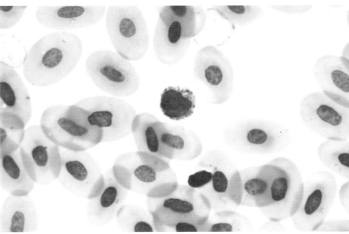

Figure 6.459 A lymphocyte (?) containing phagocytised material from an Indian python (*Python molurus*) with dermatitis (0.5×10^9/l).

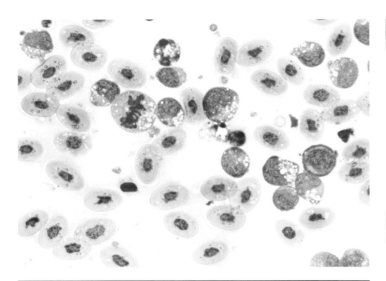

Figure 6.460 Photomicrograph of a blood smear from a bearded dragon (*Pogona vitticeps*) showing intermediate to large neoplastic lymphocytes. Immunocytochemistry demonstrated CD3 positivity consistent with acute T-cell leukaemia. Erythrocytes exhibit irregular, refractile drying artefacts (courtesy of Nicole Stacy).

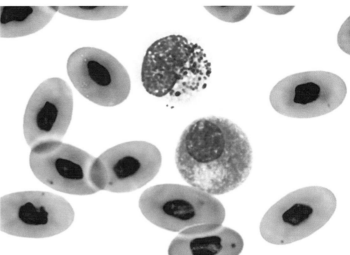

Figure 6.462 Cell above: granulocytic metamyelocyte indicating a left shift of the granulopoietic (most likely heterophilic) line, the classification of this cell into the granulopoietic cell line is only possible with careful evaluation of many cells in the blood film and detection of transient stages containing both immature granules (like in this cell) and mature brick-red ellipsoid granules (absent in this cell); cell below: plasma cell-type mononuclear cell with eccentric round nucleus, increased cytoplasmic basophilia and visible Golgi apparatus as indistinctly bordered light blue area within the cytoplasm below the nucleus in a bearded dragon (*Pogona vitticeps*) (courtesy of Helene Pendl).

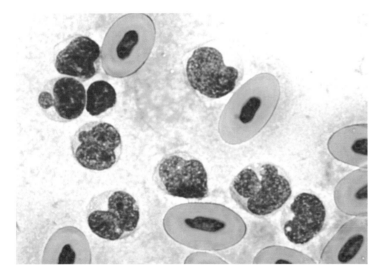

Figure 6.461 Blood film from a bearded dragon (*Pogona vitticeps*) with prominent leuko-/lymphocytosis (4.4 mm buffy coat, 98% lymphocytes in differential count) with an almost monomorphic leukocyte population consisting of atypical lymphocytes with nuclear signs of malignancy (aniso-, poly-, poikilonucleosis, indentations, amitotic-irregular divisions) suggestive of lymphoid neoplasia (e.g. lymphoid leukaemia or leukemic manifestation of lymphoma); the morphology with indentation and 'flower-like' cells raises suspicion of T-cell origin (further diagnostics: immunohistochemical stain with CD3); amorphous cloudy basophilic to amphophilic background indicating a highly proteinaceous plasma (further diagnostics: protein electrophoresis for determination of a possible monoclonal gammopathy) (courtesy of Helene Pendl and Nicole Stacy).

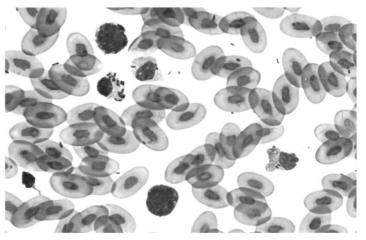

Figure 6.463 Reactive lymphocytes and heterophil in the blood smear of a bearded dragon (*Pogona vitticeps*) undergoing severe septicemia. Some phagocytic activity of bacilli can be observed (courtesy of Xavier Valls Badia).

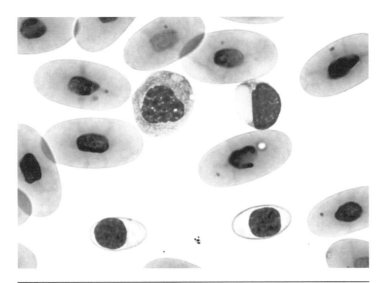

Figure 6.464 Two thrombocytes, glassy, light basophilic, intracytoplasmic inclusion bodies (ICIBs) in azurophil (left cell, small ICIB at 1 o'clock) and lymphocyte (right cell, large oval ICIB, almost completely filling the cytoplasm) highly suggestive for Boid inclusion body disease (BIBD). Arenavirus infection in a boa constrictor (*Boa constrictor*) confirmed by virology (courtesy of Helene Pendl).

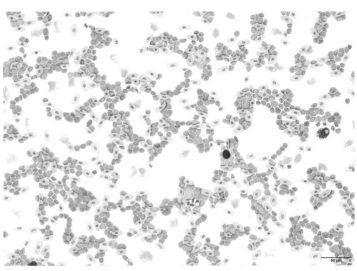

Figure 6.466 Lymphoid leukaemia in a saw shelled turtle (*Myuchelys latisternum*). This is an endemic species of Australia (courtesy of Karen Jackson).

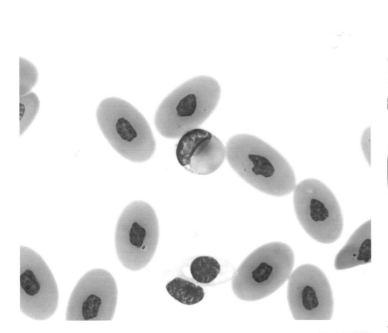

Figure 6.465 Photomicrograph of a lymphocyte with inclusion body typical of Boid inclusion body disease (BIBD) associated with Arenavirus infection in a rainbow boa (*Epicrates cenchria*). Two thrombocytes are also present (courtesy of Nicole Stacy).

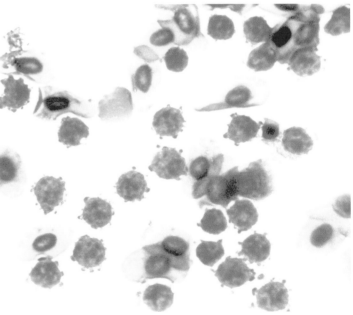

Figure 6.467 Multiple abnormal lymphocytes in the blood smear of a giant green tree frog (*Litoria infrafrenata*) with epithelial and splenic lymphoma confirmed at *post-mortem* examination (courtesy of Helen McCracken).

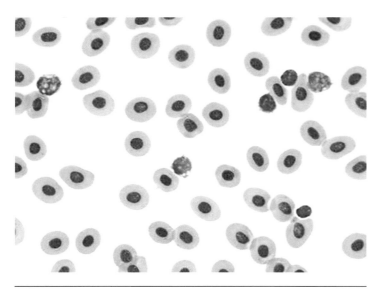

Figure 6.468 Numerous lymphocytes in the blood smear of a blue catfish (*Ictalurus furcatus*) with an ulcerative skin lesion. Haematology showed mild neutrophil with left shift and few reactive lymphocytes. Reactive lymphocytes can be vacuolated. The roundish appearance of red blood cells is normal for teleost (courtesy of Nicole Stacy).

6.1.11 Normal Monocyte Morphology

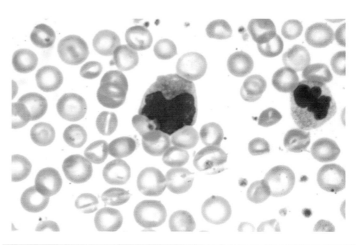

Figure 6.470 A typical monocyte and a neutrophil from a healthy cotton-headed tamarin (*Saguinus oedipus*).

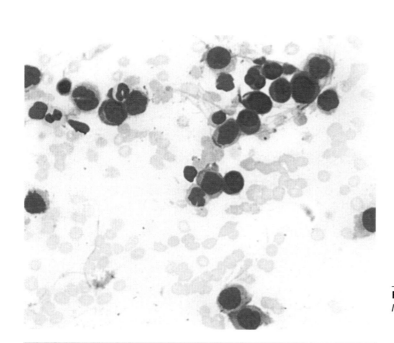

Figure 6.469 Plasma cells in the bone marrow of a case of myeloma in a domestic dog (*Canis lupus familiaris*).

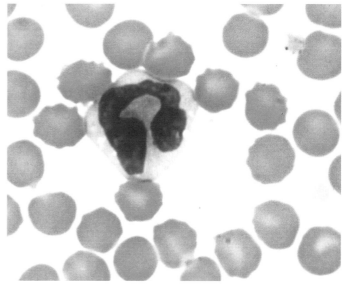

Figure 6.471 A typical monocyte from a healthy domestic dog (*Canis lupus familiaris*).

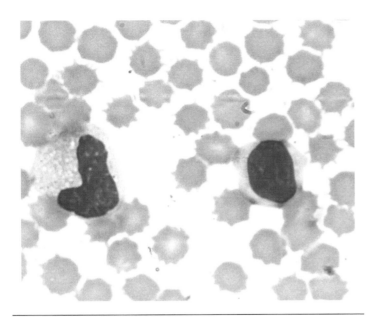

Figure 6.472 A normal monocyte (left) and lymphocyte from a healthy gaur (*Bos gaurus*). The red blood cells are crenated.

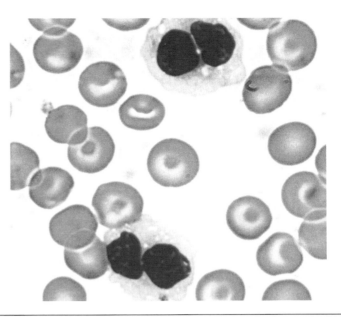

Figure 6.474 Two normal bilobed mononuclear cells from a healthy Indian elephant (*Elephas maximus indicus*). Cells of this type are found in both Indian and African elephants (*Loxodonta africana*) and are classified as monocytes on the basis of their cytochemical staining reactions.

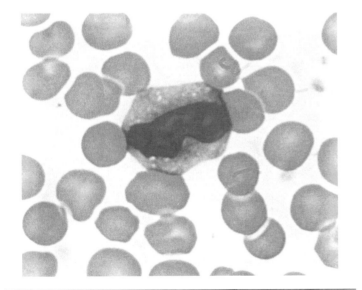

Figure 6.473 A monocyte from a healthy capybara (*Hydrochoerus hydrochaeris*).

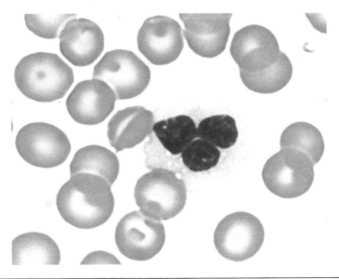

Figure 6.475 A trilobed monocyte from the elephant in the previous image.

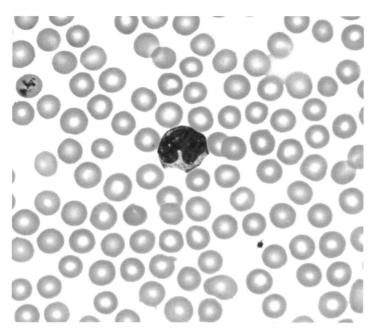

Figure 6.476 Monocyte in a blood smear from a clinically normal captive Tasmanian devil (*Sarcophilus harrissi*) (courtesy of Helen McCracken).

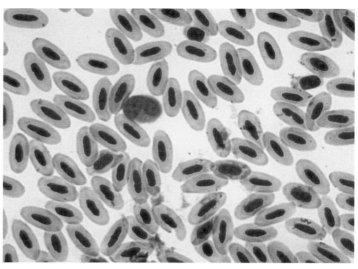

Figure 6.478 A monocyte in a blood film from a clinically normal red-breasted goose (*Branta ruficollis*) maintained in captivity (courtesy of Michelle Obrien).

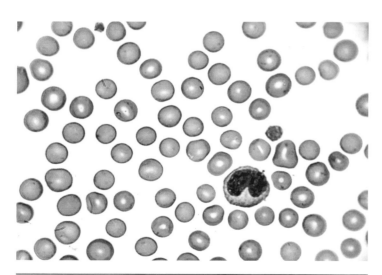

Figure 6.477 Blood smear from a captive bottlenose dolphin (*Tursiops truncatus*) showing a normal monocyte (courtesy of Guillermo Sanchez).

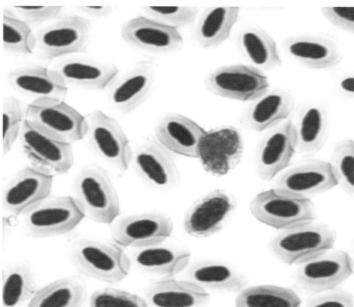

Figure 6.479 A low magnification (×40) view of the blood film from a varied honeyeater (*Lychenostomus versicolor*). A single mononuclear cell can be observed on the extreme right. This cell is thought to be a monocyte based on the morphology and size of the nucleus and the lace-like appearance of the cytoplasm. The blood film was collected as part of a study of haemoparasites of the avifauna of Australia (courtesy of Lee Peacock).

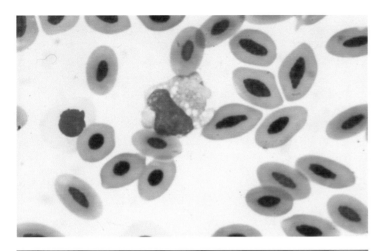

Figure 6.480 A monocyte with cytoplasmic vacuoles from a healthy ostrich (*Struthio camelus*).

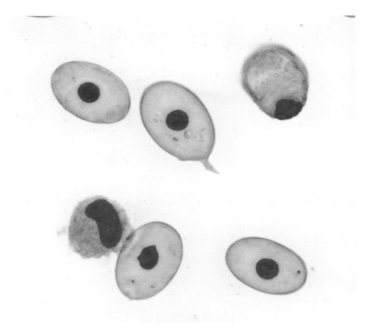

Figure 6.482 Blood smear from a broad-shelled river turtle (*Chelodina expansa*). The cell in the left lower quadrant is a monocyte. The cell in the upper right quadrant is a heterophil. The granule detail is generally not clear in intact turtle heterophils (courtesy of Helen McCracken).

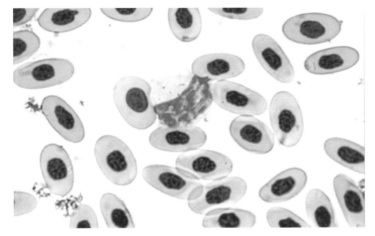

Figure 6.481 A monocyte from a healthy green iguana (*Iguana iguana*).

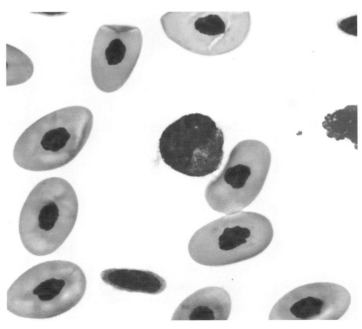

Figure 6.483 Normal monocyte in the blood smear from a baw baw frog (*Philoria frosti*) (courtesy of Helen McCracken).

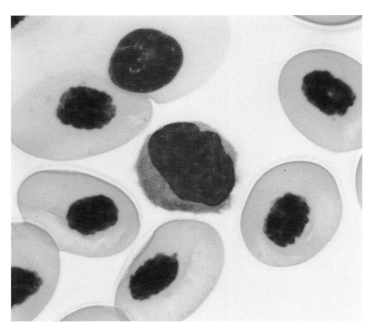

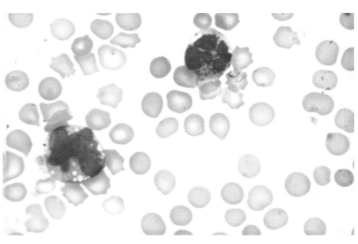

Figure 6.486 A monocyte with an irregular nucleus and cytoplasmic vacuolation from a red slender loris (*Loris tardigradus*) with a facial abscess. This animal had a high lymphocyte count and slight monocytosis (monocytes 0.9×10^9/l). A neutrophil with toxic granulation is also present.

Figure 6.484 Monocyte in a blood smear from an axolotl (*Ambystoma mexicanum*) (courtesy of Helen McCracken).

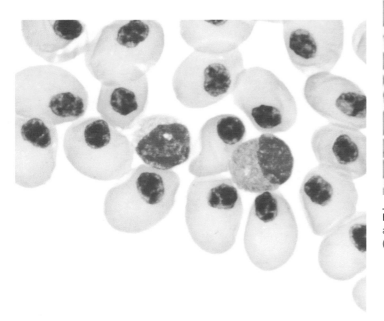

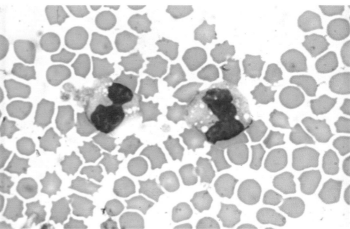

Figure 6.487 Two monocytes with cytoplasmic vacuoles from a black and white ruffed lemur (*Varecia variegata*) with chronic respiratory disease (monocyte 1.0×10^9/l).

Figure 6.485 Blood smear from a Port Jackson shark (*Heterodontus portusjacksoni*) showing one heterophil (right) and one vacuolated monocyte (left) (courtesy of Helen McCracken).

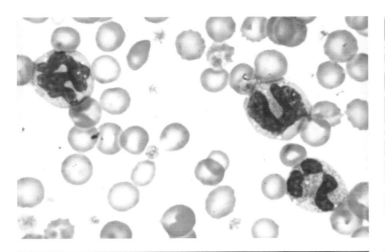

Figure 6.488 Two abnormal monocytes from a red-bellied tamarin (*Saguinus labiatus*) with an infected and necrotic tail lesion (monocytes 0.9 × 10⁹/l). One of the monocytes shows extreme nuclear malformation and the nuclei of both contain small vacuoles.

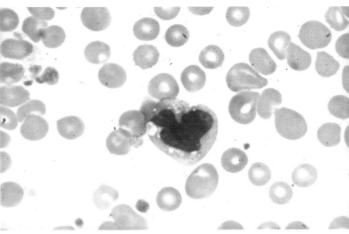

Figure 6.490 An abnormal monocyte with vacuolated cytoplasm and increased nuclear cleavage from a crab-eating macaque (*Macaca fascicularis*) with haemorrhagic anaemia (monocytes 0.7 × 10⁹/l).

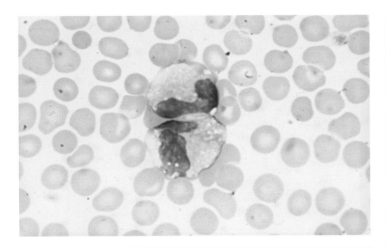

Figure 6.489 Two abnormal monocytes with cytoplasmic vacuoles from a long-haired spider monkey (*Ateles belzebuth*) with *Yersinia pseudotuberculosis* infection (monocytes 1.3 × 10⁹/l).

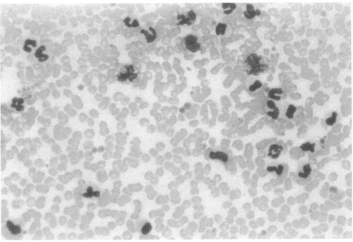

Figure 6.491 Monocytosis and neutrophilia in a dog with bacterial endocarditis (×40). The monocytes have vacuolated cytoplasm and abnormal nuclei.

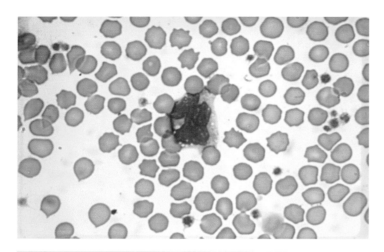

Figure 6.492 A vacuolated monocyte from a palm civet (*Paguma larvata*) with chronic infection (monocytes 1.0 × 10⁹/l).

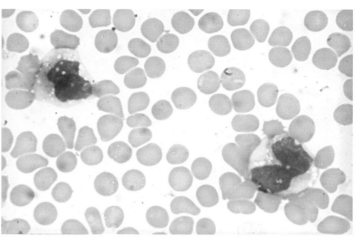

Figure 6.494 Two atypical monocytes from a gaur (*Bos gaurus*) with viral enteritis. There is vacuolation of both the cytoplasm and the nuclei (monocytes 2.0 × 10⁹/l).

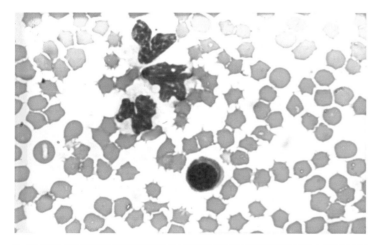

Figure 6.493 Three abnormal monocytes from a cheetah (*Acinonyx jubatus*) with *Vaccinia variola* infection. This animal had a marked monocytosis (monocytes 32.6 × 10⁹/l). The red blood cells are crenated. A polychromatic erythroblast showing evidence of retarded nuclear maturation is also present.

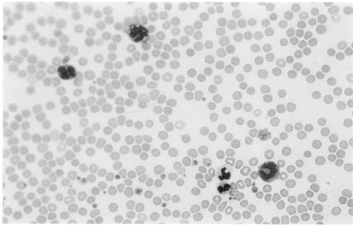

Figure 6.495 Monocytosis in a cow (*Bos taurus*) with acute mastitis (×40).

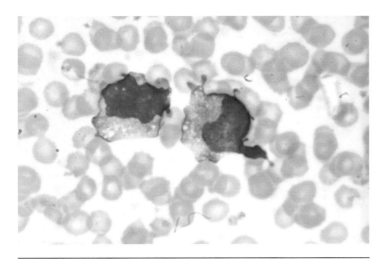

Figure 6.496 Two atypical monocytes from a guinea pig (*Cavia porcellus*) with lymphadenopathy (monocytes 2.1 × 10⁹/l).

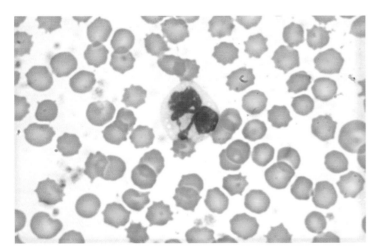

Figure 6.497 An atypical monocyte with a degenerating nucleus from a Jamaican coney (*Geocapromys brownii*) with chronic eye ulceration and a necrotic lesion of the orbit (monocytes 1.2 × 10⁹/l).

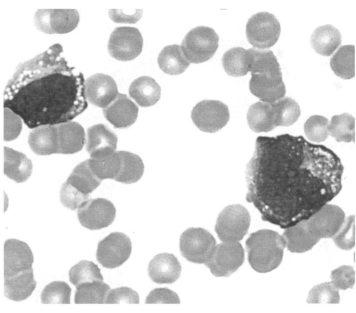

Figure 6.498 Two pathological monoblasts from a domestic dog (*Canis lupus familiaris*) with myelomonocytic leukaemia. The red blood cells show increased rouleaux formation.

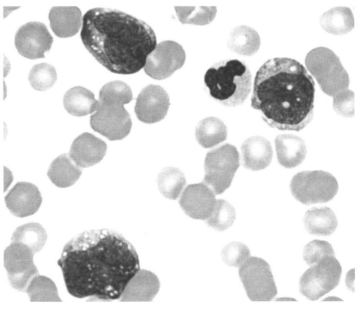

Figure 6.499 Three pathological monoblasts and a normal neutrophil from the domestic dog in the previous image.

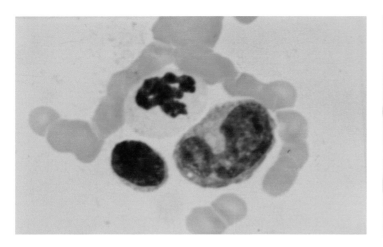

Figure 6.500 A malignant mononuclear cell in the blood of a horse (*Equus caballus*) with neoplasia.

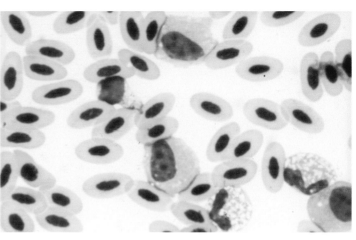

Figure 6.502 Two monocytes, two degranulated heterophils and a lymphocyte with granular cytoplasm in a sarus crane (*Grus antigone*) with *Mycobacterium avium* infection (monocytes 5.0 × 10⁹/l). The monocytes contain small vacuoles and nuclear cleavage is absent. Part of a third monocyte is also present.

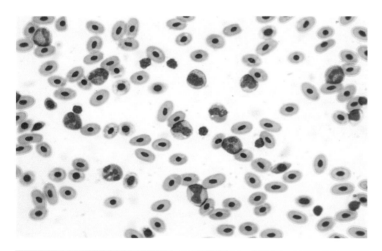

Figure 6.501 Monocytosis in an African penguin (*Spheniscus demersus*) with aspergillosis (×40, monocytes 3.1 × 10⁹/l).

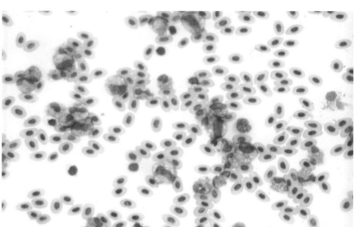

Figure 6.503 Monocytosis and heterophilia in a domestic fowl (*Gallus gallus domesticus*) with *Mycobacterium avium* infection (×40, monocytes 10.1 × 10⁹/l).

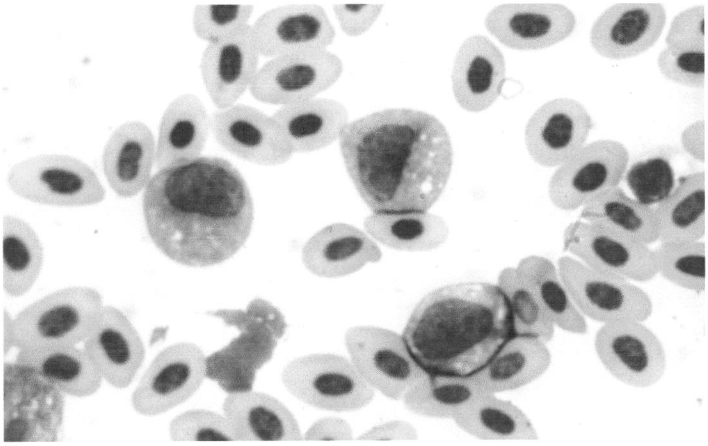

6.1.12 Normal Azurophil Morphology

Figure 6.504 A high-power view of three monocytes from the bird in the previous image. The cells have cytoplasmic vacuoles and nuclear cleavage is absent.

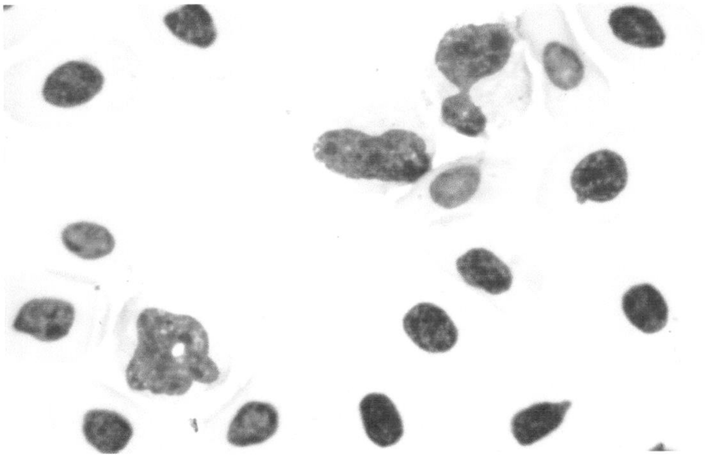

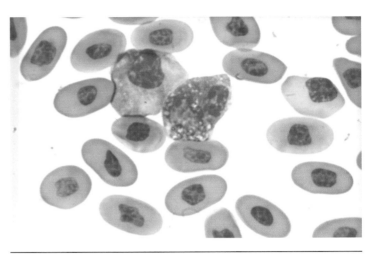

Figure 6.506 An azurophil and a heterophil from a healthy taipan snake (*Oxyuranus microlepidotus*). The azurophil is the cell on the left.

Figure 6.505 Monocytosis in an Aldabra giant tortoise (*Aldabrachelys gigantea*) suffering from anorexia and muscle wasting (monocytes 1.0 × 10⁹/l).

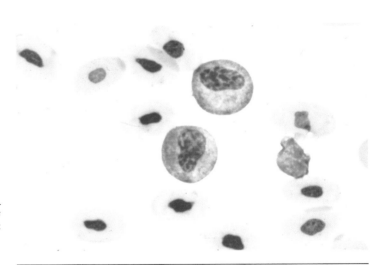

Figure 6.507 Two azurophils from a healthy reticulated python (*Malayopython reticulatus*).

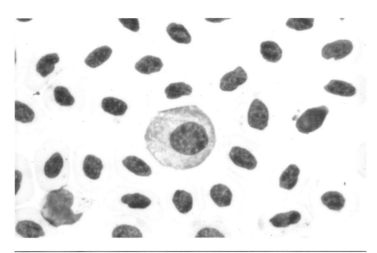

Figure 6.508 An azurophil with a large amount of cytoplasm from a healthy black-pointed tegu (*Tupinanbis teguizin*).

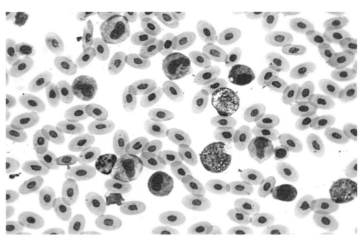

Figure 6.510 General view of a blood film showing azurophilia and heterophilia in an Eastern fox snake (*Pantherophis gloydi*) with severe progressive necrosis of the tail (×40, azurophils 20.9 × 10^9/l). The azurophils have monocytoid nuclei and a tendency for vacuolation.

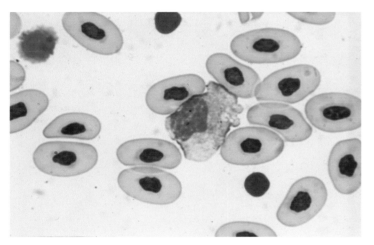

Figure 6.509 A monocytoid azurophil from a healthy green iguana (*Iguana iguana*).

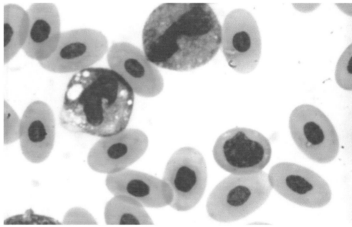

Figure 6.511 A higher magnification of two azurophils and a polychromatic erythroblast from the same snake in the previous image.

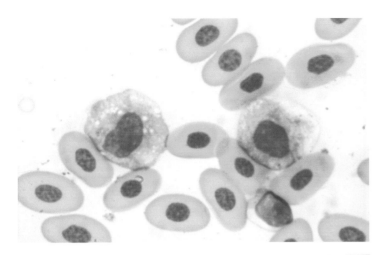

Figure 6.512 Two more large azurophils with excessive cytoplasm from the snake in the previous image. In one of the cells, the cytoplasm appears to have shrunk away from the cell membrane. A normal lymphocyte is also present.

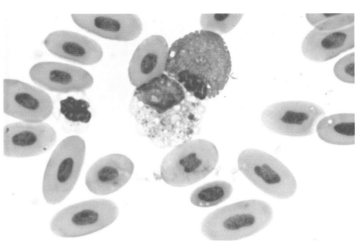

Figure 6.514 An azurophil, a heterophil and a thrombocyte from the python in the previous image. The azurophil has a scalloped cell membrane, cytoplasmic vacuolation and granulation, the latter possibly consisting of phagocytosed material.

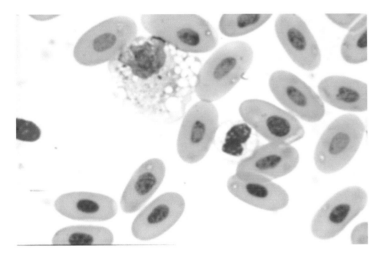

Figure 6.513 A highly vacuolated azurophil and a normal thrombocyte from a brown python (*Liasis fuscus*) with stomatitis and respiratory infection (azurophils 4.1 × 10^9/l). The cytoplasm of the azurophil contains a few abnormal granules and the cell membrane is scalloped.

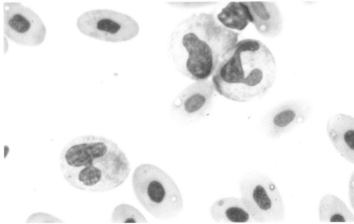

Figure 6.515 Azurophilia in a carpet python (*Morelia spilota*) with multiple abscesses, cellulitis and muscle necrosis (azurophils 5.5 × 10^9/l). The nuclei show excessive lobulation.

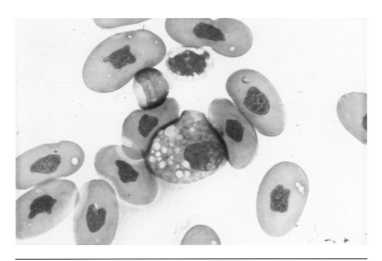

Figure 6.516 A monocytoid azurophil containing round, blue-staining cytoplasmic inclusion bodies from a Gaboon viper (*Bitis gabonica*) with wasting disease. The lymphocyte and thrombocyte in the present field are normal but the red blood cell nuclei are deformed.

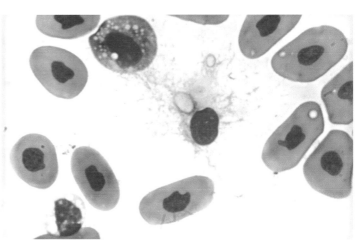

Figure 6.518 A disrupted azurophil from the snake in the previous image showing release of the blue cytoplasmic inclusion bodies.

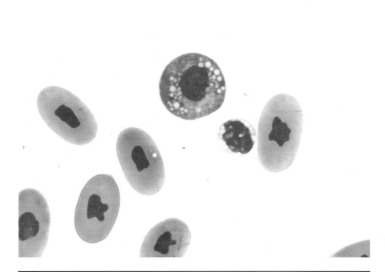

Figure 6.517 A lymphocytoid azurophil with cytoplasmic inclusions and a normal thrombocyte from the snake in the previous image.

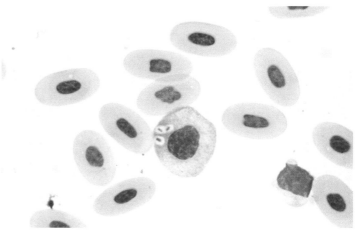

Figure 6.519 An azurophil with cytoplasmic inclusions (bacteria?) from an Indian python (*Python molurus*) with facial abscess (azurophils 1.5 × 10⁹/l).

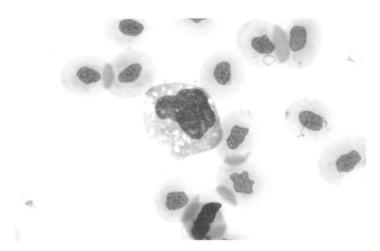

Figure 6.520 A large azurophil with an irregular nucleous and vacuolated cytoplasm from a Montpellier snake (*Malpolon monspessulanus*) suffering from ascites, skin lesions and cloacal haemorrhage. This snake had marked azurophilia (azurophils 13.4 × 10⁹/l) and was severely anaemic.

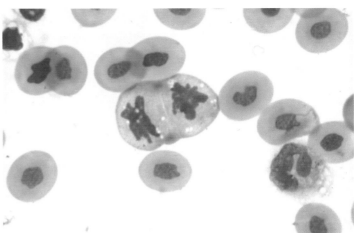

Figure 6.522 A trinucleated azurophil and an azurophil in mitosis from the snake in the previous image. The possibility that this snake was suffering from a malignancy involving the azurophils was not ruled out.

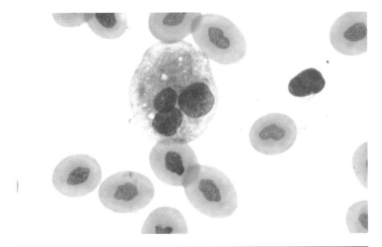

Figure 6.521 A trinucleated azurophil and an azurophil in mitosis from the snake in the previous image. The possibility that this snake was suffering from a malignancy involving the azurophils was not ruled out.

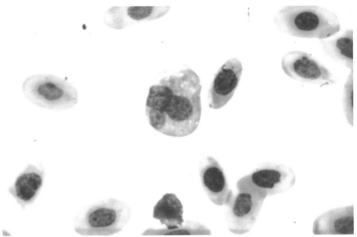

Figure 6.523 A large azurophil with a trilobed nucleus from a Chinese water dragon (*Physignathus cocincinus*) with osteomyelitis (azurophils 3.1 × 10⁹/l).

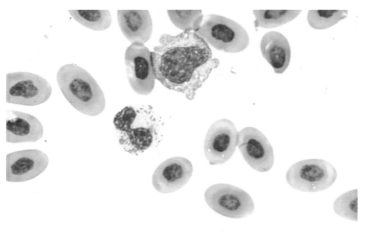

Figure 6.524 An azurophil with cytoplasmic budding and a toxic heterophil from a Chinese water dragon (*Physignathus cocincinus*) with blepharitis (azurophils 7.7 × 10⁹/l).

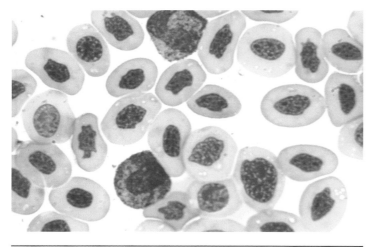

Figure 6.525 Two azurophils with intense cytoplasmic granulation from a green iguana (*Iguana iguana*) with infection associated with a fractured femur (azurophils 0.2 × 10⁹/l). The red blood cells have misshapen nuclei and one cell showing evidence of delayed nuclear maturation is present.

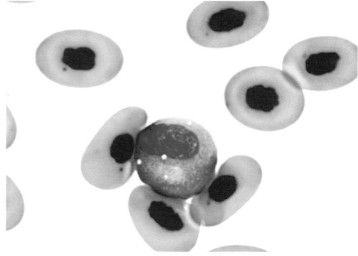

Figure 6.526 A large vacuolated mononuclear cell (azurophil or monocyte) in a blood film from a Schneider's toad or rococo toad (*Rhinella schneideri* formerly *Bufo paracnemis*). Adequate differential diagnosis cannot be achieved from a single cell and without examining the whole blood film. The rest of the film showed reactive lymphocytes, many monocyte-like cells and poorly granulated heterophils. The significance of the latter in terms of degranulation and possible toxicity could not be determined due to the lack of a reference slide from a clinically healthy animal (courtesy of Helene Pendl).

Figure 6.527 Smear from the coelomic fluid of a lumpsucker or lumpfish (*Cyclopterus lumpus*) with intranuclear microsporidia (*Nucleospora cyclopteri*), probably within a monocyte. The fish were in an experimental breeding colony and presented with weight loss and ascites in multiple animals within the group (courtesy of Mark Stidworthy).

6.1.13 Normal Variation in Platelet and Thrombocyte Morphology

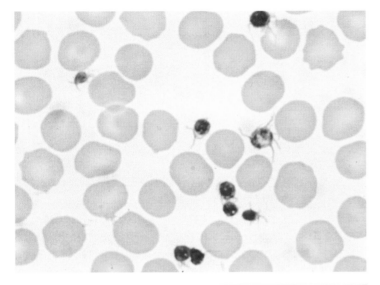

Figure 6.528 Normal platelets from a domestic dog (*Canis lupus familiaris*). Pseudopodia formation and the tendency for platelets to aggregate together are shown. These changes represent the first stage of platelet activation during sample collection.

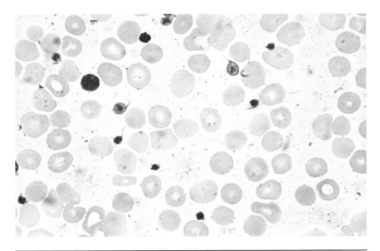

Figure 6.529 Platelet anisocytosis in a normal moose (*Alces alces*). Pseudopodia formation is evident.

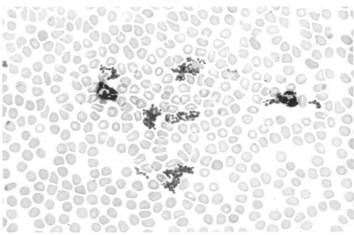

Figure 6.530 Aggregated platelets in a badly collected blood sample from a domestic goat (*Capra aegagrus hircus*) (×40).

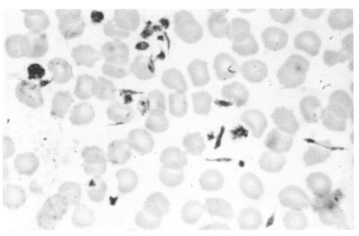

Figure 6.531 Two different morphological types of platelets from a guinea pig (*Cavia porcellus*). The filamentous platelets are unusual but apparently normal in a few species of mammals. They may represent a more easily activated subpopulation.

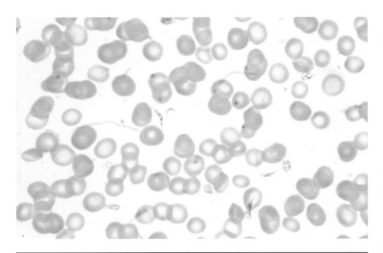

Figure 6.532 Normal and filamentous platelets from a healthy Western long-beaked echidna (*Zaglossus bruijnii*).

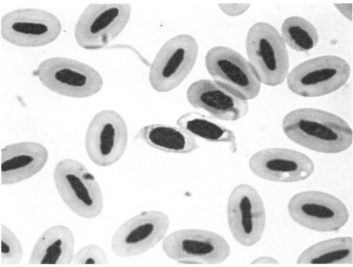

Figure 6.535 Two spindled-shaped thrombocytes from a normal kori bustard (*Ardeotis kori*).

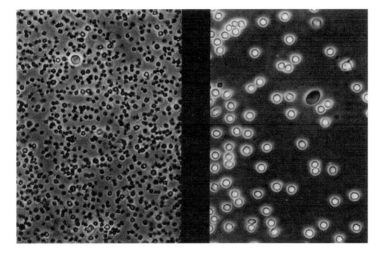

Figure 6.533 Photomicrographs at the same magnification of platelet-rich human (*Homo sapiens*) plasma (left) and thrombocytes-rich domestic fowl (*Gallus gallus domesticus*) plasma showing the difference in cell size. Thrombocytes are larger than platelets but occur in smaller number so that the packed cell volumes are similar. One red blood cell is present in the domestic fowl sample.

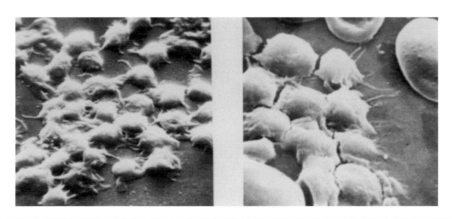

Figure 6.534 SEM of aggregated platelets from an olive baboon (*Papio anubis*) (left) and aggregated thrombocytes from a domestic turkey (*Meleagris gallopavo*) showing structural similarities.

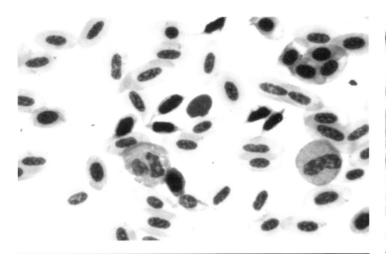

Figure 6.536 Spindled-shaped thrombocytes from a rosy flamingo (*Phoenicopterus roseus*).

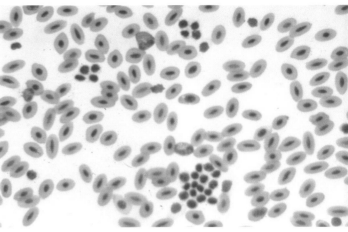

Figure 6.538 Thrombocytes aggregates from an ostrich (*Struthio camelus*) (×40).

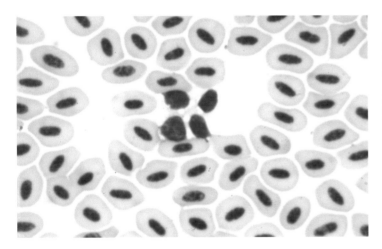

Figure 6.537 An aggregate of four normal thrombocytes from a moorhen (*Gallinula chloropus*).

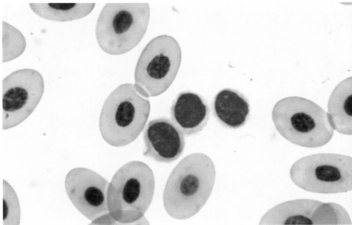

Figure 6.539 Three normal thrombocytes from a Hermann's tortoise (*Testudo hermanni*). These cells could be misidentified as lymphocytes but are smaller and have denser nuclei. The thrombocytes seen here show some evidence of aggregation.

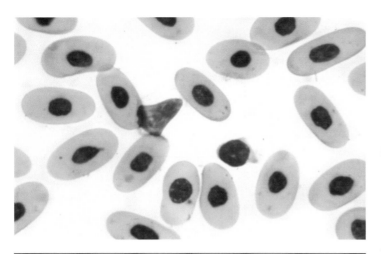

Figure 6.540 A thrombocyte and a lymphocyte from a Greek tortoise (*Testudo graeca*). In this animal the distinction between the two cell types is obvious.

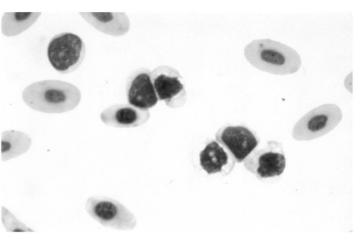

Figure 6.542 Three thrombocytes and three lymphocytes from an Indian python (*Python molurus*). The thrombocytes have smaller nuclei and less regular cytoplasm.

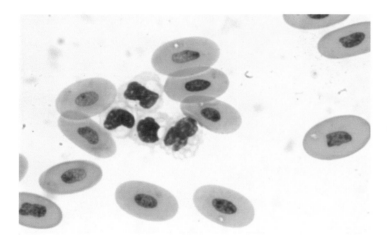

Figure 6.541 An aggregate of four thrombocytes with vacuolated cytoplasm and irregular nuclei form a brown python (*Liasis fuscus*).

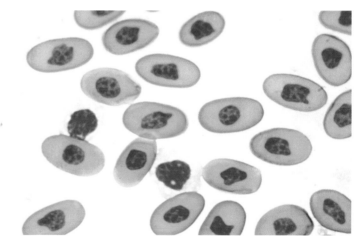

Figure 6.543 Two normal thrombocytes from a taipan snake (*Oxyuranus microlepidotus*). The cytoplasm of the cells has an irregular foamy appearance associated with partial activation.

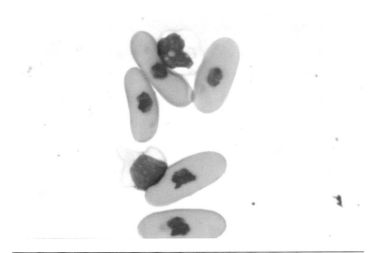

Figure 6.544 A thrombocyte and a lymphocyte from a Chinese alligator (*Alligator sinensis*). The thrombocyte is at the top of the field.

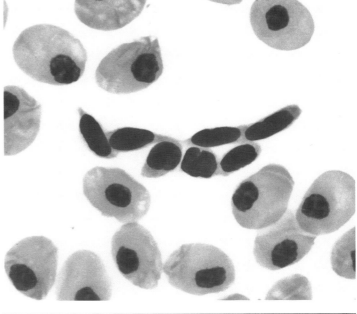

Figure 6.546 Thrombocyte clumping in a blood smear from a clinically normal Port Jackson shark (*Heterodontus portusjacksoni*) (courtesy of Helen McCracken).

6.1.14 Variations in Platelets and Thrombocytes Associated with Disease

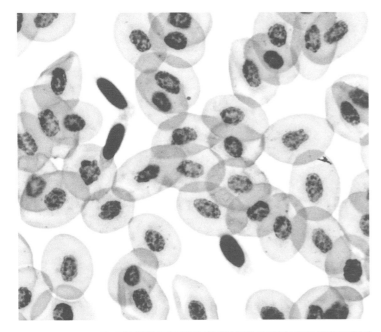

Figure 6.545 Three rather elongated thrombocytes in a blood film from a clinically normal motorbike frog (*Litoria moorei*) (courtesy of Helen McCracken).

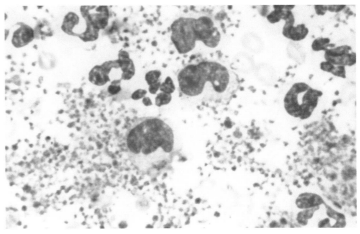

Figure 6.547 A greatly increased platelet count (thrombocytosis) in a Guaira spiny rat, also known as casiragua (*Proechimys guairae*), with infected fight wounds (platelet 1370×10^9/l). A raised platelet count is found in association with bacterial infection in many mammalian species. This animal also had neutrophilia and monocytosis.

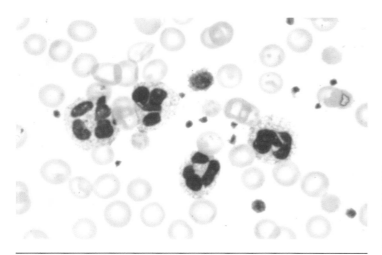

Figure 6.548 Thrombocytosis and neutrophilia in a plains vizcacha (*Lagostomus maximus*) with acute inflammatory disease (platelets 902 × 10⁹/l).

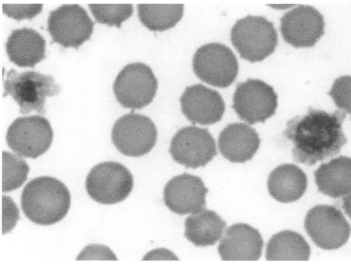

Figure 6.550 Two partially activated giant platelets from a domestic dog (*Canis lupus familiaris*) with thrombocytopenia associated with lymphoma (platelets 15 × 10⁹/l).

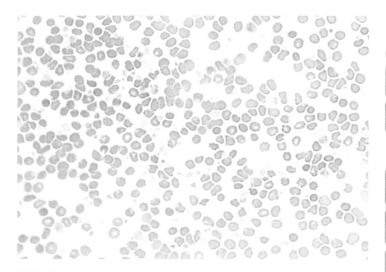

Figure 6.549 Thrombocytosis with 'shift' platelets in a domestic dog (*Canis lupus familiaris*), following rupture of a splenic haemangiosarcoma with intraperitoneal haemorrhage. The red blood cells show poikilocytosis and schistocytosis.

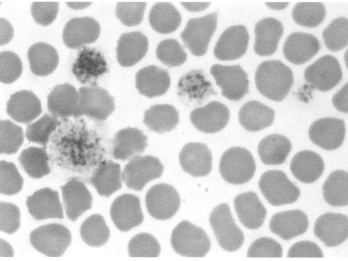

Figure 6.551 Giant platelets from a domestic dog (*Canis lupus familiaris*) with lymphoma on steroid treatment.

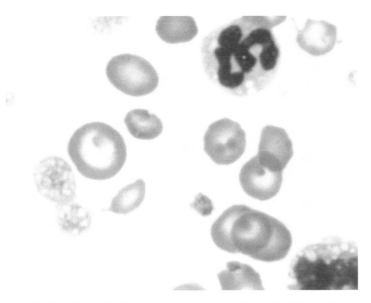

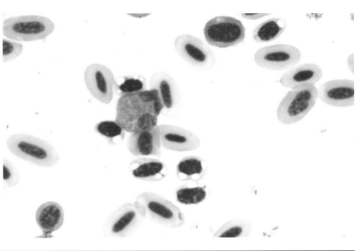

Figure 6.554 Vacuolated thrombocytes from a golden eagle (*Aquila chrysaetos*) with regenerative anaemia. This bird had a relatively high thrombocyte count (61 × 10⁹/l). The intact appearance of the thrombocyte membranes suggests that vacuolation may not be caused by activation in this case.

Figure 6.552 Giant platelets from a domestic dog (*Canis lupus familiaris*) with anaemia and a hormonal imbalance (platelets 233 × 10⁹/l). The red blood cells show anisocytosis and a neutrophil showing a right shift and cytoplasmic vacuoles are present.

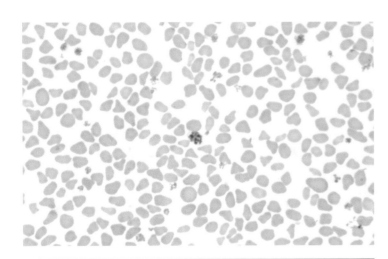

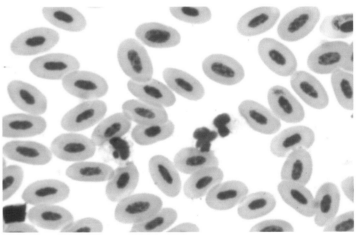

Figure 6.555 Thrombocytes with polymorphic nuclei from a grey parrot (*Psittacus erithacus*) with a respiratory infection (thrombocytes 9 × 10⁹/l).

Figure 6.553 A giant platelet with distinctive granulation from a markhor (*Capra falconeri*) with vitamin E deficiency. Triangular red blood cells occur normally as an *in vitro* artefact in this species.

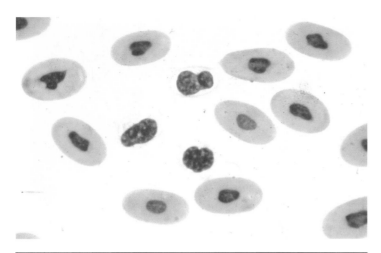

Figure 6.556 Thrombocytes with slightly polymorphic nuclei from a reticulated python (*Malayopython reticulatus*) suffering from anorexia and severe chronic stomatitis.

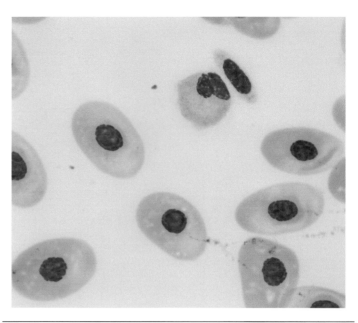

Figure 6.558 A blood film from an Eastern fiddler ray (*Trygonorrhina guaneria*). A large elongated thrombocyte can be seen on the upper right corner of the eosinophil. The eosinophil shows sub-optimal round granular definition and clarity (courtesy of Helen McCracken).

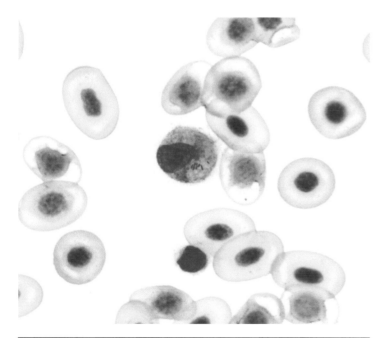

Figure 6.557 Blood smear from a spotted marsh frog (*Limnodynastes tasmaniensis*) presented with profound monocytosis with reactive monocytes, lymphopaenia and abnormal red blood cells. The microphotograph shows a monocyte with phagocytosed bacteria in the cytoplasm and a megathrombocyte at the lower end (courtesy of Helen McCracken).

6.1.15 Cellular Appearance in the Haemolymph of Invertebrates

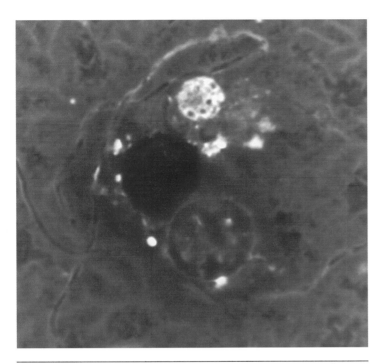

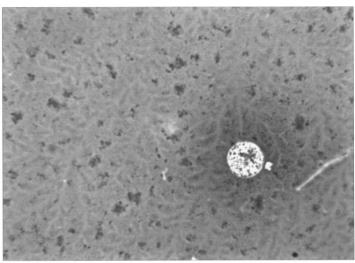

Figure 6.560 A preparation made from the haemolymph of a burgundy goliath birdeater (*Theraphosa stirmi*). One granulocyte is present. The cytoplasm is poorly stained and this makes the cell especially visible on the easily stained haemocyanin proteinaceous background (courtesy of Benjamin M. Kennedy).

Figure 6.559 The end of a smear of theraphosid haemolymph from an adult female burgundy goliath birdeater (*Theraphosa stirmi*). This is the typical appearance of cell distribution when making a slide of haemolymph as the cells produce a single cell monolayer due to its viscosity. The two heavily stained cells are plasmocytes and the poorly staining granulated cell is a granulocyte (courtesy of Benjamin M. Kennedy).

Section C

Atlas of Wild and Exotic Animal Blood Parasites

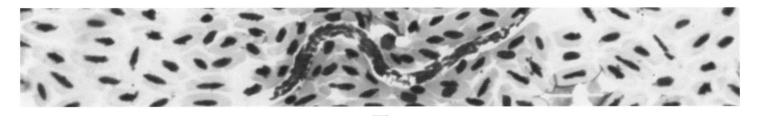

7

BLOOD PARASITES

MICHAEL A. PEIRCE

7.1 Introduction

To the inexperienced observer, there are many artefacts occurring in blood and tissue smears which can be mistaken for parasites, and the converse is also true. For diagnostic and taxonomic purposes, it is essential to avoid contamination of the smears with dust and other particles and with stain precipitates. Whenever possible, the smears should be freshly prepared from living animals as artefacts caused by disintegrating cells and contaminants are found more frequently in anticoagulated and *post-mortem* blood samples. Distortion of any parasites present may take place under these circumstances.

In this section, a wide spectrum of blood parasite species in their normal morphological forms is presented. The majority of slides were prepared from living animals with natural infections or from surveys. In a few instances, experimental material has been included for the sake of morphological clarity. A few examples are shown of parasites in *post-mortem* and anticoagulated blood samples in order to demonstrate some of the problems which can occur.

Since the publication of the previous version of this Atlas, there have been a number of taxonomic changes relating to a number of parasites and the opportunity has been taken to amend and update these where relevant.

7.2 Blood Parasites

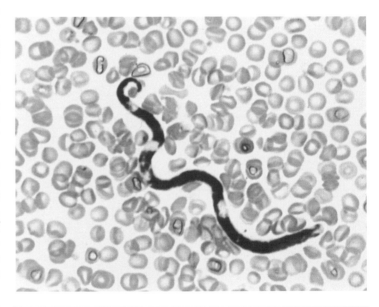

Figure 7.1 Microfilarian in blood from a newly captured capuchin monkey (*Cebus apella*) (×40).

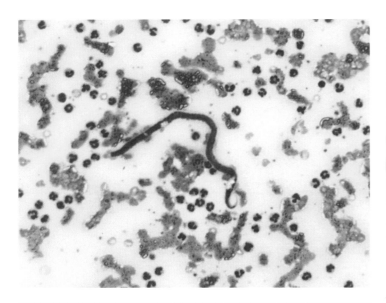

Figure 7.2 *Dirofilaria immitis* (heart worm) microfilarian from the buffy coat preparation of a domestic dog (*Canis lupus familiaris*). This organism has a cosmopolitan distribution in dogs, cats and foxes and is frequently pathogenic due to the presence of a large number of adult worms.

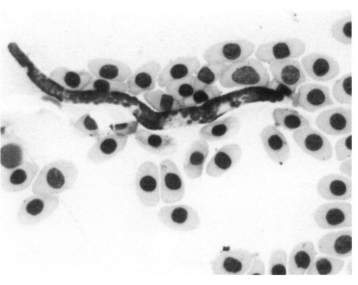

Figure 7.4 A microfilarian from a greater plated lizard (*Gerrhosaurus major*) (×40).

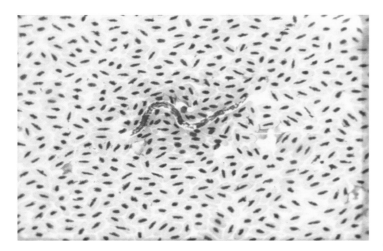

Figure 7.3 *Chandlerella sinensis* microfilarian from a red-billed blue magpie (*Urocissa erythrorhyncha*) (×40).

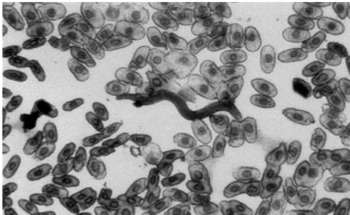

Figure 7.5 *Oswaldofilaria versterae* microfilarian from a Nile crocodile (*Crocodylus niloticus*).

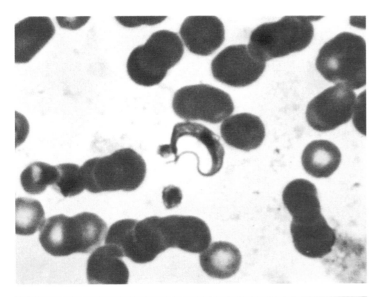

Figure 7.6 *Trypanosoma cruzi*, the cause of Chagas disease in the New World, where it has been recorded in numerous species of mammals. It may be pathogenic in domestic dogs (*Canis lupus familiaris*) and cats (*Felis catus*) but there is little evidence of this in wildlife.

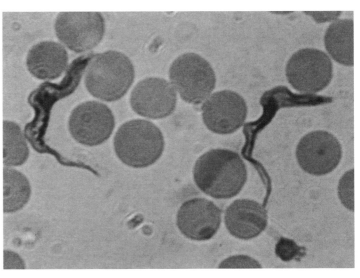

Figure 7.8 *Trypanosoma equiperdum*, the cause of 'Dourine', a venereal disease of equids. This organism is widespread and is more pathogenic in horses (*Equus caballus*) than in donkeys (*Equus asinus*) or mules (*Equus caballus* × *Equus asinus*). The parasites are found only rarely in the bloodstream.

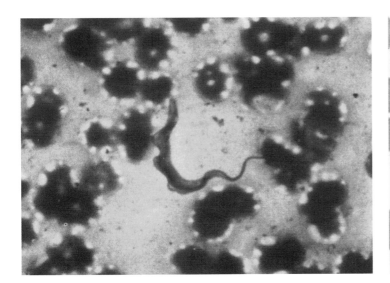

Figure 7.7 *Trypanosoma evansi*, the cause of 'Surra' in camels, equids, Asian elephants (*Elephas maximus indicus*), domestic dogs (*Canis lupus familiaris*), and other domestic and wild animals throughout a wide distribution in hot and warm-temperate climates. Pathogenicity depends upon the virulence of the parasite strain.

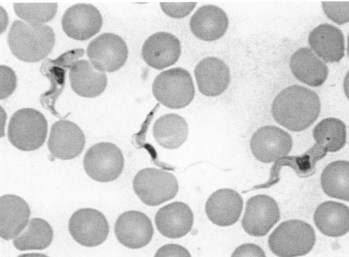

Figure 7.9 *Trypanosoma brucei*, a polymorphic salivarian trypanosome of Africa where it is highly pathogenic and frequently fatal in domestic mammals, particularly horses (*Equus caballus*). The parasite is found in a large number of wild mammal reservoir hosts, especially antelopes.

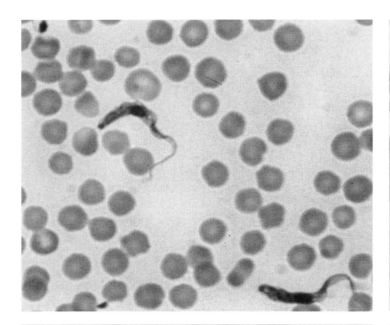

Figure 7.10 *Trypanosoma vivax*, the cause of 'Souma' in Africa and, to a lesser extent, in central South America, the West Indies and Mauritius, where it is pathogenic in domestic mammals, particularly in ungulates. The reservoir hosts are wild mammals, especially antelopes.

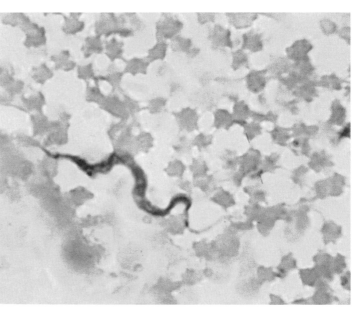

Figure 7.12 *Trypanosoma theileri*, a parasite with a worldwide distribution in cattle, also occurs in European bison (*Bison bonasus*) and in African antelopes and generally considered non-pathogenic.

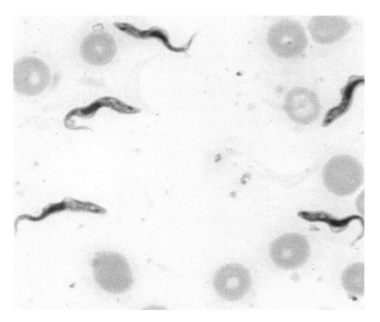

Figure 7.11 *Trypanosoma congolense*, an African species highly pathogenic in domestic mammals and common in wild ruminants. Pathogenicity depends upon the virulence of the parasite strain.

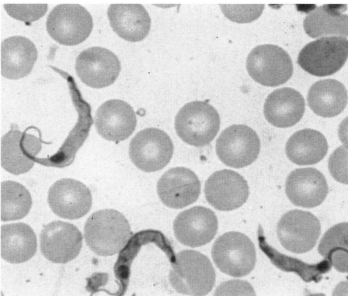

Figure 7.13 *Trypanosoma lewisi*, a parasite with a worldwide distribution in rats (*Rattus* spp.) that is generally considered to be non-pathogenic.

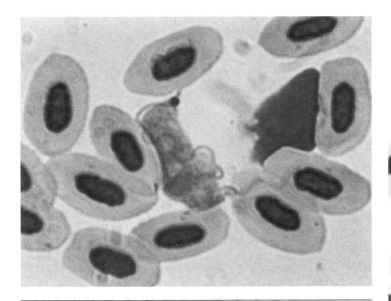

Figure 7.14 *Trypanosoma everetti* from a blue tit (*Cyanistes caeruleus*). This is a small trypanosome, occurring in a wide range of avian hosts in Europe, Africa and probably Asia. It is not known to be pathogenic.

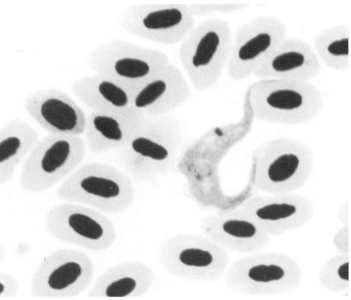

Figure 7.16 *Trypanosoma bouffardi* from a blue waxbill (*Uraeginthus angolensis*). This is a non-pathogenic species occurring in African passerines.

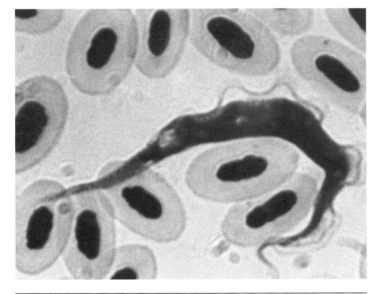

Figure 7.15 *Trypanosoma pycnonoti* from a white-eared bulbul (*Pycnonotus leucotis*). This is a non-pathogenic species occurring in African bulbuls.

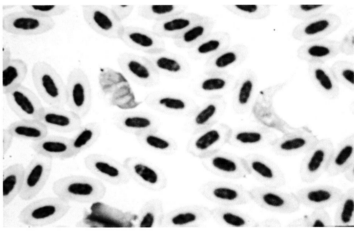

Figure 7.17 *Trypanosoma everetti* and *T. bouffardi* mixed infection from a blue waxbill (*Uraeginthus angolensis*).

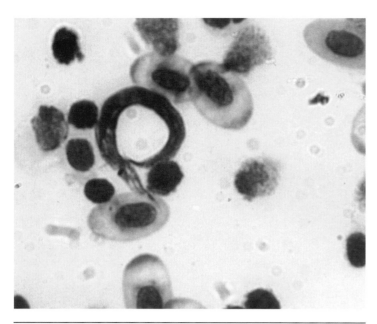

Figure 7.18 *Trypanosoma avium* from a tawny owl (*Strix aluco*).

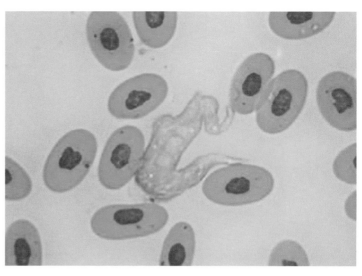

Figure 7.20 *Trypanosoma constrictor* from a rainbow boa (*Epicrates cenchria*) showing degenerating morphology.

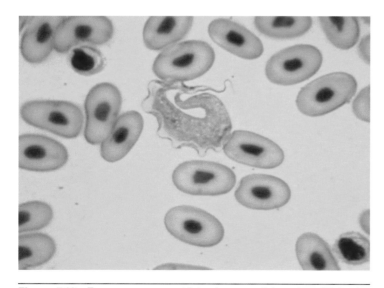

Figure 7.19 *Trypanosoma constrictor* from a rainbow boa (*Epicrates cenchria*).

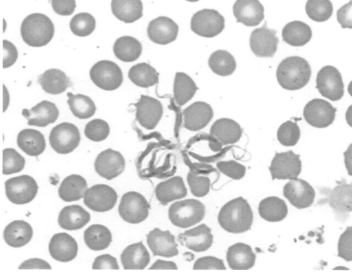

Figure 7.21 *Trypanosoma nabiasi* from a rabbit (*Oryctolagus cuniculus*) showing a cluster of parasites.

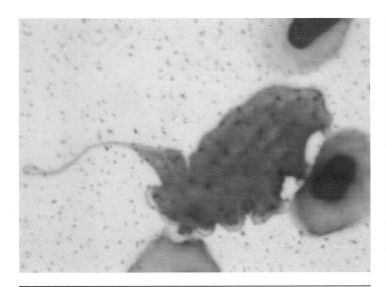

Figure 7.22 *Trypanosoma rotatorium* from a pool frog (*Pelophylax lessonae*).

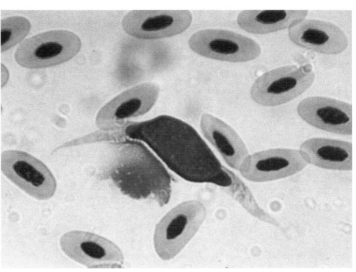

Figure 7.24 *Leucocytozoon danilewskyi* macrogametocyte in an erythrocyte from little owl (*Athene noctua*). This parasite occurs worldwide in Strigiformes but is not known to be pathogenic.

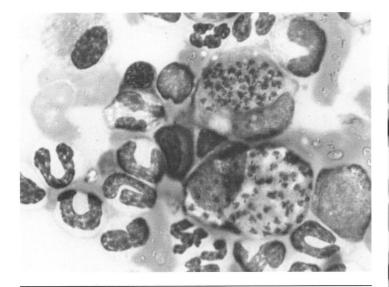

Figure 7.23 *Leishmania donovani* in the white cells of a domestic dog (*Canis lupus familiaris*) in which it is pathogenic, although rarely fatal (bone marrow smear).

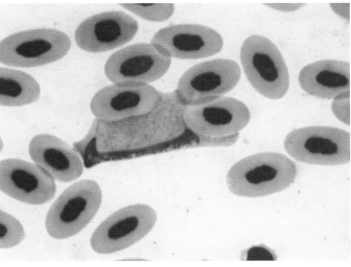

Figure 7.25 A distorted microgametocyte of *L. danilewskyi* in an EDTA blood sample from a tawny owl (*Strix aluco*).

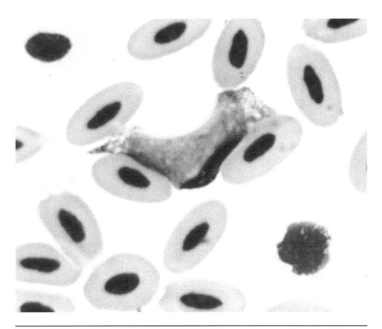

Figure 7.26 *Leucocytozoon* sp., macrogametocyte from a Mikado pheasant (*Syrmaticus mikado*).

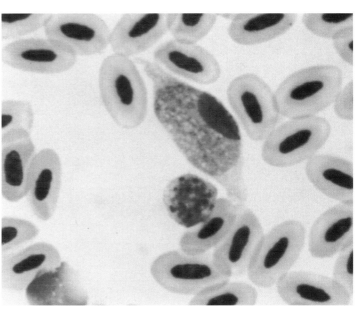

Figure 7.28 *L. balmorali* macrogametocyte (elongate form) from a puffback shrike (*Dryoscopus cubla*).

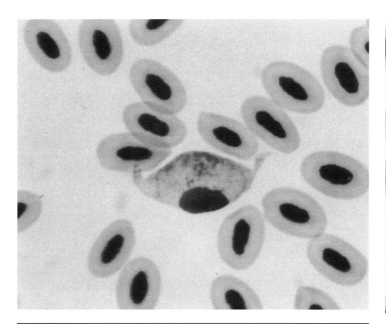

Figure 7.27 *Leucocytozoon balmorali* microgametocyte (elongate form) in a puffback shrike (*Dryoscopus cubla*) erythrocyte. This organism is not known to be pathogenic.

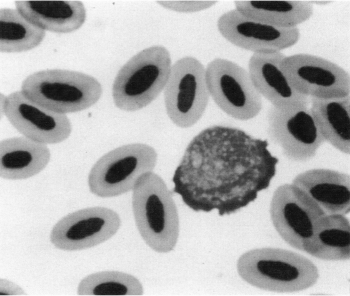

Figure 7.29 *L. balmorali* macrogametocyte (round form) from a puffback shrike (*Dryoscopus cubla*).

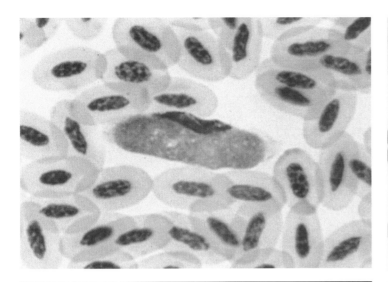

Figure 7.30 *Leucocytozoon toddi* macrogametocyte from a goshawk (*Accipiter gentilis*). This organism is not known to be pathogenic, although high parasitaemias frequently occur in chicks of some Accipitridae. It has a worldwide distribution.

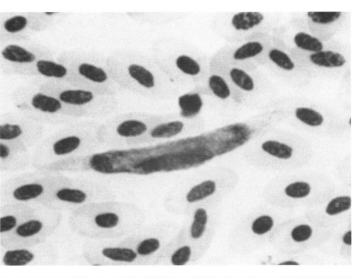

Figure 7.32 *Leucocytozoon naevei* macrogametocyte from a yellow-necked spurfowl (*Pternistis leucoscepus*). This parasite is common in African Phasianidae but there is no evidence of pathogenicity.

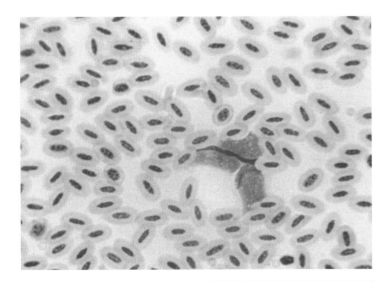

Figure 7.31 Three distorted *L. toddi* macrogametocytes from a goshawk (*Accipiter gentilis*).

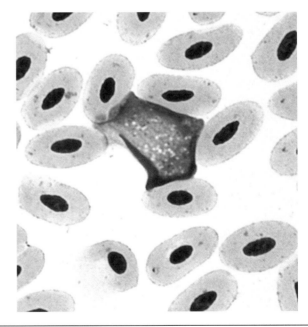

Figure 7.33 *Leucocytozoon tawaki* macrogametocyte from a macaroni penguin (*Eudyptes chrysolophus*).

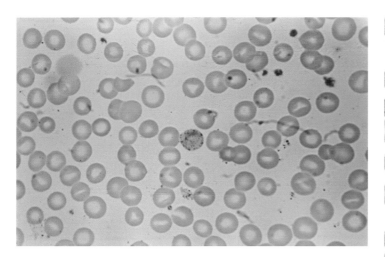

Figure 7.34 *Plasmodium brasilianum* macrogametocyte from a newly imported squirrel monkey (*Saimiri sciureus*). This is a pathogenic parasite occurring in New World monkeys.

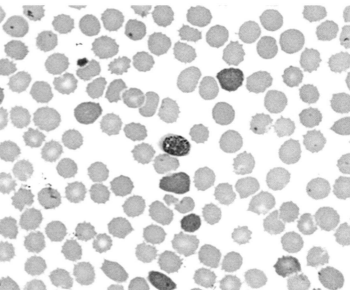

Figure 7.36 *Polychromophilus melanipherus* microgametocyte from a West African cave bat (*Miniopterus inflatus*).

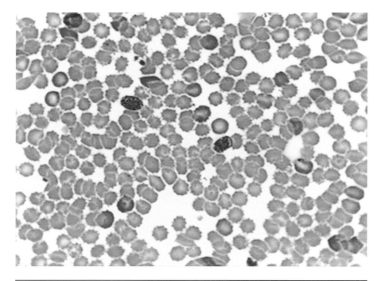

Figure 7.35 *Polychromophilus melanipherus* macrogametocytes from a West African cave bat (*Miniopterus inflatus*).

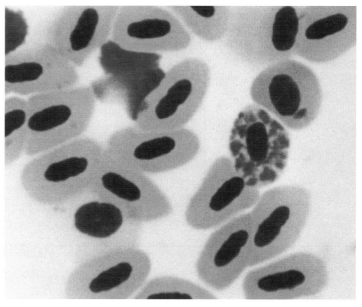

Figure 7.37 *Plasmodium circumflexum* schizont from a blue waxbill (*Uraeginthus angolensis*). This organism has a cosmopolitan distribution in passerines and is also found in grouse, geese and waders. It is mildly pathogenic in its natural hosts.

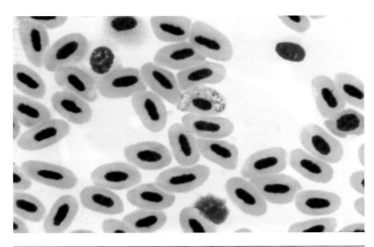

Figure 7.38 *Plasmodium circumflexum* microgametocyte from a tawny owl (*Strix aluco*).

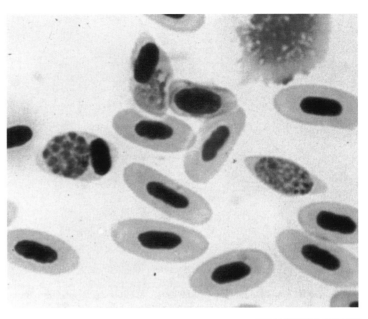

Figure 7.40 *Plasmodium relictum*, a parasite with worldwide distribution in many avian families, in some of which it is pathogenic and often fatal.

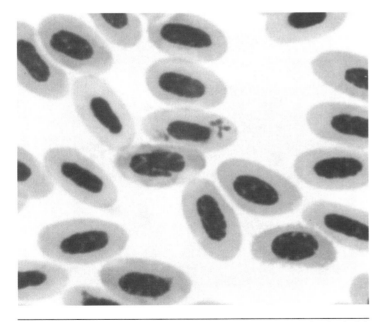

Figure 7.39 *Plasmodium rouxi* schizont from a dark plain-backed pipit (*Anthus leucophrys*). The parasite is common in Africa and Asia and has been recorded less often in Europe and the US. It occurs more frequently in passerines than in other avian orders and may be mildly pathogenic in its natural hosts.

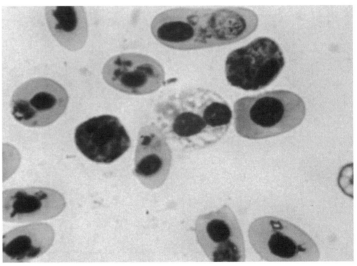

Figure 7.41 *Plasmodium gallinaceum* from a domestic fowl (*Gallus gallus domesticus*). The natural host in Asia is the jungle fowl (*Gallus gallus*) in which it is non-pathogenic. In domestic fowls it may be highly pathogenic.

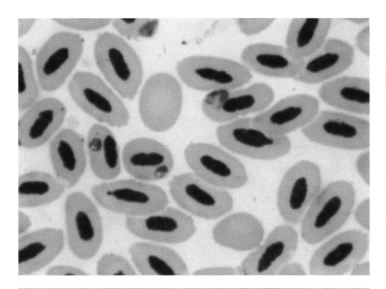

Figure 7.42 *Plasmodium* sp., trophozoites from the *post-mortem* examination of a wattled starling (*Creatophora cinerea*). There are two erythoplastids in the field.

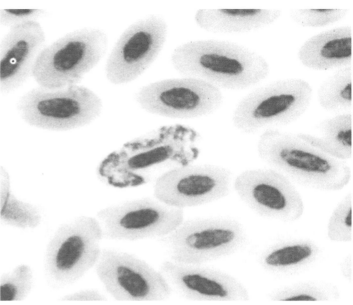

Figure 7.44 *Haemoproteus nisi* macrogametocyte from a sparrowhawk (*Accipiter nisus*). Note the reddish-purple volutin granules in the parasite cytoplasm. This species is not known to be pathogenic.

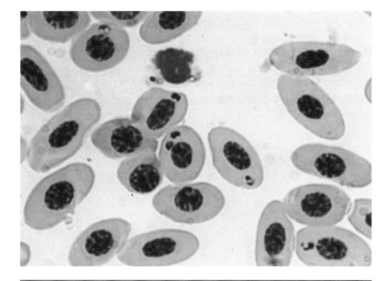

Figure 7.43 *Plasmodium* sp., trophozoites in an EDTA sample from a snowy owl (*Bubo scandiacus*).

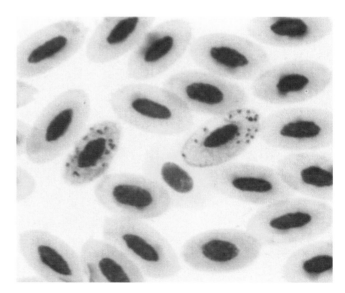

Figure 7.45 Macrogametocyte and microgametocyte of *H. nisi* from a sparrow hawk (*Accipiter nisus*).

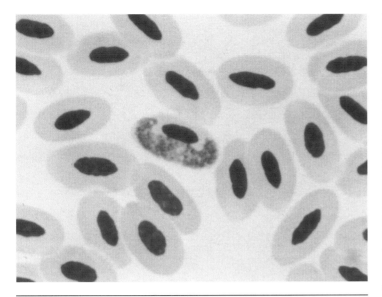

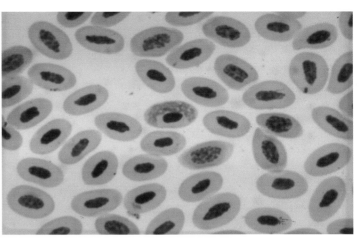

Figure 7.48 *Haemoproteus handai* macrogametocytes from a ring-necked parakeet (*Psittacula krameri*). The parasite occurs in Asian and Australasian psittacines and is not known to be pathogenic, although high parasitaemia may cause anaemia.

Figure 7.46 *Haemoproteus columbae* macrogametocyte from a Cape turtle dove (*Streptopelia capicola*). This parasite has a cosmopolitan distribution in Columbiformes and is reported to be pathogenic in young pigeons.

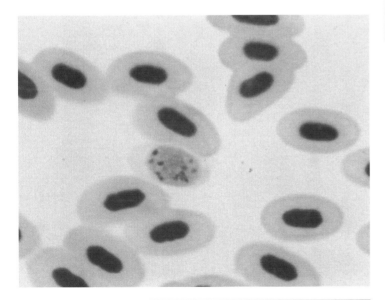

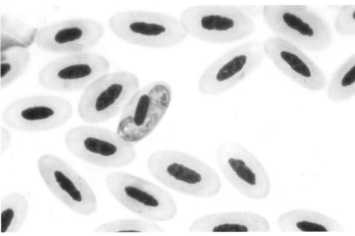

Figure 7.49 *Haemoproteus antigonis* from a crowned crane (*Balearica regulorum*).

Figure 7.47 *Haemoproteus sequeirae* macrogametocyte in an enucleated erythrocyte of a scarlet-chested sunbird (*Chalcomitra senegalensis*). This is a non-pathogenic species occurring in Nectariniidae.

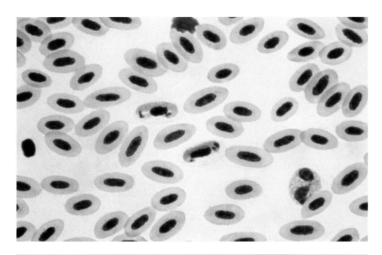

Figure 7.50 Macrogametocyte and microgametocyte of *Haemoproteus psittaci* from an African grey parrot (*Psittacus erithacus*).

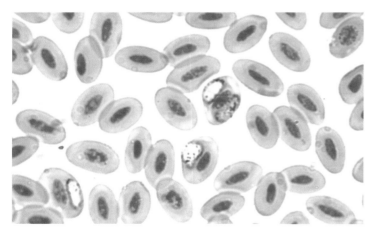

Figure 7.51 Distorted *Haemoproteus crumeniferus* parasite in an EDTA blood sample from a healthy marabou stork (*Leptoptilos crumenifer*).

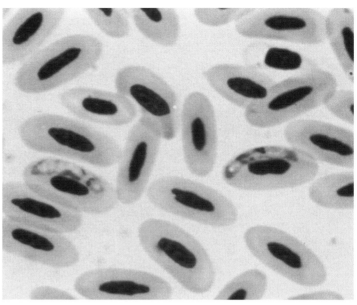

Figure 7.52 Macrogametocyte and microgametocyte of *Haemoproteus balearicae* from a crowned crane (*Balearica regulorum*). This organism is not known to be pathogenic.

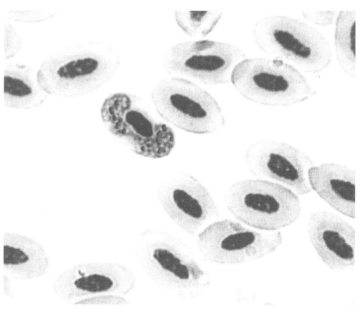

Figure 7.53 Macrogametocyte and microgametocyte of *Haemoproteus syrnii* from a tawny owl (*Strix aluco*).

Figure 7.54 Macrogametocytes of *Haemoproteus passeris* from a house sparrow (*Passer domesticus*).

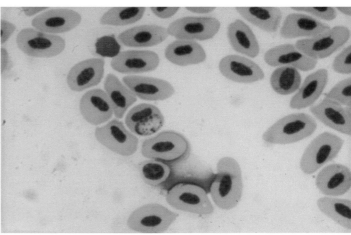

Figure 7.56 Distorted *Haemoproteus* sp., in an EDTA blood sample from a tawny owl (*Strix aluco*).

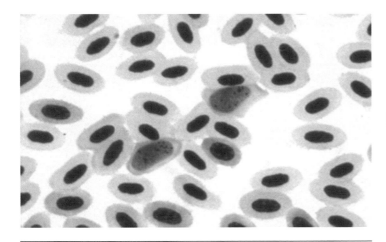

Figure 7.55 Macrogametocytes of *Haemoproteus enucleator* from a kingfisher (*Alcedinidae*).

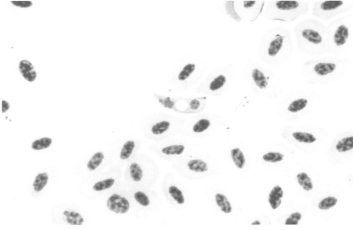

Figure 7.57 *Haemoproteus* sp., ookinete in a 48-hour-old EDTA blood sample from a tawny owl (*Strix aluco*). This stage normally occurs in the vector but may be seen occasionally in blood samples maintained at ambient temperatures for sufficient time to permit development.

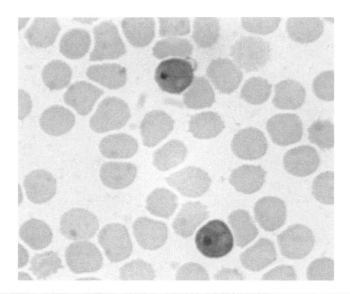

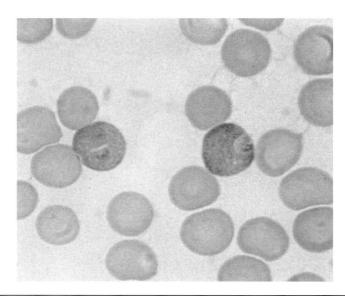

Figure 7.58 *Hepatocystis epomophori* macrogametocyte and microgametocyte from a Zambian fruit bat (*Epomophorus gambianus*). This parasite is not known to be pathogenic.

Figure 7.60 Mature gametocytes of *H. kochi* from a vervet money (*Chlorocebus pygerythrus*).

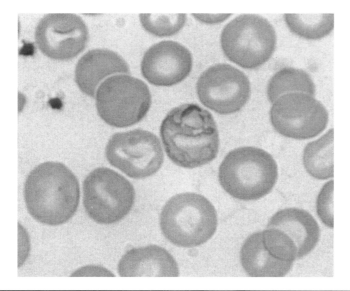

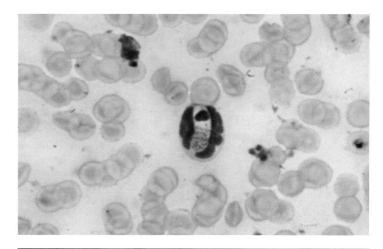

Figure 7.61 *Hepatozoon sciuri* in a white blood cell from a grey squirrel (*Sciurus carolinensis*). The parasite is non-pathogenic.

Figure 7.59 *Hepatocystis kochi* immature gametocyte from a vervet monkey (*Chlorocebus pygerythrus*). This parasite occurs in guenons and other African monkeys and is mildly pathogenic but non-fatal.

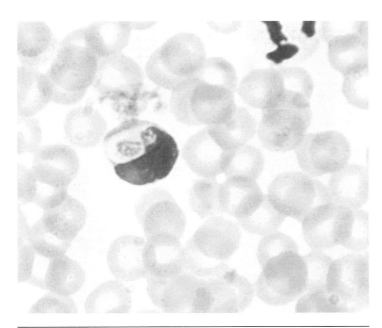

Figure 7.64 *Hepatozoon* sp., from a giant panda (*Ailuropoda melanoleuca*).

Figure 7.62 *Hepatozoon canis* in a white blood cell from a domestic dog (*Canis lupus familiaris*). The parasite has a worldwide distribution in dogs and wild canids. In dogs it is usually pathogenic and often fatal.

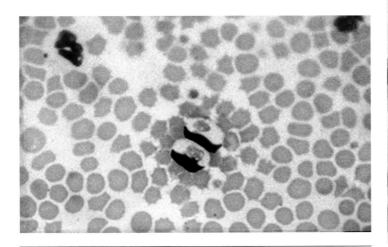

Figure 7.63 *Hepatozoon* sp., gametocytes from a cheetah (*Acinonyx jubatus*).

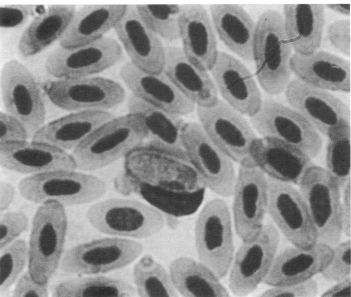

Figure 7.65 *Hepatozoon albatrossi* in a mononuclear white blood cell from a wandering albatross (*Diomedea exulans*). The parasite is not known to be pathogenic.

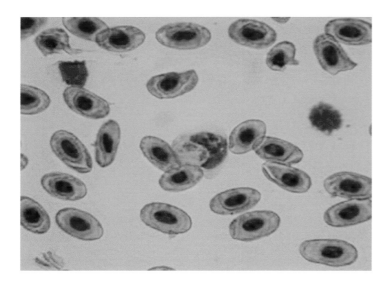

Figure 7.66 *Hepatozoon kiwii* from a brown kiwi (*Apteryx mantelli*).

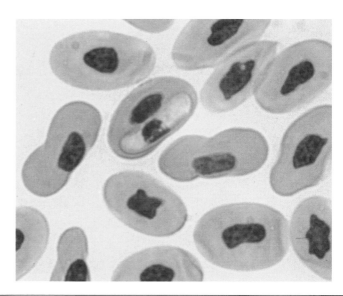

Figure 7.68 *Hepatozoon sebai* from an African rock python (*Python sebae*). The parasite is not known to be pathogenic.

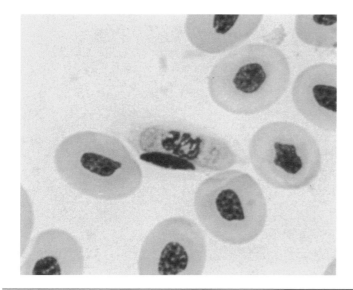

Figure 7.67 *Hepatozoon boodoni* in a Zambian house snake (*Boaedon fuliginosus*) red blood cell. The parasite is not known to be pathogenic.

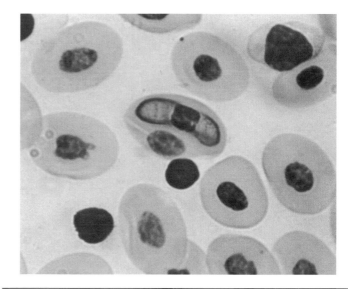

Figure 7.69 *Hepatozoon musotae* from a Seychelles house snake (*Boaedon geometricus*). This parasite also occurs in other species of house snake on mainland Africa. It is not known to be pathogenic.

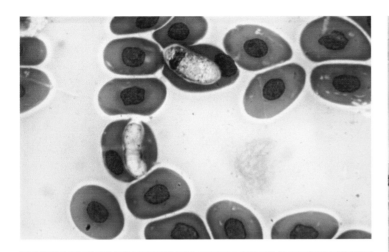

Figure 7.70 *Hepatozoon serpentium* in a *post-mortem* blood sample from a green anaconda (*Eunectes murinus*).

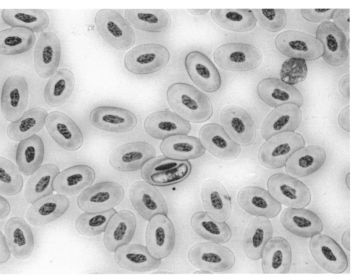

Figure 7.72 *Hepatozoon guttata* from a corn snake (*Pantherophis guttatus*).

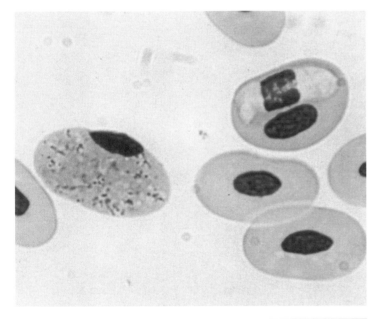

Figure 7.71 *Hepatozoon najae* and *Haemoproteus mesnili* from a Mozambique spitting cobra (*Naja mossambica*). This haemogregarine is widely distributed in African cobras but is not known to be pathogenic.

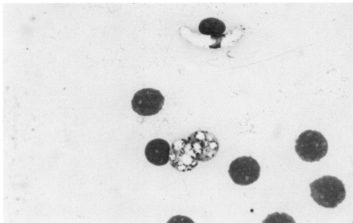

Figure 7.73 Distorted *Hepatozoon* and *Haemoproteus* parasites in a hae-molysed EDTA blood sample from a yellow-footed tortoise (*Chelonoidis denticulatus*).

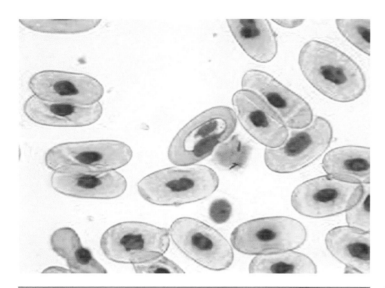

Figure 7.74 *Hepatozoon caimani* from a caiman (*Caimaninae*).

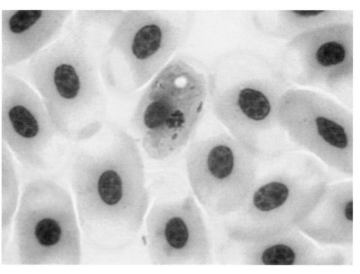

Figure 7.76 *Haemoproteus testudinalis* from a leopard tortoise (*Stigmochelys pardalis*).

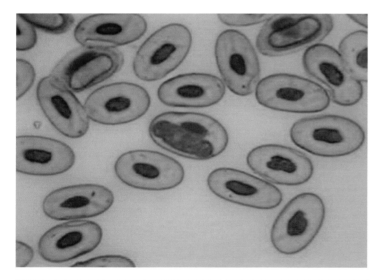

Figure 7.75 *Haemocystidium apigmentada* and *Hepatozoon* sp., from a spiny-tailed lizard (*Uromastyx aegyptia*).

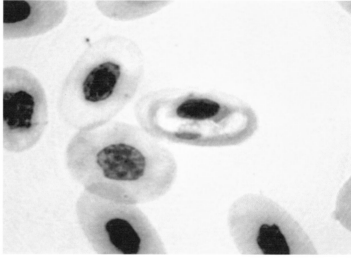

Figure 7.77 *Hepatozoon fitzsimonsi* from an angulate tortoise (*Chersina angulata*).

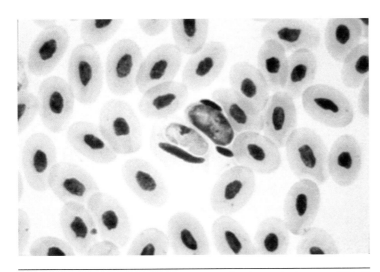

Figure 7.78 *Karyolysus latus* macrogametocyte and microgametocyte from a sand lizard (*Lacerta agilis*). Not known to be pathogenic.

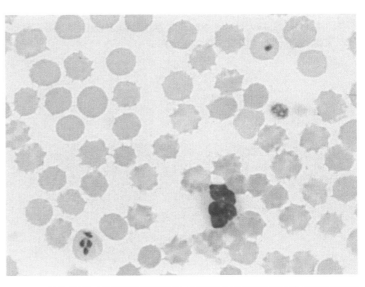

Figure 7.80 *Babesia gibsoni* from a domestic dog (*Canis lupus familiaris*). The parasite is common in dogs in the Indian sub-continent, China and North Africa and is pathogenic, often fatal. It also occurs in wild canids in which it is non-pathogenic.

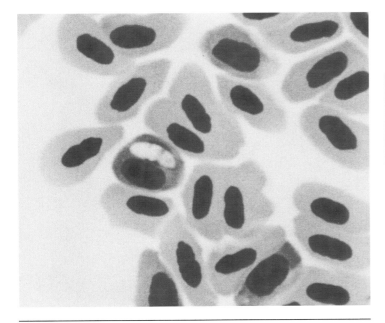

Figure 7.79 *Lankesterella* sp., in a basophilic erythroblast from a house sparrow (*Passer domesticus*). The parasite may be pathogenic in young birds.

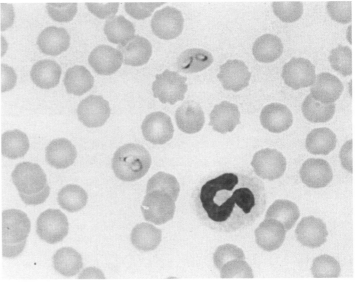

Figure 7.81 *Babesia canis* chronic infection in a domestic dog (*Canis lupus familiaris*). This organism has a cosmopolitan distribution in the dog and wild carnivores and is pathogenic, often fatal.

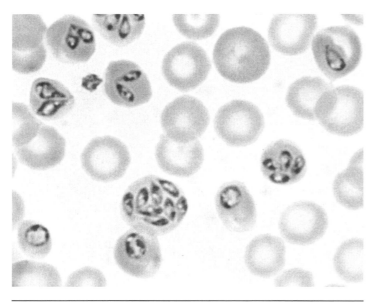

Figure 7.82 *B. canis* acute infection in a domestic dog (*Canis lupus familiaris*) showing multiple divisions in the erythrocytes, leading to severe anaemia.

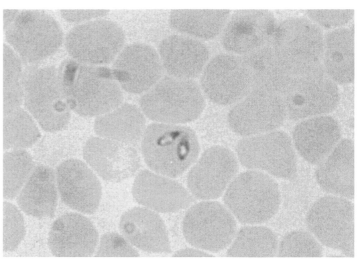

Figure 7.84 *Babesia caballi* from a horse (*Equus caballus*). Dividing pyriform merozoites are present. The organism is widely distributed in horses, donkeys and mules and is less pathogenic than *T. equi* although it can be fatal.

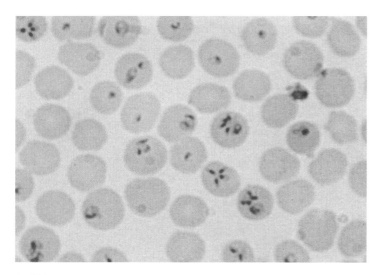

Figure 7.83 *Theileria equi* from a horse (*Equus caballus*). This organism is widely distributed in equids including zebras. In most equids it is pathogenic and frequently fatal although it is not known to be fatal in zebras.

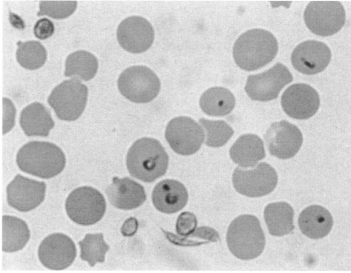

Figure 7.85 Acute infection with *B. caballi* from a horse (*Equus caballus*) showing some parasites becoming pyknotic.

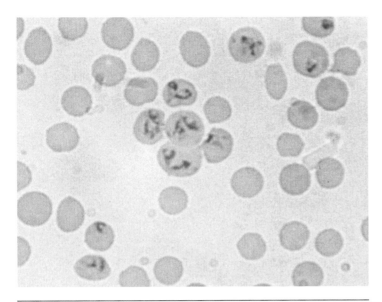

Figure 7.86 *Babesia major* from cattle (*Bos taurus*). This organism is widely distributed in cattle in South America, southern Europe, Great Britain, Russia and parts of North and West Africa. It is also found in bison (*Bison bonasus*). Pathogenicity is usually mild, except in bison in which it can be fatal.

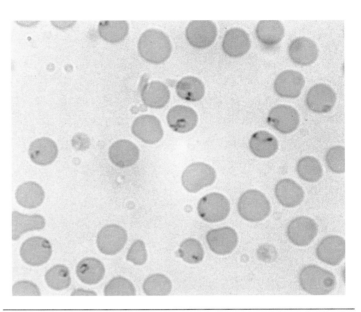

Figure 7.88 *Babesia divergens* from cattle (*Bos taurus*). The distribution of this parasite is mainly confined to Northern Europe where it occurs in cattle and probably also in deer. It is pathogenic and sometimes fatal, with haemoglobinuria giving rise to the classical 'Red water fever'.

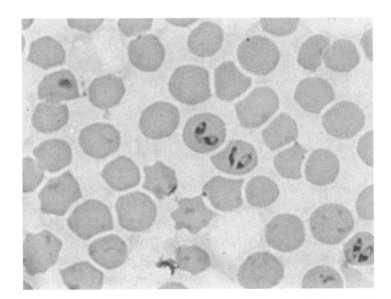

Figure 7.87 *Babesia bigemina* from cattle (*Bos taurus*). This parasite is widely distributed in tropical and sub-tropical regions and is pathogenic with a high mortality in acute cases.

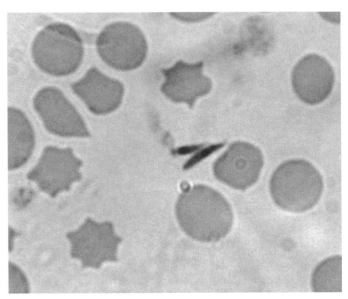

Figure 7.89 *B. divergens* vermicules in cattle (*Bos taurus*) blood from a natural tick-transmitted infection. It is very rare to observe these stages in the blood.

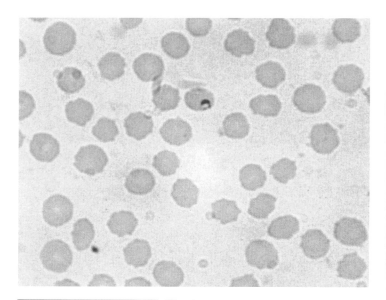

Figure 7.90 *Babesia ovis* from a sheep (*Ovis aries*). This parasite is widely distributed in sheep and goats (*Capra aegagrus hircus*) and is pathogenic but rarely fatal.

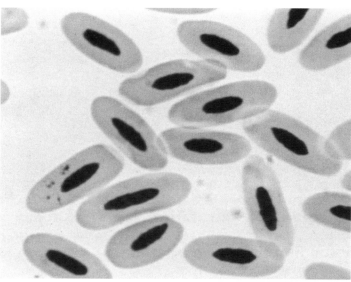

Figure 7.92 *Babesia balearicae* from a crowned crane (*Balearica regulorum*). Parasites of the genus *Babesia* are rare in birds, although several species have been described.

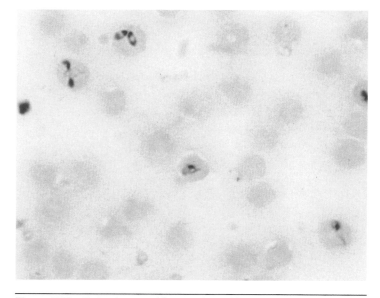

Figure 7.91 *Babesia motasi* from a sheep (*Ovis aries*). Also widely distributed in sheep and goats (*Capra aegagrus hircus*) but, compared with *B. ovis*, is larger, more pathogenic and frequently fatal in the acute form.

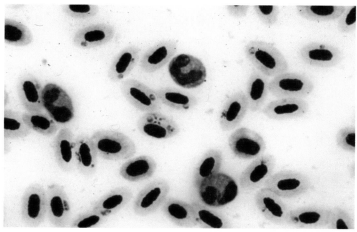

Figure 7.93 *Babesia shortti* from a kestrel (*Falco tinnunculus*) showing typical Maltese-cross shaped schizonts. This species often shows a high parasitaemia and can be pathogenic.

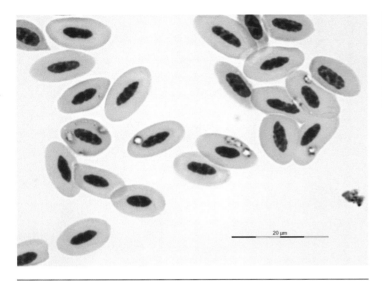

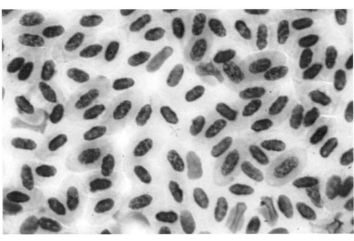

Figure 7.96 *Babesia rustica* from a red-rumped swallow (*Cecropis dau-rica*) showing a cruciform schizont.

Figure 7.94 *Babesia shortti* from a saker falcon (*Falco cherrug*) showing trophozoites and schizont precursors.

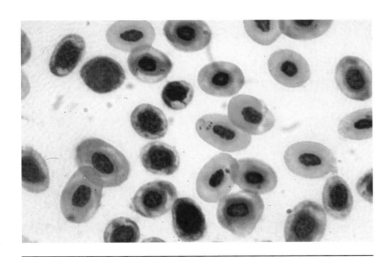

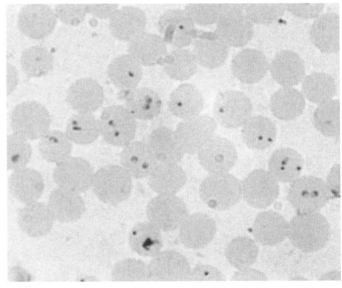

Figure 7.95 *Babesia peircei* from an African penguin (*Spheniscus demersus*) showing a fan-shaped schizont frequently found in this species. Known to be pathogenic.

Figure 7.97 *Theileria parva* (East Coast fever) from cattle (*Bos taurus*). This intracellular organism is pathogenic with a high mortality in East African cattle but is non-pathogenic in Cape buffalo (*Syncerus caffer*).

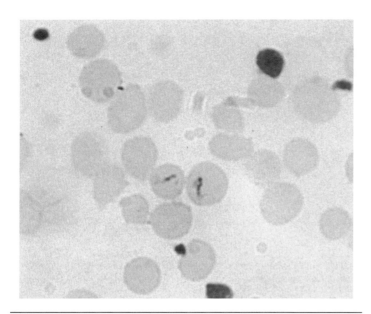

Figure 7.98 *Theileria orientalis*, the cause of benign bovine theileriosis, has a wide distribution but most strains are only slightly pathogenic and non-fatal. *T. orientalis* is morphologically indistinguishable from *T. mutans* which is now considered to be a strictly African species.

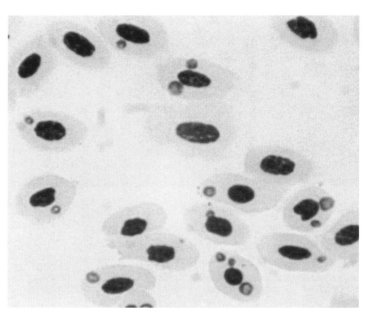

Figure 7.100 *Aegyptianella pullorum* from a domestic fowl (*Gallus gallus domesticus*). This organism occurs in the red blood cells of domestic ducks (*Anas platyrhynchos domesticus*), geese (*Anser* sp.) and turkeys (*Meleagris gallopavo*) and may be highly pathogenic, particularly to newly introduced stock. Its distribution follows that of the fowl tick (*Argas persicus*) by which it is transmitted.

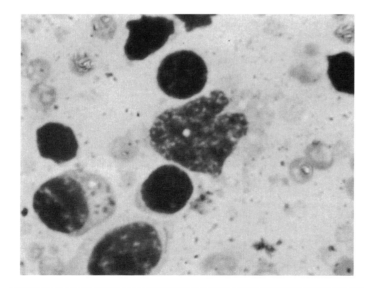

Figure 7.99 Schizont of *Theileria lestoquardi* (*T. hirci*) in a lymphocyte from a sheep (*Ovis aries*) (lymph node smear). The parasite is widely distributed in sheep (*Ovis aries*) and goats (*Capra aegagrus hircus*) and is highly pathogenic, usually fatal, in adult animals.

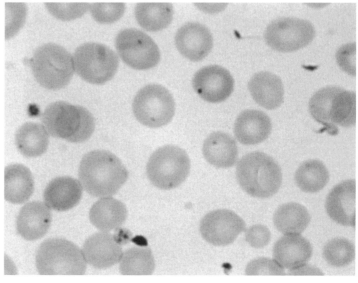

Figure 7.101 *Anaplasma marginale* from an eland (*Taurotragus oryx*). This intracellular parasite has a wide distribution in cattle (*Bos taurus*) and wild ungulates. In cattle, pathogenicity increases with age and, in animals older than three years, the disease is often fatal.

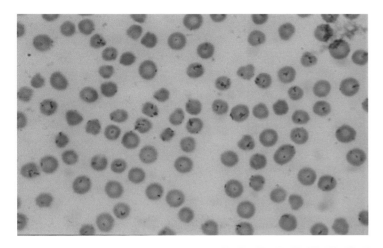

Figure 7.102 *Eperythrozoon ovis* from a sheep (*Ovis aries*). This intracellular organism has a wide distribution in sheep and goats (*Capra aegagrus hircus*) in which it is occasionally pathogenic.

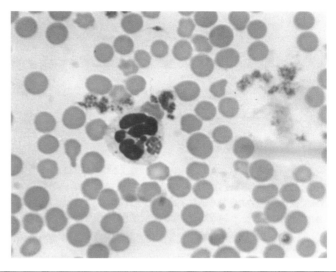

Figure 7.104 *Ehrlichia* (= *Cytoecetes*) *phagocytophilia* from cattle (*Bos taurus*), the cause of bovine and ovine tick-borne fever in Europe. The parasite occurs as a morula in leucocytes. The disease is generally of low pathogenicity and is rarely fatal, although it may cause abortion and drop in milk yield.

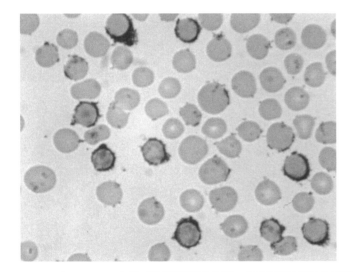

Figure 7.103 Bovine eperythrozoonosis. This has a worldwide distribution in cattle (*Bos taurus*) and may be mildly pathogenic although rarely fatal.

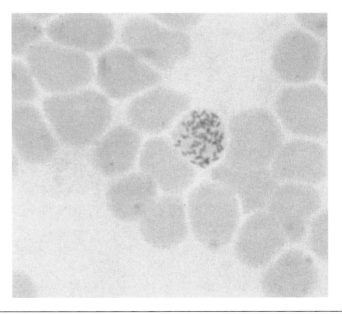

Figure 7.105 *Grahamella* sp., from a giant rat (*Cricetomys gambianus*). This organism occurs commonly in many species of small mammals worldwide. It is not known to be pathogenic.

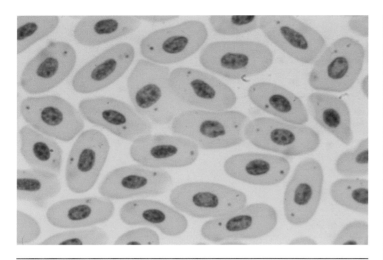

Figure 7.106 *Pirhemocyton* sp., from an Indian python (*Python molurus*). A widely distributed parasite in reptiles. The morphology is variable and the parasitaemia can be 100%, but not known to be pathogenic.

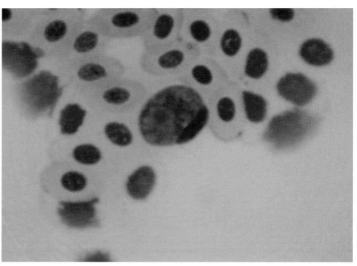

Figure 7.108 *Eimeria reichenowi* mature trophozoite in blood from a Eurasian crane (*Grus grus*). Clinically, the cause of disseminated visceral coccidiosis (DVC).

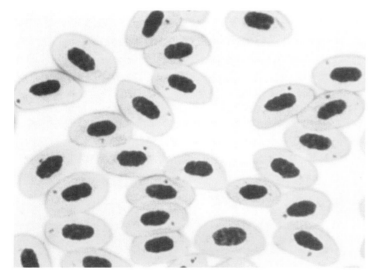

Figure 7.107 Rickettsia-like organisms, morphologically resembling *Pirhemocyton*, in the blood of an olive thrush (*Turdus abyssinicus*). Similar organisms are frequently encountered in avian hosts.

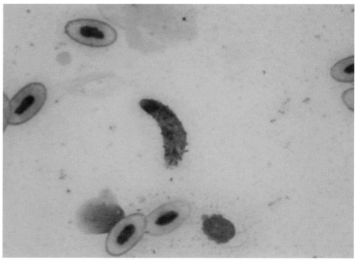

Figure 7.109 DVC from a Eurasian crane (*Grus grus*) showing an extra-cellular large sporozoite.

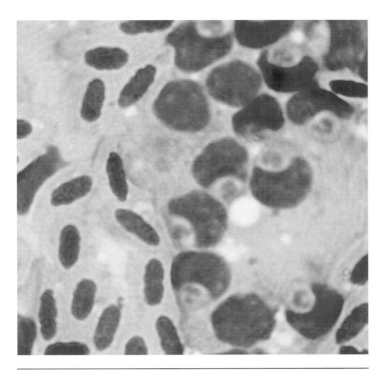

Figure 7.110 *Isospora normanlevinei* from a cirl bunting (*Emberiza cirlus*). Merozoites in leucocytes from a spleen impression smear.

Section D

Appendix

APPENDIX
NORMAL HAEMATOLOGY REFERENCE VALUES IN SELECTED WILD AND EXOTIC ANIMALS

Mammals

Haematology Reference Values in Ferrets (*Mustela putorius furo*) (Female and Male Animals Combined)

Parameter	Reference Range	
Haematocrit (l/l)	0.4–0.7	
Haemoglobin (g/l)	13.85–20.94	
Erythrocytes (t/l)	7.4–13.0	
Leukocytes (g/l)	3.0–16.7	
Thrombocytes s (g/l)	171.7–1280.6	
MCV (fl)	49.6–60.6	
MCH (mmol/l)	17.8–20.9	
MCHC (fmol/l)	1.0–1.2	
Parameter	(g/l)	(%)
Monocytes	0.0–0.5	0.0–6.5
Lymphocytes	0.6–10.5	12.6–80.6
Band neutrophils	0.0–0.1	0.0–1.2
Segmented neutrophils	0.9–7.4	17.2–81.9
Eosinophils	0.0–0.7	0.0–5.7
Basophils	0.0–0.2	0.0–1.4

Extracted from: Hein J, Speyer F, Hartmann K, Sauter-Louis C (2012). Reference ranges for laboratory parameters in ferrets. *Vet Rec*, 171: 218.

Haematology Reference Values in New Zealand White Rabbits (*Oryctolagus cuniculus*)

Parameters	Adult Male	Adult Female	1–3 Months
RBC ($\times 10^6$/μl)	5.46–7.94	5.11–6.51	5.15–6.48
PCV (%)	33–50	31.0–48.6	38.1–44.1
Hgb (g/dl)	10.4–17.4	9.8–15.8	10.7–13.9
MCV (fl)	58.5–66.5	57.8–65.4	66.2–80.3
MCH (pg)	18.7–22.7	17.1–23.5	19.5–22.7
MCHC (%)	33–50	28.7–35.7	24.2–32.6
Platelets ($\times 10^3$/μl)	304–656	270–630	
WBC ($\times 10^3$/μl)	5.5–12.5	5.2–10.6	4.1–9.79
Neutrophils (%)	38–54	36.4–50.4	18.8–46.4
Lymphocytes (%)	28–50	31.5–52.1	44.6–77.8
Eosinophils (%)	0.5–3.5	0.8–3.2	0–2.4
Basophils (%)	2.5–7.5	2.4–6.2	0.1–4.5
Monocytes (%)	4–12	6.6–13.4	0–13.1

Extracted from: Moore DM, Zimmermann K, Smith SA (2015). Hematological assessment in pet rabbits. *Vet Clin Exot Anim*, 18:9–19.

Haematology Reference Values in Guinea Pigs (*Cavia porcellus*)

Parameter	Adult Male	Adult Female	2–90 Days Old
RBC (×10⁶/µl)	4.36–6.84	3.35–6.15	4.06–6.02
PCV (%)	37–47	40.9–49.9	33.8–48.8
Hgb (g/dl)	11.6–17.2	11.4–17.0	10.13–15.1
MCV (fl)	71–83	86.1–95.9	77.5–88.7
MCH (pg)	24.2–27.2	23.1–26.3	–
MCHC (%)	29.7–38.9	28.2–34.4	28.3–32.4
Platelets (×10³/µl)	260–740	266–634	–
WBC (×10³/µl)	5.5–17.5	5.2–16.4	2.66–10.1
Neutrophils (%)	28–56	20.3–41.9	14.8–42.6
Lymphocytes (%)	40.0–62.5	46.4–80.4	52.6–83.2
Eosinophils (%)	1–7	0–7	0.1–3.6
Basophils (%)	0–1.7	0–0.8	0–0.58
Monocytes (%)	3.3–5.3	1.0–2.6	0–0.37

Extracted from: Zimmerman K, Moore DM, Smith SA (2015). Hematological assessment in pet guinea pigs (*Cavia porcellus*): Blood sample collection and blood cell identification. *Vet Clin Exot Anim*, 18:33–40.

Haematology Reference Values in Chinchillas (*Chinchilla laniger*)*

Parameter	Mean ± Standard Error of Mean	95% Confidence Limit
PCV (%)	42.6 ± 1.2	39.2–45.9
Hgb (g/dl)	15.6 ± 0.8	13.6–17.6
RBC (×10⁶/µl)	4.3 ± 0.1	4.1–4.6
MCV (fl)	101.3 ± 3.9	90.4–112.3
MCH (pg)	3.6 ± 0.3	3.0–4.3
MCHC (g/dl)	40.4 ± 1.3	36.7–44.1
WBC (×10³/µl)	11.1 ± 1.3	8.0–14.3
Lymphocytes (%)	40.2 ± 3.9	30.9–49.6
Monocytes (%)	6.3 ± 1.2	3.6–9.1
Segmented neutrophils (%)	46.9 ± 3.9	37.6–56.2
Bastonated neutrophils (%)	2.1 ± 0.7	0.4–3.8
Basophils (%)	1.7–0.7±	0.9–3.4
Eosinophils (%)	2.6 ± 1.1	0.2–5.2

*Blood sample collected by cardiac puncture following anaesthesia
Extracted from: De Silva OT, Kreutz LC, Barcellos LJG, Borella J, SoSo AB, Souza C (2005). Reference values for chinchilla (*Chinchilla laniger*) blood cells and serum biochemical parameters. *Cienc Rural*, 35(3):602–606.

Haematology Reference Values in Rats, Mice, Hamsters and Gerbils

Parameters	Rat Male	Rat Female	Mouse Male	Mouse Female	Hamster Male	Hamster Female	Gerbil Male	Gerbil Female
RBC (×10⁶/ml)	8.15–9.75	6.76–9.2	6.9–11.7	6.86–11.3	4.7–10.3	3.96–9.96	7.1–8.6	8.0–9.4
PCV (%)	44.4–50.4	37.6–50.6	33.1–49.9	39.7–44.5	47.9–57.1	39.2–58.8	42–49	43–50
Hgb (g/dL)	13.4–15.8	11.5–16.1	11.1–11.5	10.7–11.1	14.4–19.2	13.1–18.9	12.1–13.8	13.1–16.9
MCV (fL)	49.8–57.8	50.9–65.5	47.5–50.5	47–52	64.8–77.6	64–76	46.6–60	46.64–60.04
MCH (pg)	14.3–18.3	15.6–19	11.7–12.7	11.1–12.7	19.9–24.9	20.2–25.8	16.1–19.4	16.3–19.4
MCHC (%)	26.2–35.4	26.5–36.1	23.2–31.2	22.3–29.5	27.5–36.5	27.8–37.4	30.6–33.3	30.6–33.3
Platelets (×10³/ml)	150–450	160–460	157–412	170–410	367–573	300–490	432–710	540–632
WBC (×10³/ml)	8.0–11.8	6.6–12.6	12.5–15.9	12.1–13.7	5.02–10.2	6.48–10.6	4.3–12.3	5.6–12.8
Neutrophils (%)	6.2–42.6	4.4–49.2	13.2–21.6	15.7–18.5	17.1–27.1	22.8–35.2	9.3–23.6	10.7–25.8
Lymphocytes (%)	57.6–83.2	50.2–84.5	62.4–82.8	65.9–77.9	54.7–92.3	50.9–84.9	68–76.8	58.9–78.1
Eosinophils (%)	0.1–0.63	0–1.96	1.37–2.81	2.05–2.77	0.26–1.54	0.22–1.18	0–1.6	0–2.3
Basophils (%)	0–0.6	0–0.4	0.22–0.82	0.13–0.85	0–5	0–2.1	0–1.6	0–0.8
Monocytes (%)	0–0.65	0–1.81	2.22–2.47	0.98–1.11	0.9–4.1	0.4–4.4	0–6.5	1.7–6.2

Extracted from: Lindstrom NM, Moore DM, Zimmerman K, Smith SA (2015). Hematologic assessment in pet rats, mice, hamsters, and gerbils: blood sample collection and blood cell identification. *Vet Clin North Am Exot Anim Pract*, 18(1):21–32.

Haematology Reference Values in Free-Ranging Lions (*Panthera leo*)

Parameter	n	Mean ± Standard Deviation	Reference Range
Hgb (g/dL)	324	11.8 ± 1.47	8.9–14.6
Hct (%)	324	35.5 ± 4.43	26.8–44.1
RBC (×10⁶/µl)	44 (M)	6.7 ± 0.81	5.1–8.3
	41 (F)	6.7 ± 0.89	5.0–8.4
	69 (M)	7.2 ± 0.78	5.7–8.7
	104 (F)	6.8 ± 0.83	5.2–8.5
MCV (fl)	113 (M)	51.3 ± 2.4	46.6–55.9
	145 (F)	52.3 ± 2.2	48.0–56.7
MCH (pg)	113 (M)	16.7 ± 0.94	14.8–18.5
	145 (F)	17.3 ± 0. 94	15.5–19.1
MCHC (g/dl)	113 (M)	32.6 ± 1.51	29.6–35.5
	145 (F)	33.1 ± 1.35	30.4–35.7
WBC (×10³/µl)	183	19.0 ± 4.73	9.7–28.2

Extracted from: Maas M, Keet DF, Nielsen M (2013). Hematologic and serum chemistry reference intervals for free-ranging lions (*Panthera leo*). Res Vet Sci, 95(1):266–268.

Haematology Reference Values in Indian Elephants (*Elephas maximus*)

Parameters	Juvenile (n = 11)	Males (n = 14)	Non-Pregnant Females (n = 11)	Pregnant Females (n = 5)
Erythrocytes (millions/mm³)	2.42 ± 0.44*	2.47 ± 0.42	2.40 ± 0.51	1.84 ± 0.43
Haemoglobin g/dl)	11.12 ± 2.42	10.24 ± 2.40	10.72 ± 2.29	9.98 ± 1.98
PCV (%)	34.7 ± 3.49	34.8 ± 4.84	34.8 ± 4.64	29.8 ± 2.05
MCV (µm³)	144.8 ± 17.97	142.0 ± 12.89	146.9 ± 13.06	168.6 ± 2.05
MCH (µµg)	46.67 ± 10.09	41.93 ± 9.87	44.49 ± 10.22	56.93 ± 16.15
MCHC (%)	32.13 ± 6.85	29.69 ± 7.27	31.06 ± 6.92	33.77 ± 7.72
Leucocytes (thousands/mm³)	11.9 ± 4.20	8.78 ± 2.02	9.81 ± 4.20	12.4 ± 8.20
Neutrophils (%)	32.1 ± 5.08	34.2 ± 3.39	32.3 ± 5.42	44.1 ± 4.1
Eosinophils (%)	3.8 ± 2.09	6.2 ± 2.71	6.6 ± 3.04	1.9 ± 1.38
Basophils (%)	0.6 ± 0.67	0.7 ± 0.96	0.9 ± 0.79	0.5 ± 0.84
Lymphocytes (%)	59.0 ± 5.74	52.8 ± 5.14	56.2 ± 4.42	50.6 ± 4.61
Monocytes (%)	4.50 ± 1.89	6.07± 3.11	3.90 ± 3.63	2.90 ±2.68

*Mean ± Standard Deviation
Extracted from: Nirmalan G, Nair SG, Simon KJ (1967). Hematology of the Indian elephant (*Elephas maximus*). Can J Physiol Pharmacol, 45(6):985–991.

Haematology Reference Values in Black Rhinoceros (*Diceros bicornis*)

Parameter	Sample Size	Value	Range
Haemoglobin (g/l)	29	158.1 ± 18.2	108–192
Red cell count (×10^{12}/l)	29	5.08 ± 0.59	3.89–6.38
Haematocrit (l/l)	29	0,445 ± 0.049	0.308–0.513
Red cell distribution	(Width) 20	14.4 ± 1.88	10.3–17.0
Mean cell volume (fl)	29	87.2 ± 5.8	75.3–97.1
Mean cell haemoglobin (pg)	29	31.2 ± 2.35	27.9–36.1
White cell count	30	13.98 ± 4.77	5.6–26.2
Neutrophils	30	8.39 ± 3.30	1.8–18.5
Lymphocytes		3.86 ± 1.88	1.01–9.49
Monocytes		1.05 ± 0.76	0.37–3.86
Eosinophils		0.69 ± 0.56	0.09–2.36
Platelets	18	223.5 ± 67.6	126–410
ESR	29	23 ± 17.5	2–78
White cell count	30	13.98 ± 4.77	5.6–26.2

Extracted from: Paul B, Du Toit R, Lloyd S, Mandisodza A (1988). Haematological studies on wild black rhinoceros (*Diceros bicornis*) evidence of an unstable haemoglobin. *J Zool*, 214(3):399–405.

Haematology Reference Values in White Rhinoceros (*Ceratotherium simum*)

Parameter	Number	Mean ± Standard Deviation	Minimum - maximum
WBC (×10^3/ml)	142	15.36 ± 4.85	8.00–33.40
Neutrophils (×10^3/ml)	115	7.51 ± 2.42	3.12–16.10
Band neutrophils (×10^3/ml)	115	0.01 ± 0.06	0.00–0.45
Lymphocytes (×10^3/ml)	115	4.27 ± 1.97	0.97–12.69
Monocytes (×10^3/ml)	115	1.86 ± 1.04	0.00–5.4
Eosinophils (×10^3/ml)	115	2.33 ± 1.27	0.38–7.01
Basophils (×10^3/ml)	115	0.02 ± 0.06	0.00–0.37
RBC (×10^6/ml)	80	7.35 ± 1.24	3.94–10.95
Hgb (g/dl)	80	14.94 ± 2.54	8.50–20.70

Parameter	Number	Mean ± Standard Deviation	Minimum - maximum
PCV (%)	80	44.46 ± 6.68	24.80–58.70
MCV (fl)	80	60.61 ± 5.19 0	35.00–71.00
MCH (pg/cell)	80	20.72 ± 4.08	14.30–29.60
MCHC (g/dl)	80	33.89 ± 5.19	25.30–43.50
Platelets (×10^3/ml)	80	587.55 ± 180.24	101.00–900.00

Extracted from: Miller M, Buss P, Wanty R, Parsons S, van Helden P, Olea-Popelka F (2015). Baseline hematologic results for free-ranging white rhinoceros (*Ceratotherium simum*) in Kruger National Park, South Africa. *J Wildl Dis*, 51(4):916–922.

Haematology Reference Values in Chimpanzees (*Pan troglodytes*)

Parameter	No. of Animals	No. of Tests	Mean ± Standard Deviation	Range
RBC (×10^6)	98	211	5.02 ± 0.48	3.05–6.23
WBC (×10^6)	98	211	12.7 ± 6.2	4.8–50.3
Haematocrit (%)	98	211	43.9 ± 3.68	29–58
Haemoglobin (g %)	98	211	13.9 ± 1.16	7.9–18.9
Platelets (×10^3)	98	211	268.1 ± 113.12	60–975
Reticulocytes (%)	98	211	0.21 ± 0.28	0–1.7
Sedimentation rate (mm/h)	98	211	21.7 ± 15.1	0–53
MCV (μm^3)	98	211	87.8 ± 8.0	64.9–109.1
MCH (ng)	98	211	27.8 ± 2.2	20.3–35.0
MCHC (%)	98	211	31.8 ± 0.1 0	27.0–36.
Seg. neutrophils				
%	98	211	64.48 18.29	13–94
Absolute (×10^3)	98	211	8.72 6.01	1–44.3
Band neutrophils				
%	98	211	0.58 1.09	0–7
Absolute (×10^3)	98	211	0.08 ± 0.21	0–2
Lymphocytes				
%	98	211	30.32 ± 17.33	4–69
Absolute (×10^3)	98	211	3.40 ± 2.0	0.7–20.7

(*Continued*)

(*Continued*)

Parameter	No. of Animals	No. of Tests	Mean ± Standard Deviation	Range
Monocytes				
%	98	211	2.01 ± 1.65	0–10
Absolute (×10³)	98	211	0.24 ± 0.23	0–1.5
Eosinophils				
%	98	211	2.12 ± 2.66	0–17
Absolute (×10³)	98	211	0.24 ± 0.31	0–1.7
Basophils				
%	98	211	0.22 ± 0.50	0–3
Absolute (×10³)	98	211	0.02 ± 0.06	0–0.5

Extracted from: McClure HM, Keeling ME, Guilloud NB (1972). Hematologic and blood chemistry data for the chimpanzee (*Pan troglodytes*). *Folia Primatol*, 18:444–462.

Haematology Reference Values in Captive Killer Whales (*Orcinus orca*)

Parameter		No. of Animals Studied Male	Females	No. of Samples	Tendency	Range
Haemoglobin (g/dl)	High*	1	1	142	17.1	16–19
	Low**	5	7	550	15.0	13–16
Haematocrit (%)	High	1	1	114	49	44–55
	Low	5	7	581	43	37–49
	High	1	1	143	43	38–50
	Low	5	7	575	3.8	3.2–4.3
MCV (µm³)		6	8	682	112	94–123
MCH (µµg)		6	8	667	39	35–45
MCHC (%)		6	8	684	35	32–38
WBC (mm³)		6	7	851	7,800	4,500–11,000
Neutrophils	%	6	8	907	74	34–86
	Absolute				5.77	
Bands	%	6	8	897	0	0
lymphocytes	%	6	8	893	21	8–32
	Absolute				1,638	
Monocytes	%	6	8	906	3	0–6
	Absolute				234	
Eosinophils	%	6	8	822	2	0–8
	Absolute				136	
Basophils		6	8	667	0	0

Extracted from: Cornell LH (1983). Hematology and clinical chemistry in the killer whale (*Orcinus orca*). *J Wildl Dis*, 19(3):259–264.

Haematology Reference Values in Atlantic Bottle Nosed Dolphins (*Tursiops truncatus*) n = 59

Parameter	Mean ± Standard Deviation	Range
WBC (×10³/µl)	10.31 ± 2.58	5.8–19.5
RBC (×10⁶/µl)	3.61 ± 0.28	2.8–4.8
Haemoglobin (g/dl)	14.47 ± 1.11	11.3–18.2
MCV (fl)	112.2 ± 6.46	96–126
MCH (pg)	40.41 ± 2.55	33–45
MCHC (g/dl)	35.93 ± 1.01	32–38
Basophils (×10³/µl)	0.04 ± 0.08	0–0.3
Segmented neutrophils		
Relative (%)	44.55 ± 9.89	25.35–68.37
Absolute (×10³/µl)	4.6 ± 1.65	1.8–12.7
Bands		
Relative (%)	0.04 ± 0.31	0–2.41
Absolute (×10³/µl)	0–0.03	0–0.2
Lymphocytes		
Relative (%)	19.29 ± 7.8	2.04–47.33
Absolute (×10³/µl)	1.96 ± 0.93	0.2–6.2
Monocytes		
Relative (%)	3.29 ± 2.24	0–10.69
Absolute (×10³/µl)	0.35 ± 0.31	0–1.6
Eosinophils		
Relative (%)	32.4 ± 9.28	13.68–52.75
Absolute (×10³/µl)	3.35 ± 1.27	1.3–7.1
Platelets (×10³/µl)	167.12 ± 41.68	73–281
Fibrinogen (mg/dl)	138.14 ± 89.22	50–400

Extracted from: Goldstein JD, Reese E, Reif JS, Varela RA, McCulloch SD, Defran RH1. Fair PA, Bossart GD (2006). Hematologic, biochemical, and cytologic findings from apparently healthy Atlantic bottlenose dolphins (*Tursiops truncatus*) inhabiting the Indian River Lagoon, Florida, USA. *Journal of Wildlife Diseases*, 42(2):447–454.

Haematology Reference Values in California Sea Lions (*Zalophus californianus*) n = 66

Parameter	Mean	95% Confidence Interval
RBC (×10⁶/mm3)	4.18	4.06–4.29
WBC (×10³/mm3)	9.5	8.8–10.2
Hgb (g/dL)	15.5	15.0–15.9
MCV (fl)	106	105–107

Parameter	Mean	95% Confidence Interval
Platelets (×10³/mm³)	299	267–336
Neutrophils (×10³/mm³)	5.0	4.5–5.6
Bands (×10³/mm³)	0.1	0.1–0.2
Lymphocytes (×10³/mm³)	2.8	2.5–3.0
Monocytes (×10³/mm³)	0.2	0.2–0.3
Eosinophils (×10³/mm³)	1.3	1.1–1.5

Extracted from: Williams KM (2013). Clinical values of blood variables in wild and stranded California sea lions (*Zalophus californianus*) and blood sample storage stability. A thesis presented to the Faculty of the Division of Science and Environmental Policy California State University, Monterey Bay, and Moss Landing Marine Laboratories. In partial fulfilment of the requirements for the Degree Master of Science in Marine Science. California, USA.

Haematology Reference Values in Przewalski's Horses (*Equus przewalskii*)

Parameter	Mean ± Standard Deviation	Minimum–Maximum	n
RBC (×10¹²/l)	6.6 ± 1.2	4.92–9.82	21/35
Haematocrit (l/l)	0.41 ± 0.05	0.3–0.52	23/37
Haemoglobin (g/l)	132.07 ± 22.14	91.55–187.15	17/32
MCV (fl)	63.61 ± 10.2	40.15–83.61	20/35
MCH (pg)	19.6 ± 3.7	12.43–29.61	16/30
MCHC (g/l)	314.44 ± 52.16	179.51–445.6	17/32
WBC (×10⁹/l)	6.68 ± 1.55	4.3–10.2	21/36
Bands (×10⁹/l)	0.26 ± 0.14	0–0.73	21/35
Segments (×10⁹/l)	3.08 ± 1.05	1.57–7.24	21/36
Lymphocytes (×10⁹/l)	2.92 ± 0.80	1.68–4.84	21/36
Monocytes (×10⁹/l)	0.34 ± 0.24	0.05–0.98	21/36
Eosinophils (×10⁹/l)	0.09 ± 0.16	0–0.62	21/36

Extracted from: Tomenendalova J, Vodicka R, Uhrikova I, Doubek J (2014). Determination of haematological and biochemical parameters of Przewalski's horses (*Equus przewalskii*). *Veterinarni Medicina*, 59:11–21.

Haematology Reference Values in Roe Deer (*Capreolus capreolus*)

Parameter	Mean ± SE	Range	n
Platelet (×10³/µl)	332.63 ± 18.52	166–602	33
RBC (×10¹²/l)	11.76 ± 0.14	9.6–13.8	35

(Continued)

Parameter	Mean ± SE	Range	n
Hgb (g/l)	185.91 ± 1.90	150–204	35
Hct (l/l)	0.52 ± 0.005	0.43–0.59	35
MCV (fl)	44.94 ± 0.44	39–50	35
MCHC (g/l)	352.4 ± 1.39	332–374	35
Reticulocytes (×10³/l)	11.68 ± 1.66	1–34	35
Reticulocytes (%)	0.1 ± 0.01	0.01–0.3	35
WBC (×10³/l)	4.91 ± 0.23	2.8–8.1	35
Segmented neutrophils (×10³/l)	2.38 ± 0.21	0.7–5.6	32
Banded neutrophils (×10³/l)	0.006 ± 0.004	0–0.1	32
Eosinophils (×10³/l)	0.04 ± 0.01	0–0.2	35
Basophils (×10³/l)	0.05 ± 0.01	0–0.3	35
Lymphocytes (×10³/l)	2.40 ± 0.12	1–4.1	35
Monocytes (×10³/l)	0.08 ± 0.01	0–0.3	35

Extracted from: Küker S, Huber N, Evans A, Kjellander P, Bergvall UA, Jones KL, Arnemo JM (2015). Hematology, serum chemistry, and serum protein electrophoresis ranges for free-ranging roe deer (*Capreolus capreolus*) in Sweden. *J Wildl Dis*, 51(1):269–273.

Haematology Reference Values in Tammar Wallabies (*Macropus eugenii*)

Parameter	Mean ± Standard Error of Mean
Haemoglobin (g/l)	161.27 ± 1.79 (n = 30 females)
	165.86 ± 2.26 (n = 7 males)
Erythrocytes (×10¹²/l)	6.29 ± 0.08
Haematocrit (l/l)	0.47 ± 003
MCV (fl)	75.57 ± 0.49
MCH (pg)	25.84 ± 0.17
MCHC 9g/l)	340.6 ± 1.04 (n = 30 females)
	435.74 ± 1.46 (n = 7 males)
Leucocytes (×10⁹/l)	6.11 ± 0.27 (n = 38)
Neutrophils (×10⁹/l)	1.64 ± 0.14
Lymphocytes (×10⁹/l)	3.87 ± 0.22
Monocytes (×10⁹/l)	0.2 ± 0.04
Eosinophils (×10⁹/l)	0.12 ± 0.02
Platelets (×10⁹/l)	254.71 ± 19.5

Extracted from: Mckenzie S, Deane EM, Burnett L (2002). Haematology and serum biochemistry of the Tammar wallaby, *Macropus eugenii*. *Comp Clin Path*, 11:229–237.

Haematology Reference Values in Pig-Tailed Macaques (*Macaca nemestrina*)

	Male n = 26		Female n = 4	
Parameter	Mean	Range	Mean	Range
Body weight (kg)	8.6	5.9–11.7	4.4	2.0–5.7
Leucocytes (10³/mm³)	11.78	8.50–16.71	12.10	8.50–17.60
Lymphocytes (%)	44.9	37.0–51.2	48.5	41.2–54.2
Monocytes (%)	2.3	0.5–8.0	1.0	0.5–2.6
Neutrophils (%)	50.1	44.8–58.3	48.8	42.0–55.8
Eosinophils (%)	2.0	0.0–5.1	1.4	0.5–2.0
Basophils (%)	0.7	0.0–3.0	0.3	0.2–1.0
Erythrocyte (10³/mm³)	5.95	4.75–7.01	5.62	5.13–6.33
Haematocrit (%)	41.8	31.0–47.5	41.2	38.0–42.6
Haemoglobin (g/dl)	11.3	8.5–14.2	11.5	9.8–12.2
MCV (µm3)	70.4	60.8–85.4	71.9	66.2–79.5
MCH (µµg)	19.0	16.3–24.4	20.4	17.1–22.9
MCHC (%)	27.1	22.1	278	25.8–28.1

Extracted from: Rahlmann DF, Pace N, Barnstein NJ (1967). Hematology of the pig-tailed monkey (*Macaca nemestrina*). *Folia Primt*, 5:280–284.

Birds*

Haematology Reference Values in Gentoo Penguins (*Pygoscelis papua*), King Penguins (*Aptenodytes patagonicus*), Rockhopper Penguins (*Eudyptes chrysocome*), Black-Footed Penguins (*Spheniscus demersus*) and Humboldt Penguins (*Spheniscus humboldti*) in Captivity

Parameter	Gentoo (n = 29)	King (n = 13)	Rockhopper (n = 28)	Black-Footed (n = 54)	Humboldt's (n = 34)
Hb (g/dl)	15.7 ± 1.9* (12.7–19.2)**	16.9±4.2 (10.2–22.0)	17.9 ± 1.1 (16.1–19.2)	16.8 ± 1.6 (13.4–19.5)	17.6 ± 2.5 14.5–21.3)
RBC (×10^{12}/l)	1.61 ± 014 (1.39–1.76)	1.58 ± 0.30 (1.18–2.02)	1.94 ± 0.20 (1.69–2.30)	1.74 ± 0.20 (1.32–2.12)	2.05 ± 0.31 (1.46–2.46)
PCV (l/l)	0.45 ± 0.04 (0.37–0.51)	0.45 ± 0.08 (0.32–0.55)	0.46 ± 0.03 (0.41–0.50)	0.44 ± 0.04 (0.36–0.51)	0.48 ± 0.04 (0.41–0.54)
MCV (fl)	258 ± 31 (215–301)	288 ± 18 (270–301)	234 ± 18 (210–267)	254 ± 11 (232–273)	228 ± 17 (194–267)
MCH (pg)	95.0 ± 8.8 (81.4–110.5)	108.0 ± 12.5 (86.4–120.9)	91.7 ± 7.7 (80.3–104.0)	95.1 ± 4.5 (87.2–104.3)	85.3 ± 5.2 (76.0–94.3
MCHC (g/dl)	37,7 ± 1.7 (34.8–40.0)	37.5 ± 3.8 (31.9–41.5)	39.1 ± 0.8 (37.5–40.0)	37.8 ± 1.4 (35.4–40.0)	38.0 ± 1.7 (35.0–40.6)
WBC (×10^9/l)	8.2 ± 4.1 (3.2–16.1)	4.3 ± 1.4 (2.8–6.7)	4.7 ± 1.4 (3.0–7.7)	9.3 ± 3.5 (3.5–16.3)	15.9 ± 5.1 (5.6–25.8)
Heterophils (×10^9/l)	3.9 ± 1.3 (2.2–6.0)	2.3 ± 1.0 (1.4–3.9)	3.6 ± 1.1 (1.5–5.2)	8.1 ± 2.3 (5.0–12.3)	11.9 ± 4.5 (4.1–17.9)
Lymphocytes (×10^9/l)	2.2 ± 1.0 (0.6–4.8)	2.0 ± 0.6 (1.3–2.8)	1.6 ± 0.8 (0.3–2.5)	3.1 ± 1.4 (0.8–5.2)	2.7 ± 1.8 (1.0–5.0)
Monocytes (×10^9/l)	0	0	<0.1 (0.0–0.1)	0	<0.5 (0.0–0.5)
Eosinophils (×10^9/l)	0	0	<0.3 (0.0–0.3)	<0.1 (0.0–0.2)	<0.2 (0.0–0.02)
Basophils (×10^9/l)	0	0	<0.1 (0.0–0.1)	<0.1 (0.0–0.3)	<0.5 (0.0–0.5)
Thrombocytes (×10^9/l)	-	-	7.3 ± 4.2 (4–15)	11 ± 5 (5–19)	9.5 ± 4.9 (7–20)
Fibrinogen (g/l)	3.2 ± 0.8 (2.1–4.2)	2.4	2.9 ± 0.9 (2.0–2.7)	2.9 ± 0.4 (2.2–3.7)	3.5 ± 0.9 (2.0–5.3)

*Mean ± Standard Error of Mean ** (Minimum–Maximum)
Extracted from: Hawkey CM, Samour HJ (1988). The value of clinical hematology in exotic birds. In: Jacobson ER, Kollias GV Jr, editors. *Exotic Animals: Contemporary Issues in Small Animal Practice*. Churchill Livingstone, London, UK. 109–141.

*All the values expressed for birds have been obtained by counting and/or measuring the different parameters using manual methods.

Haematology Reference Values in American White Pelicans (*Pelecanus erythrorhynchos*) and Brown Pelicans (*Pelecanus occidentalis*)

Parameter	White Pelican (n = 10)	Brown Pelican (n = 5)
Hb (g/dl)	13.0 ± 1.8* (9.8–16.6)**	14.5 (14.3–14.8)
RBC (×10^{12}/l)	2.3 ± 0.3 (1.9–2.7)	2.7 (2.6–2.8)
PCV (l/l)	0.39 ± 0.04 (0.33–0.45)	0.46 (0.43–0.49)
MCV (fl)	166 ± 9 (152–182)	168 (166–173)
MCH (pg)	52.3 ± 2.9 (46.0–59.3)	53.4 (51.2–56.8)
MCHC (g/dl)	32.6 ± 2.0 (28.4–35.2)	31.7 (30.4–32.9)
WBC (×10^9/l)	9.5 ± 3.4 (5.0–15.0)	11.9 (6.6–19.4)
Heterophils (×10^9/l)	7.2 ± 3.0 (4.2–9.3)	6.7 (4.0–9.5)
Lymphocytes (×10^9/l)	3.4 ± 0.7 (2.7–4.5)	4.0 (2.5–7.0)
Monocytes (×10^9/l)	(0.0–0.2)	(0.0–0.2)
Eosinophils (×10^9/l)	(0.0–0.3)	(0.0–0.2)
Basophils (×10^9/l)	(0.1–1.6)	(0.0–0.2)
Thrombocytes (×10^9/l)	29 ± 6 (21–38)	(17–38)
Fibrinogen (g/l)	0.9 ± 0.4 (0.3–1.5)	2.9 (2.6–3.1)

*Mean ± Standard Error of Mean ** (Minimum–Maximum)
Extracted from: Hawkey CM, Samour HJ (1988). The value of clinical hematology in exotic birds. In: Jacobson ER, Kollias GV Jr, editors. *Exotic Animals: Contemporary Issues in Small Animal Practice*. Churchill Livingstone, London, UK. 109–141.

Haematology Reference Values in American Rosy Flamingos (*Phoenicopterus ruber*), Chilean Flamingos (*Phoenicopterus chilensis*), Greater Flamingos (*Phoenicopterus roseus*) and Lesser Flamingos (*Phoeniconaias minor*)

Parameter	American Rosy (n = 36)	Chilean (n = 24)	Greater (n = 9)	Lesser (n = 10)
Hb (g/dl)	17. ± 1.4* (13.8–19.1)**	16.2 ± 1.1 (14.1–18.1)	17.3 (15.9–19.6)	16.8 ± 1.6 (15.2–19.5)
RBC (×10^{12}/l)	2.6 ± 0.2 (2.3–3.0)	2.7 ± 0.2 (2.4–3.0)	2.6 (2.3–2.8)	2.7 ± 0.1 (2.3–2.9)
PCV (l/l)	0.47 ± 0.04 (0.40–0.54)	0.46 ± 0.03 (0.41–0.51)	0.50 (0.47–0.57)	0.51 ± 0.03 (0.46–0.54)
MCV (fl)	184 ± 7 (168–196)	171 ± 6 (161–183)	193 (170–207)	188 ± 6 (179–195)
MCH (pg)	66.4 ± 4.0 (59.0–73.5)	60.6 ± 2.1 (57.3–64.8)	66.2 (57.6–70.0)	62.0 ± 5.4 (55.4–70.4)
MCHC (g/dl)	35.9 ± 2.3 (31.7–39.8)	35.6 ± 0.9 (33.3–37.9)	34.4 (33.5–35.2)	33.0 ± 2.5 (30.8–37.5)
WBC (×10^9/l)	5.1 ± 1.7 (2.4–8.7)	4.9 ± 2.5 (1.6–9.0)	2.4 (0.9–3.4)	6.1 ± 2.0 (3.8–8.5)
Heterophils (×10^9/l)	2.4 ± 0.9 (1.0–4.4)	2.4 ± 1.7 (0.4–4.8)	1.2 (0.2–3.0)	4.6 ± 1.8 (1.7–6.9)
Lymphocytes (×10^9/l)	1.6 ± 0.6 (0.7–3.0)	1.8 ± 0.6 (0.8–2.7)	0.9 (0.4–1.6)	1.2 ± 0.6 (0.5–2.4)
Monocytes (×10^9/l)	(0.0–0.5)	0	(0.0–0.2)	(0.0–0.4)
Eosinophils (×10^9/l)	(0.0–0.6)	(0.0–0.7)	(0.0–0.4)	0
Basophils (×10^9/l)	(0.0–0.5)	(0.0–0.4)	0.0–0.4)	(0.0–0.3)
Thrombocytes (×10^9/l)	14 ±6 (4–29)	15 ± 6 (6–33)	4 (2–7)	16 ± 8 (3–23)
Fibrinogen (g/l)	2.7 ± 0.6 (1.7–3.7)	2.3 ± 0.6 (1.3–3.6)	2.6 (1.5–3.3)	2.3 ± 0.4 (1.4–2.9)

*Mean ± Standard Error of Mean ** (Minimum–Maximum)
Extracted from: Hawkey CM, Samour HJ (1988). The value of clinical hematology in exotic birds. In: Jacobson ER, Kollias GV Jr, editors. *Exotic Animals: Contemporary Issues in Small Animal Practice*. Churchill Livingstone, London, UK. 109–141.

Haematology Reference Values in Night Herons (*Nycticorax nycticorax*), Marabou Storks (*Leptoptilos crumenifer*), White Storks (*Ciconia ciconia*), Maguari Stork (*Ciconia maguari*), Australian White Ibises (*Threskiornis moluccus*) and Scarlet Ibises (*Eudocimus ruber*)

Parameters	Night Heron (n = 22)	Marabou Stork (n = 8)	White Stork (n = 16)	Maguari Stork (n = 5)	White Ibis (n = 8)	Scarlet Ibis (n = 15)
Hb (g/dl)	15.1 ± 1.4* (12.9–17.8)**	15.8 ± 0.9 (14.8–17.1)	15.9 ± 1.1 (14.4–17.7)	16.1 (13.8–17.8)	18.2 ± 1.8 (14.6–19.5)	15.34 ± 1.0 (13.3–17.1)
RBC (×10^{12}/l)	2.6 ± 0.2 (2.3–2.9)	2.3 ± 0.1 (2.2–2.5)	2.4 ± 0.1 (2.1–2.7)	2.3 (2.2–2.7)	3.0 ± 0.2 (2.6–3.3)	3.2 ± 0.3 (2.6–3.8)
PCV (l/l)	0.45 ± 0.03 (0.40–0.50)	0.47 ± 0.04 (0.42–0.52)	0.45 ± 0.03 (0.41–0.48	0.46 (0.42–0.50)	0.48 ± 0.04 (0.39–0.51)	0.49 ± 0.05 (0.41–0.53)
MCV (fl)	178 ± 8 (162–194)	202 ± 12 (175–212)	189 ± 6 (172–195)	195 (186–210)	159 ± 7 (150–170)	153 ± 7 (142–164)
MCH (pg)	59.5 ± 3.4 (53.2–64.0)	67.3 ± 3.4 (62.1–70.9)	67.2 ± 2.9 (60.2–69.9)	69.0 (61.3–75.7)	60.5 ± 3.6 (55.5–67.2)	48.8 ± 2.9 (45.2–53.4)
MCHC (g/dl)	33.6 ± 1.7 (30.3–35.8)	33.5 ± 1.8 (31.1–35.9)	35.3 ± 1.5 (31.0–36.9)	35.3 (32.9–36.2)	38.1 ± 1.1 (36.8–39.4)	31.5 ± 1.3 (29.2–33.7)
WBC (×10^9/l)	9.9 ± 2.8 (5.8–15.2)	19.5 ± 4.1 (14.4–23.3)	10.8 ± 3.1 (7.0–14.3)	9.8 (7.2–15.5)	6.7 ± 2.8 (2.1–10.0)	7.1 ± 3.2 (2.6–12.6)
Heterophils (×10^9/l)	7.1 ± 2.4 (3.7–11.5)	12.7 ± 4.1 (7.6–18.7)	9.2 ± 3.2 (5.1–14.9)	5.9 (2.0–11.5)	5.7 ± 2.6 (1.4–8.8)	4.5 ± 2.5 (1.6–8.5)
Lymphocytes (×10^9/l)	2.5 ± 0.9 (1.4–4.2)	4.1 ± 0.9 (2.5–5.3)	0.8 ± 0.4 (0.2–1.6)	2.6 (1.4–3.3)	1.0 ± 0.3 (0.5–1.5)	2.0 ± 0.7 (0.8–3.0)
Monocytes (×10^9/l)	(0.0–0.9)	(0.0–2.3)	(0.0–0.3)	(0.0–0.7)	0	0
Eosinophils (×10^9/l)	(0.0–1.1)	(0.2–4.1)	(0.0–0.7)	(0.7–2.2)	(0.0–0.3)	(0.0–0.8)
Basophils (×10^9/l)	0	(0.0–0.7)	(0.0–0.5)	(0.0–0.8)	(0.0–0.3)	(0.0–0.7)
Thrombocytes (×10^9/l)	16 ± 4 (8–25)	16 ± 3 (12–19)	19 ± 8 (8–32)	10 (8–11)	33 ± 12 (18–48)	22 ± 8 (11–35)
Fibrinogen (g/l)	1.8 ± 0.6 (1.1–3.1)	3.2 ± 0.6 (2.6–4.4)	2.3 ± 0.4 (1.7–3.2)	1.7 (1.3–2.1)	2.3 ± 0.3 (1.9–2.7)	2.6 ± 0.6 (1.9–3.7)

*Mean ± Standard Error of Mean ** (Minimum–Maximum)
Extracted from: Hawkey CM, Samour HJ (1988). The value of clinical hematology in exotic birds. In: Jacobson ER, Kollias GV Jr, editors. *Exotic Animals: Contemporary Issues in Small Animal Practice*. Churchill Livingstone, London, UK. 109–141.

Haematology Reference Values in Turkey Vultures (*Cathartes aura*), Egyptian Vultures (*Neophron percnopterus*), Buzzards (*Buteo buteo*), Golden Eagles (*Aquila chrysaetos*), Caracaras (*Caracara cheriway*) and Secretary Birds (*Sagittarius serpentarius*)

Parameters	Turkey Vulture (n = 10)	Egyptian Vulture (n = 4)	Buzzard (n = 6)	Golden Eagle (n = 4)	Caracara (n = 9)	Secretary Bird (n = 4)
Hb (g/dl)	16.3* (15.7–17.3)**	14.8 (13.3–16.5)	12.9 (11.6–14.6)	13.8 (12.1–15.2)	16.2 (13.1–20.6)	16.7 (15.2–18.6)
RBC (×10^{12}/l)	2.7 (2.4–2.9)	2.3 (1.9–2.6)	2.4 (2.2–2.7)	2.4 (1.9–2.7)	2.8 (2.5–3.3)	2.18 (2.0–2.3)
PCV (l/l)	0.54 (0.51–0.58)	0.43 (0.37–0.46)	0.38 (0.34–0.42)	0.41 (0.35–0.47)	0.46 (0.38–0.59)	0.46 (0.42–0.50)
MCV (fl)	204 (194–224)	190 (183–206)	158 (151–171)	174 (160–184)	165 (149–173)	208 (201–216)

(Continued)

Parameters	Turkey Vulture (n = 10)	Egyptian Vulture (n = 4)	Buzzard (n = 6)	Golden Eagle (n = 4)	Caracara (n = 9)	Secretary Bird (n = 4)
MCH (pg)	61.7 (58.6–65.0)	67.7 (65.2–72.9)	53.8 (48.8–57.5)	58.9 (56.3–62.7)	57.8 (51.6–62.4)	76.7 (73.4–80.8)
MCHC (g/dl)	30.2 (28.6–32.0)	35.2 (35.0–35.5)	33.9 (31.4–36.0)	35.2 (34.0–36.0)	35.2 (34.0–36.0)	36.9 (35–37.6)
WBC (×10^9/l)	20.1 (10.5–31.9)	7.6 (4.7–10.6)	9.1 (4.6–13.9)	6.8 (3.3–11.6)	6.8 (3.3–11.6)	8.1 (6.8–10.0)
Heterophils (×10^9/l)	11.8 (6.7–19.8)	4.0 (1.2–5.5)	5.5 (2.3–8.8)	4.2 (0.6–5.9)	4.2 (0.6–5.9)	5.3 (3.0–9.0)
Lymphocytes (×10^9/l)	3.3 (0.8–5.6)	2.5 (1.5–3.4)	1.7 (1.1–2.4)	2.4 (0.9–5.6)	2.4 (0.9–5.6)	2.4 (0.8–4.2)
Monocytes (×10^9/l)	(0.0–0.4)	(0.0–0.4)	0	(0.0–0.6)	(0.0–0.6)	(0.0–0.4)
Eosinophils (×10^9/l)	(1.5–7.5)	(0.3–1.4)	(0.1–3.1)	(0.0–0.3)	(0.0–0.3)	(0.0–0.2)
Basophils (×10^9/l)	(0.0–2.3)	0	(0.0–0.6)	(0.0–0.3)	(0.0–0.3)	(0.0–0.4)
Thrombocytes (×10^9/l)	14 (7–22)	13 (6–15)	27 (18–36)	27 (18–35)	27 (18–35)	9 (7–10)
Fibrinogen (g/l)	–	1.6 (1.0–1.9)	2.3 (1.3–3.3)	2.4 (1.2–3.8)	2.4 (1.2–3.8)	2.7 (2.0–3.3)

*Mean ± Standard Error of Mean ** (Minimum–Maximum)
Extracted from: Hawkey CM, Samour HJ (1988). The value of clinical hematology in exotic birds. In: Jacobson ER, Kollias GV Jr, editors. *Exotic Animals: Contemporary Issues in Small Animal Practice*. Churchill Livingstone, London, UK. 109–141.

Haematology Reference Values in Crowned Cranes (*Balearica* spp), Demoiselle Cranes (*Grus virgo*), Manchurian Cranes (*Grus japonensis*), Sarus Cranes (*Grus antigone*) and Blue Cranes (*Anthropoides paradisea*)

Parameter	Crowned (n = 33)	Demoiselle (n = 11)	Manchurian (n = 11)	Sarus (n = 9)	Blue (n = 8)
Hb (g/dl)	15.6 ± 1.9* (11.9–18.8)**	14.9 ± 1.0 (13.1–16.2)	14.3 ± 1.3 (12.6–16.8)	15.5 (13.8–17.7)	15.0 (13.4–16.8)
RBC (×10^12/l)	2.8 ± 0.3 (2.4–3.1)	2.7 ± 0.2 (2.3–3.0)	2.2 ± 0.3 (1.9–2.7)	2.2 (2.0–2.5)	2.4 (2.2–2.7)
PCV (l/l)	0.47 ± 0.03 (0.44–0.52)	0.43 ± 1.1 (0.39–0.47)	0.42 ± 0.04 (0.38–0.50)	0.45 (0.42–0.49)	0.42 (0.39–0.46)
MCV (fl)	171 ± 7 (156–182)	162 ± 7 (154–172)	191 ± 9 (180–204)	203 (196–209)	174 (158–190)
MCH (pg)	64.3 ± 3.0 (59.8–70.2)	55.6 ± 2.4 (51.5–60.0)	64.8 ± 4.5 (56.0–68.8)	69.6 (60.0–77.0)	61.9 (57.3–65.1)
MCHC (g/dl)	36.2 ± 2.3 (34.5–39.2)	34.3 ± 1.1 (32.6–36.2)	34.3 ± 1.5 (32.7–37.2)	34.5 (30.6–37.0)	35.5 (33.1–39.5)
WBC (×10^9/l)	11.1 ± 2.8 (6.3–15.6)	5.3 ± 1.7 (2.9–8.6)	9.5 ± 1.9 (5.7–11.6)	9.4 (3.5–12.2)	9.1 (2.9–16.9)
Heterophils (×10^9/l)	8.2 ± 3.0 (4.1–13.3)	3.8 ± 1.4 (1.7–6.6)	6.7 ± 1.6 (4.5–9.3)	6.5 (1.4–9.5)	4.5 (1.0–10.0)
Lymphocytes (×10^9/l)	1.6 ± 0.6 (0.6–2.7)	0.8 ± 0.4 (0.4–1.5)	1.9 ± 0.8 (0.5–2.9)	2.1 (1.2–3.0)	2.5 (1.1–4.2)

(Continued)

Parameter	Crowned (n = 33)	Demoiselle (n = 11)	Manchurian (n = 11)	Sarus (n = 9)	Blue (n = 8)
Monocytes (×10⁹/l)	(0.0–0.3)	(0.0–0.4)	0	(0.0–0.6)	(0.0–0.8)
Eosinophils (×10⁹/l)	(0.0–1.3)	(0.0–0.9)	(0.0–1.2)	(0.1–0.7)	(0.4–2.1)
Basophils (×10⁹/l)	(0.1–0.8)	(0.0–0.3)	(0.0–0.9)	(0.0–0.9)	(0.0–0.5)
Thrombocytes (×10⁹/l)	(5.8–1.8)	12 ±10 (4–32)	13 ± 2 (11–15)	15 (11–21)	18 (11–27)
Fibrinogen (g/l)	–	2.5 ± 0.8 (1.4–3.7)	2.7 ± 0.4 (2.3–3.6)	2.7 (1.6–5.0)	3.5 (2.7–4.5)

*Mean ± Standard Error of Mean ** (Minimum–Maximum)
Extracted from: Hawkey CM, Samour HJ (1988). The value of clinical hematology in exotic birds. In: Jacobson ER, Kollias GV Jr, editors. *Exotic Animals: Contemporary Issues in Small Animal Practice*. Churchill Livingstone, London, UK. 109–141.

Haematology Reference Values in Barn Owls (*Tito alba*), European Eagle Owls (*Bubo bubo*), African Eagle Owls (*Bubo africanus*), Spectacled Owls (*Pulsatrix perspicillata*), Tawny Owls (*Strix aluco*) and Boobook Owls (*Ninox boobook*)

Parameter	Barn owl (n = 10)	European Eagle Owl (n = 14)	African Eagle Owl (n = 6)	Spectacled Owl (n = 4)	Tawny Owl (n = 14)	Boobook Owl (n = 5)
Hb (g/dl)	14.2 ± 1.5* (12.7–16.4)**	14.2 ± 1.5 (11.7–16.8)	15.8 (13.9–19.9)	14.2 (12.4–16.3)	14.6 ± 1.1 (12.9–16.4)	15.1 (14.4–15.9)
RBC (×10¹²/l)	2.7 ± 0.3 (2.2–3.0)	1.9 ± 0.2 (1.4–2.3)	2.4 (2.1–2.8)	1.6 (1.4–1.8)	2.5 ± 0.2 (2.0–2.9)	2.5 (2.4–2.9)
PCV (l/l)	0.46 ± 0.03 (0.42–0.51)	0.39 ± 0.04 (0.31–0.45)	0.45 (0.41–0.53)	0.42 (0.37–0.45)	0.40 ± 0.03 (0.36–0.47)	0.42 (0.40–0.45)
MCV (fl)	176 ± 22 (145–216)	207 ± 17 (178–239)	189 (171–214)	261 (245–267)	158 ± 9 (147–177)	172 (165–175)
MCH (pg)	51.1 ± 5.7 (44.9–60.7)	75.1 ± 8.1 (67.1–87.1)	66.4 (58.9–76.2)	87.8 (86.1–89.1)	56.8 ± 4.8 (49.8–66.6)	61.5 (60.8–61.6)
MCHC (g/dl)	31.8 ± 2.2 (28.9–34.9	36.3 ± 2.0 (33.8–38.4)	35.1 (33.9–36.6)	33.7 (32.3–36.2)	36.3 ± 0.9 (34.9–38.0)	36.0 (34.8–37.3)
WBC (×10⁹/l)	16.6 ± 4.2 (11.5–22.3	10.8 ± 4.0 (5.3–18.6)	6.2 (4.7–8.0)	9.6 (6.9–11.1)	6.7 ± 3.3 (2.4–11.8)	6.4 (3.7–11.2)
Heterophils (×10⁹/l)	8.9 ± 3.0 (5.2–12.5)	6.9 ± 3.2 (2.6–11.8)	3.0 (1.3–5.2)	4.9 (2.8–7.6)	3.4 ± 2.0 (1.1–7.2)	4.6 (2.3–9.1)
Lymphocytes (×10⁹/l)	5.0 ± 1.7 (2.5–7.5)	3.8 ± 1.9 (1.9–6.7)	2.3 (1.9–3.2)	4.3 (2.7–7.3)	3.3 ± 1.4 (0.9–5.1)	1.4 (0.9–1.7)
Monocytes (×10⁹/l)	(0.0–1.0)	0	0	0	(0.0–0.3)	(0.0–0.5)
Eosinophils (×10⁹/l)	(0.0–2.5)	(0.0–1.6)	(0.0–1.0)	(0.0–0.6)	(0.0–1.9)	(0.0–0.5)
Basophils (×10⁹/l)	(0.0–0.9)	(0.0–0.6)	(0.0–0.6)	(0.0–0.4)	(0.0–0.9)	(0.0–0.2)
Thrombocytes (×10⁹/l)	33 ± 15 (14–58)	15 ± 3 (9–17)	22 (14–29)	18	17 ± 5 (10–24)	–
Fibrinogen (g/l)	2.7 ± 0.5 (1.9–3.3)	3.3 ± 0.9 (1.4–5.0)	5.2 (3.6–7.7)	7.0 (6.4–8.8)	3.6 ± 0.7 (2.6–5.3)	2.8 (1.6–3.8)

*Mean ± Standard Error of Mean ** (Minimum–Maximum)
Extracted from: Hawkey CM, Samour HJ (1988). The value of clinical hematology in exotic birds. In: Jacobson ER, Kollias GV Jr, editors. *Exotic Animals: Contemporary Issues in Small Animal Practice*. Churchill Livingstone, London, UK. 109–141.

Haematology Reference Values in the Little Corellas (*Cacatua sanguinea*), Greater Sulphur-Crested Cockatoos (*Cacatua g. galerita*) and Galah Cockatoos (*Eolophus roseicapilla*)

Parameter	Little Corella (n = 11)	Greater Sulphur-Crested Cockatoo (n = 25)	Galah Cockatoo (n = 55)
Hb (g/dl)	17.0 ± 1.2* (15.4–19.00)**	15.7 ± 1.0 (13.8–17.1)	16.7 ± 1.3 (14.0–18.8)
RBC (×10^{12}/l)	2.9 ± 0.3 (2.5–3.4)	2.7 ± 0.2 (2.4–3.0)	3.6 ± 0.2 (3.1–3.9)
PCV (l/l)	0.53 ± 0.04 (0.47–0.60)	0.45 ± 0.03 (0.41–0.49)	0.54 ± 0.03 (0.49–0.60)
MCV (fl)	188 ± 1.1 (181–200)	165 ± 9 (145–187)	149 ± 8 (136–164)
MCH (pg)	60.5 ± 2.4 (56.6–63.1)	57.6 ± 2.1 (53.8–60.6)	45.9 ± 2.9 (43.5–51.3)
MCHC (g/dl)	32.1 ± 2.4 (28.7–36.9)	34.9 ± 1.3 (33.3–37.6)	31.0 ± 1.7 (27.5–33.9)
WBC (×10^9/l)	7.3 ± 2.8 (4.2–11.8)	6.4 ± 2.9 (1.4–10.7)	6.3 ± 3.1 (1.6–11.9)
Heterophils (×10^9/l)	5.2 ± 2.7 (2.8–10.6)	3.7 ± 1.7 (1.0–6.6)	4.6 ± 2.5 (0.6–9.2)
Lymphocytes (×10^9/l)	1.5 ± 0.9 (0.5–3.9)	1.9 ± 0.8 (1.0–3.6)	1.2 ± 0.5 (0.5–2.0)
Monocytes (×10^9/l)	(0.0–0.5)	(0.0–0.2)	(0.0–0.1)
Eosinophils (×10^9/l)	(0.0–0.7)	(0.0–0.2)	(0.0–0.2)
Basophils (×10^9/l)	(0.0–0.8)	(0.0–0.9)	(0.0–0.8)
Thrombocytes (×10^9/l)	12 ± 7 (5–24)	13 ± 7 (7–24)	–
Fibrinogen (g/l)	2.0 ± 0.4 (1.5–2.8)	1.4 ± 0.3 (0.9–2.0)	1.8 ± 0.7 (0.8–3.5)

*Mean ± Standard Error of Mean ** (Minimum–Maximum)
Extracted from: Hawkey CM, Samour HJ (1988). The value of clinical hematology in exotic birds. In: Jacobson ER, Kollias GV Jr, editors. *Exotic Animals: Contemporary Issues in Small Animal Practice.* Churchill Livingstone, London, UK. 109–141.

Haematology Reference Values for Blue and Gold Macaws (*Ara ararauna*), Green-Winged Macaws (*Ara chloropterus*), Hyacinth Macaws (*Anodorhynchus hyacinthinus*) and military Macaws (*Ara militaris*)

Parameters	Blue and Gold Macaw (n = 27)	Green-Winged Macaw (n = 24)	Hyacinth Macaw (n = 6)	Military Macaw (n = 5)
Hb (g/dl)	16.8 ± 1.2* (14.8–18.9)**	16.2 ± 1.2 (14.7–18.8)	16.2 (15.4–18.0)	16.6 (14.4–19.6)
RBC (×10^{12}/l)	3.2 ± 0.2 (2.7–3.5)	3.1 ± 0.2 (2.8–3.3)	3.2 (2.9–3.4)	2.9 (2.7–3.1)
PCV (l/l)	0.46 ± 0.03 (0.41–0.51)	0.48 ± 0.05 (0.41–0.56)	0.48 (0.43–0.53)	0.47 (0.44–0.50)
MCV (fl)	146 ± 7 (132–157)	158 ± 9 (141–174)	150 (145–156)	159 (154–160)

(Continued)

Parameters	Blue and Gold Macaw (n = 27)	Green-Winged Macaw (n = 24)	Hyacinth Macaw (n = 6)	Military Macaw (n = 5)
MCH (pg)	52.3 ± 2.2 (49.4–56.4)	53.4 ± 2.4 (50.5–57.7)	50.0 (45.2–54.6)	55.9 (40.6–59.6)
MCHC (g/dl)	36.6 ± 1.7 (34.7–39.8)	33.9 ± 3.0 (29–38.3)	33.5 (30.2–37.3)	35.3 (34.2–37.0)
WBC ($\times 10^9$/l)	8.5 ± 3.6 (4.5–15.4)	10.6 ± 4.7 (6.0–19.3)	7.3 (5.6–8.9)	14.8 (12.6–17.8)
Heterophils ($\times 10^9$/l)	5.0 ± 1.5 (2.3–8.0)	4.1 ± 0.6 (3.7–5.8)	5.7 (3.3–6.8)	13.1 (10.0–15.3)
Lymphocytes ($\times 10^9$/l)	2.0 ± 0.6 (0.9–3.3)	3.9 ± 1.1 (2.30–5.5)	1.3 (0.8–2.0)	1.6 (0.7–2.5)
Monocytes ($\times 10^9$/l)	(0.0–0.3)	(0.0–0.1)	0	0
Eosinophils ($\times 10^9$/l)	0	0	0	0
Basophils ($\times 10^9$/l)	(0.0–0.2)	(0.0–0.2)	(0.2–0.3)	0
Thrombocytes ($\times 10^9$/l)	21 ± 6 (11–34)	11 ± 3 (8–15)	19 (7–29)	24 (19–30)
Fibrinogen (g/l)	1.9 ± 0.6 (1.0–3.2)	2.0 ± 0.5 (1.1–2.7)	1.8 (1.3–2.4)	1.3 (1.0–2.0)

*Mean ± Standard Error of Mean ** (Minimum–Maximum)
Extracted from: Hawkey CM, Samour HJ (1988). The value of clinical hematology in exotic birds. In: Jacobson ER, Kollias GV Jr, editors. *Exotic Animals: Contemporary Issues in Small Animal Practice*. Churchill Livingstone, London, UK. 109–141.

Haematology Reference Values in Red-Fronted Macaws (*Ara rubrogenys*), Scarlet Macaws (*Ara macao*), Severe Macaws (*Ara severus*) and Yellow-Naped Macaws (*Primolius auricollis*)

Parameter	Red-Fronted Macaw (n = 7)	Scarlet Macaw (n = 7)	Severe Macaw (n = 4)	Yellow-Naped Macaw (n = 7)
Hb (g/dl)	16.4* (14.4–19.6)**	17.3 (15.8–18.4)	16.7 (16.0–17.1)	17.4 (15.0–19.2)
RBC ($\times 10^{12}$/l)	3.3 (2.6–3.5)	3.0 (2.7–3.2)	3.3 (3.1–3.6)	3.5 (2.9–3.9)
PCV (l/l)	0.46 (0.41–0.51)	0.48 (0.46–0.52)	0.49 (0.45–0.53)	0.50 (0.43–0.53)
MCV (fl)	142 (136–158)	160 (143–175)	149 (145–156)	144 (133–150)
MCH (pg)	50.8 (40.6–59.6)	57.6 (51.1–64.2)	51.4 (47.9–55.2)	50.6 (48.0–53.8)
MCHC (g/dl)	35.3 (29.3–38.4)	35.9 (32.6–38.5)	34.5 (32.1–35.6)	35.0 (32.9–36.9)
WBC ($\times 10^9$/l)	5.9 (3.0–8.1)	10.2 (6.4–15.4)	7.8 (4.2–10.2)	10.4 (5.0–16.9)
Heterophils ($\times 10^9$/l)	4.1 (1.8–6.2)	8.0 (4.9–12.8)	5.2 (3.0–6.7)	8.2 (4.0–15.2)
Lymphocytes ($\times 10^9$/l)	1.7 (1.0–2.6)	1.6 (1.2–2.2)	2.2 (0.5–3.6)	2.2 (0.9–4.8)
Monocytes ($\times 10^9$/l)	(0.0–0.2)	0	(0.0–0.3)	(0.0–0.3)

(Continued)

Parameter	Red-Fronted Macaw (n = 7)	Scarlet Macaw (n = 7)	Severe Macaw (n = 4)	Yellow-Naped Macaw (n = 7)
Eosinophils (×10⁹/l)	0	0	(0.0–0.5)	0
Basophils (×10⁹/l)	(0.0–0.2)	(0.0–0.8)	(0.0–0.1)	(0.0–0.4)
Thrombocytes (×10⁹/l)	13 (4–28)	22 (17–30)	12 (10–13)	18 (10–30)
Fibrinogen (g/l)	1.5 (1.3–2.8)	1.7 (1.0–2.2)	1.8 (1.6–2.1)	1.4 (1.0–1.9)

Extracted from: Hawkey CM, Samour HJ (1988). The value of clinical hematology in exotic birds. In: Jacobson ER, Kollias GV Jr, editors. *Exotic Animals: Contemporary Issues in Small Animal Practice*. Churchill Livingstone, London, UK. 109–141.

Haematology Reference Values in Grey Parrots (*Psittacus erithacus*), Amazon Parrots (*Amazona* spp.) and Keas (*Nestor notabilis*)

Parameter	Grey Parrot	Amazon Parrot	Kea
Hb (g/dl)	15.5* (14.2–17.0)**	15.5 ± 1.3 (13.8–17.9)	13.4 ± 2.2 (10.6–16.9)
RBC (×10¹²/l)	3.3 ± 0.2 (3.0–3.6)	2.9 ± 0.3 (2.6–3.5)	2.6 ± 0.3 (2.3–3.1)
PCV (l/l)	0.48 ± 0.03 (0.43–0.51)	0.51 ± 0.03 (0.44–0.56)	0.40 ± 0.04 (0.34–0.46)
MCV (fl)	145 ± 6 (137–155)	173 ± 11 (156–194)	154 ± 17 (137–186)
MCH (pg)	47.2 ± 3.0 (41.9–52.8)	52.6 ± 5.0 (44.7–58.6)	51.2 ± 9.0 (41.6–68.1)
MCHC (g/dl)	32.5 ± 2.0 (28.9–34.0)	31.6 ± 2.8 (28.9–35.8)	33.2 ± 2.6 (30.4–37.0)
WBC (×10⁹/l)	7.0 ± 2.3 (3.3–10.3)	4.6 ± 1.4 (2.3–6.5)	16.0 ± 3.8 (12.1–22.6)
Heterophils (×10⁹/l)	4.9 ± 1.7 (1.8–7.3)	2.9 ± 0.7 (1.6–3.8)	13.8 ± 3.5 (9.4–20.1)
Lymphocytes (×10⁹/l)	1.4 ± 0.4 (0.7–2.1)	1.7 ± 0.8 (0.6–2.8)	1.9 ± 0.6 (1.1–2.7)
Monocytes (×10⁹/l)	(0.0–0.3)	(0.0–0.1)	0
Eosinophils (×10⁹/l)	0	(0.0–0.1)	(0.0–0.5)
Basophils (×10⁹/l)	(0.0–0.8)	(0.0–0.2)	(0.0–0.6)
Thrombocytes (×10⁹/l)	22 ± 9 (11–42)	32 ± 12 (10–67)	16 ± 5 (11–24)
Fibrinogen (g/l)	2.2 ± 0.5 (1.5–2.8)	2.2 ± 0.5 (1.4–3.0)	1.5 ± 0.2 (1.1–1.8)

*Mean ± Standard Error of Mean ** (Minimum–Maximum)
Extracted from: Hawkey CM, Samour HJ (1988). The value of clinical hematology in exotic birds. In: Jacobson ER, Kollias GV Jr, editors. *Exotic Animals: Contemporary Issues in Small Animal Practice*. Churchill Livingstone, London, UK. 109–141.

Haematology Reference Values in Indian Peafowl (*Pavo cristatus*) (n = 69)

Parameter	P$_{2.5}$–P$_{97.5}$*	Mean	Standard Deviation **	Range
RBC ($\times10^{12}$/l)	1.88–2.82	2.25	0.28	1.79–3.51
Hb (g/dl)	11.97–17.18	13.87	1.26	11.60–17.90
PCV (%)	33.23–44.78	38.13	2.95	32.0–46.0
MCV (fl)	144.44–195.19	170.46	13.58	131.0–196.8
MCH (pg)	51.39–71.83	61.89	5.16	50.90–73.70
MCHC (g/dl)	34.26–38.76	36.32	1.05	34.10–38.90
WBC ($\times10^{9}$/l)	4.77–18.31	9.52	3.54	3.80–18.90
Heterophils ($\times10^{9}$/l)	1.32–9.08	4.52	2.29	1.01–13.20
Lymphocytes ($\times10^{9}$/l)	1.57–9.36	4.68	1.94	1.27–10.20
Monocytes ($\times10^{9}$/l)	0.00–0.048	0.08	0.13	0.00–0.50
Eosinophils ($\times10^{9}$/l)	0.00–0.52	0.16	0.16	0.00–0.74
Basophils ($\times10^{9}$/l)	0.00–0.37	0.07	0.09	0.00–0.40
Thrombocytes ($\times10^{9}$/l)	16.37–56.30	31.60	10.73	15.4–69.20
Fibrinogen (g/l)	1.00–3.89	2.20	0.70	1.00–4.00

Extracted from: Samour J, Naldo J, Rahman H, Sakkir M (2010). Haematology and plasma chemistry values in captive Indian peafowl (*Pavo cristatus*). *J Avian Med Surg*, 24(2): 99–106.

Haematology Reference Values in Saker Falcons (*Falco cherrug*), Peregrine Falcons (*Falco peregrinus*) and Gyr Falcons (*Falco rusticolus*)

Parameter	Saker Falcon (n = 25)	Peregrine Falcon (n = 48)	Gyr Falcons (n = 25)
RBC ($\times10^{12}$/l)	2.65 ± 0.08 (2.05–3.90)	3.49 ± 0.21 (2.76–4.05)	3.91 ± 0.14* (3.1–5.12)
Hb (g/dl)	15.93 ± 0.38 (13.3–21.2)	14.82 ± 1.32 (11.6–19.1)	18.85 ± 0.23 (16.0–21.2)
Hct (l/l)	0.47 ± 0.59 (0.42–0.53)	0.40 ± 0.38 (0.26–0.58)	51.36 ± 0.90 (44–59)
MCV (fl)	183.16 ± 3.84 (135.8–219.5)	117.51 ± 7.70 (100.8–176.0)	135.83 ± 3.59 (106.18–162.36)
MCH (pg)	60.74 ± 1.42 (50.62–78.94)	—	49.44 ± 1.32 (39.17–59.67)
MCHC (g/dl)	33.28 ± 0.63 (28.33–40)	—	36.41 ± 0.16 (35.47–37.84)
WBC ($\times10^{9}$/l)	5.7 ± 0.31 (2.8–8.4)	12.56 ± 3.06 (7.6–21.2)	7.3 ± 0.38 (4.2–10.8)
Heterophils ($\times10^{9}$/l)	4.14 ± 0.24 (2.18–5.96)	4.52 ± 1.2 (1.38–7.53)	4.67 ± 0.34 (2.31–8.85)

(Continued)

Parameter	Saker Falcon (n = 25)	Peregrine Falcon (n = 48)	Gyr Falcons (n = 25)
Lymphocytes ($\times 10^9$/l)	1.33 ± 0.09 (0.52–2.29)	5.52 ± 1.36 (1.75–7.53)	1.43 ± 0.10 (0.48–2.36)
Monocytes ($\times 10^9$/l)	0.21 ± 0.03 (0.04–0.64)	0.25 ± 0.03 (0.12–0.62)	0.42 ± 0.05 (0.03–0.9)
Eosinophils ($\times 10^9$/l)	0	2.3 ± 0.9 (1–4.77)	0.27 ± 0.04 (0.0–0.68)
Basophils ($\times 10^9$/l)	0.08 ± 0.01 (0–0.32)	–	0.05 ± 0.02 (0.0–0.29)
Thrombocytes ($\times 10^9$/l)	0.41 ± 0.03 (0.17–0.76)	2.97 ± 1.2 (1.25–7.15)	22.57 ± 1.04 (12.67–29. 93)
Fibrinogen (g/l)	2.82 ± 0.14 (1.78–4.7)	–	3.61 ± 0.21 (1.72–5.63)

Saker falcons and gyr falcons: Mean ± Standard Error of Mean (Minimum–Maximum)
Peregrine falcons: Mean ± Standard Deviation (Minimum–Maximum)
Extracted from: Samour JH, D'Aloia M-A, Howlett JC (1996). Normal haematology of the saker falcon (*Falco cherrug*). *Comp Haem Int*, 6:50–52.
Extracted from: Dötlinger HS, Bird DM (1995) Haematological parameters in captive peregrine falcons (*Falco peregrinus*). *Falco* Newsletter No. 4, the Middle East Falcon Research Group, National Avian Research Centre, United Arab Emirates.
Extracted from: Samour JH, John SK, Naldo JL (2005). Normal haematology values in gyr falcons (*Falco rusticolus*) in Saudi Arabia. *Vet Rec*, 157:844–847.

Haematology Values in the Stone Curlews (*Burhinus oedicnemus*)

Parameter	Stone Curlew (n = 18)
RBC ($\times 10^{12}$/l)	2.86 ± 0.04* (2.59–3.27)**
HB (g/dl)	14.45 ± 0.23 (12.2–16.6)
Hct (l/l)	0.47 ± 0.7 (0.44–0.58)
MCV (fl)	167.3 ± 2.7 (149.9–196.2)
MCH (pg)	50.7 ± 0.83 (43.72–57.14)
MCHC (g/dl)	30.34 ± 0.38 (27.72–35.51)
WBC ($\times 10^9$/l)	7.88 ± 0.66 (2.45–12.6)

(Continued)

Parameter	Stone Curlew (n = 18)
Heterophils ($\times 10^9$/l)	5.99 ± 0.70 (0.99–11.59)
Eosinophils ($\times 10^9$/l)	0.68 ± 0.19 (0.00–2.78)
Basophils ($\times 10^9$/l)	0.19 ± 0.05 (0.00–0.87)
Lymphocytes ($\times 10^9$/l)	0.54 ± 0.08 (0.20–1.38)
Monocytes ($\times 10^9$/l)	0.46 ± 0.05 (0.00–0.99)
Thrombocytes ($\times 10^9$/l)	8.93 ± 0.91 (3.4–18.2)
Fibrinogen (g/l)	3.31 ± 0.13 (2.1–4.1)

*Mean ± Standard Error of Mean ** (Minimum–Maximum)
Extracted from: Samour JH, Howlett JC, Silvanose C, Bailey TA, Wernery U (1998). Normal haematology and blood chemistry of captive adult stone curlews (*Burhinus oedicnemus*). *Comp Haem Int*, 8:219–224.

Haematology Values in Female and Male Houbara Bustards (*Chlamydotis macqueenii*)

Parameter	Females (n = 20)	Males (n = 14)
RBC (×10⁹/l)	2.49 ± 0.08 * (2.09–3.36) **	2.53 ± 0.09 (1.95–3.15)
Hb (g/dl)	15.15 ± 0.19 (13.3–16.3)	14.72 ± 0.14 (13.7–15.7)
Hct (l/l)	0.47 ± 0.6 (0.43–0.55)	0.47 ± 0.98 (0.42–0.51)
MCV (fl)	192.91 ± 4.94 (152.6–227.9)	189.7 ± 8.85 (146.3–259.1)
MCH (pg)	61.98 ± 1.94 (42.3–75.1)	58.9 ± 2.44 (46.6–74.3)
MCHC (g/dl)	31.89 ± 0.4 (27.7–35.3)	31.16 ± 0.54 (26.1–34.1)
WBC (×10⁹/l)	5.81 ± 0.25 (3.2–7.85)	5.81 ± 0.29 (4.25–7.6)
Heterophils (×10⁹/l)	3.83 ± 0.24 (1.7–5.7)	3.64 ± 0.24 (1.99–4.82)
Lymphocytes (×10⁹/l)	1.82 ± 0.16 (0.8–3.3)	1.84 ± 0.15 (0.97–3.24)
Monocytes (×10⁹/l)	0.13 ± 0.06 (0.0–1.15)	0.15 ± 0.03 (0.0–0.42)
Eosinophils (×10⁹/l)	0.06 ± 0.02 (0.0–0.39)	0.07 ± 0.01 (0.0–0.23)
Basophils (×10⁹/l)	0.06 ± 0.02 (0.0–0.34)	0.07 ± 0.02 (0.0–0.26)
Thrombocytes (×10⁹/l)	6.85 ± 0.86 (2.09–18.45)	6.82 ± 0.59 (2.76–9.88)
Fibrinogen (g/l)	2 ± 0.14 (0.68–3.06)	1.87 ± 0.26 (0.8–4.8)

*Mean ± Standard Error of Mean ** (Minimum–Maximum)
Extracted from: Samour JH, Howlett JC, Hart MG, Bailey TA, Naldo J, D'Aloia M-A (1994). Normal haematology of the houbara bustard (*Chlamydotis undulata macqueenii*). *Comp Haem Int*, 4:198–202.

Haematology Reference Values in Kori Bustards (*Ardeotis kori*)

Parameter	Mean + Standard Error of Mean (Minimum–Maximum)	n
RBC (×10¹²/1)	2.30+0.06 (1.74–2.95)	28
Hb (g/dl)	14.10+0.16 (11.9–15.9)	28
Hct (1/1)	0.47+0.05 (0.395–0.525)	28
MCV (fl)	208.5+5.1 (161.9–275.4)	28
MCH (pg)	62.4+ 1.6 (48.0–84.6)	28
MCHC (g/dl)	30.0+0.4 (29.7–34.9)	28
WBC (×10⁹/1)	7.29+0.42 (3.05–12.85)	28
Heterophils (×10⁹/1)	3.98+0.32 (0.95–9.25)	28
Lymphocytes (×10⁹/1)	2.21+0.24 (0.41–5.45)	28
Monocytes (×10⁹/1)	0.60+0.07 (0.0–1.57)	28
Eosinophils (×10⁹/1)	0.35+0.05 (0.0–1.15)	27
Basophils (×10⁹/1)	0.20+0.03 (0.0–0.80)	28
Thrombocytes (×10⁹/1)	5.5+0.7 (1.49–18.0)	25
Fibrinogen (g/1)	2.42+0.10 (1.42–4.5)	27

Extracted from: Howlett JC, Samour JH, D'Aloia M-A, Bailey TA, Naldo J (1995). Normal haematology of captive adult kori bustards (*Ardeotis kori*). *Comp Haem Int*, 5:102–105.

Age-Related Haematology Values in Captive Masai Ostriches (*Struthio camelus massaicus*) (n = 24)

Parameter	Months											
	1	2	3	4	5	6	7	8	9	10	11	12
RBC (×10^{12}/l)	1.8±0.04*	1.8±0.04	1.6±0.02	1.7±0.04	1.8±0.03	1.8±0.03	1.9±0.04	1.9±0.03	2.1±0.04	1.9±0.03	2.1±0.03	2.1±0.04
	1.4–2.2**	1.6–2.3	1.4–1.8	1.3–2.2	1.4–2.0	1.5–2.1	1.6–2.5	1.7–2.3	1.7–2.5	1.7–2.2	1.7–2.6	1.7–2.6
Hb (g/dl)	12.7±0.1	11.89±0.2	10.7±0.1	11.5±0.2	11.3±0.1	11.9±0.1	13.3±0.2	14.1±0.1	14.9±0.2	15.2±0.1	15.5±0.2	15.3±0.2
	11.1–13.8	10.3–14.4	9.7–11.9	10.3–13.6	10.1–12.9	10.6–13.4	12.0–14.8	12.1–15.5	13.4–16.3	14.1–16.2	13.9–17.5	13.7–17.2
PCV (%)	35.9±0.5	34.1±0.4	31.9±0.3	33.4±0.4	34.2±0.4	35.1±0.4	38.4±0.7	40.9±0.4	43.6±0.5	44.4±0.4	45.1±0.5	44.6±2.4
	30.6–40.8	30.5–38.9	29.0–34.9	29.1–37.0	31.1–37.9	32.0–39.4	26.0–42.9	38.0–43.9	39.1–47.4	40.5–47.9	40.1–50.7	40.1–49.9
MCV (fl)	197.1±3.2	185.5±2.9	194.0±2.2	196.8±3.9	192.3±3.3	198.9±3.2	194.9±4.2	213.7±3.08	203.6±4.9	222.8±2.9	219.0±2.8	212.5±3.2
	161.6–225.1	153.6–210.4	178.3–209.9	166.3–241.9	171.71–231.9	169.5–226.4	143.9–228.8	179.8–237.4	125.3–232.8	201.4–261.2	196.1–247.8	186.8–241.8
MCH (pg)	70.1±1.3	64.6±0.9	65.2±0.7	67.4±1.2	63.1±0.9	67.9±1.2	67.4±1.3	74.6±1.1	71.1±1.2	76.32±1.04	75.1±0.9	72.9±1.1
	55.3–81.9	56.8–74.7	58.6–71.6	59.2–84.5	55.9–74.3	58.8–80.0	52.5–77.3	62.4–83.4	60.9–81.3	69.31–90.69	67.4–86.2	63.6–83.8
MCHC (g/dl)	35.4±0.3	34.9±0.5	33.6±0.3	34.3±0.3	32.8±0.2	34.1±0.3	34.8±0.7	34.9±0.2	34.3±0.1	34.2±0.2	34.3±0.2	34.3±0.2
	32.9–37.9	31.6–39.9	31.8–37.2	31.9–37.9	30.6–35.2	31.6–36.8	32.3–48.5	33.5– −36.6	32.7–35.6	32.3–35.9	32.7–35.5	32.6–36.0
WBC (×10^9/l)	11.9±0.7	10.9±0.8	11.3±1.1	11.8±0.9	14.3±0.9	16.7±0.8	23.9±1.2	17.6±0.9	24.2±1.2	24.6±1.5	24.5±1.3	23.0±1.1
	5.4–18.9	5.5–19.6	3.8–28.1	4.9–20.8	6.1–22.1	9.2–27.5	17.1–38.6	9.6–28.5	13.1–34.7	14.4–42.2	17.3–39.2	12.55–32.9
Het (×10^9/l)	9.0±0.6	8.7±0.7	8.7±0.9	9.4±0.7	11.0±0.8	12.9±0.8	18.9±1.2	14.1±0.1	19.4±1.2	18.5±1.4	18.7±1.3	17.4±0.9
	3.9–15.5	3.1–16.5	3.2–21.4	3.8–16.6	4.5–19.2	6.5–23.5	12.2–33.9	12.1–15.5	8.9–30.8	9.7–36.5	11.1–32.5	8.5–26.0
Lymp (×10^9/l)	2.7±0.1	2.03±0.1	2.2±0.2	1.9±0.2	3.1±0.2	3.2±0.2	4.6±0.3	4.02±0.2	4.4±0.2	5.3±0.2	4.9±0.3	4.2±0.2
	1.3–4.6	0.9–3.4	0.6–5.1	0.7–3.8	1.2–5.3	1.6–5.0	2.5–7.1	2.3–5.6	2.4–6.6	3.8–8.0	2.4–7.9	2.7–6.3
Mon (×10^9/l)	0.02±0.01	0.04±0.01	0.04±0.01	0.1±0.03	0.04±0.01	0.1±0.04	0.3±0.05	0.3±0.03	0.2±0.04	0.3±0.04	1.15±0.2	0.7±0.09
	0.0–0.2	0.0–0.2	0.0–0.2	0.0–0.8	0.0–0.2	0.0–0.6	0.0–1.1	0.09–0.7	0.0–0.7	0.00–0.9	0.2–2.9	0.01–1.5
Eos (×10^9/l)	0.01±0.0	0.0±0.0	0.02±0.01	0.09±0.02	0.05±0.01	0.2±0.02	0.2±0.05	0.2±0.03	0.2±0.04	0.2±0.03	0.45±0.06	0.37±0.06
	0.0–0.2	0.0–0.07	0.0–0.3	0.0–0.3	0.0–0.2	0.0–0.4	0.0–0.8	0.0–0.7	0.0–0.8	0.00–0.5	0.2–1.6	0.0–1.02
Bas (×10^9/l)	0.2±0.02	0.1±0.02	0.2±0.07	0.2±0.04	0.09±0.02	0.2±0.04	0.1±0.03	0.07±0.02	0.05±0.02	0.3±0.05	0.3±0.03	0.3±0.1
	0.0–0.4	0.0–0.4	0.0–1.4	0.0–0.7	0.0–0.4	0.0–0.8	0.0–0.5	0.0–0.4	0.0–0.4	0.00–0.8	0.0–0.8	0.0–1.0
Throm (×10^9/l)	17.5±1.0	15.7±1.1	14.4±1.1	13.9±0.7	15.1±0.7	17.7±1.0	27.2±1.4	22.3±1.0	26.7±1.7	26.8±1.1	23.1±1.3	24.6±1.5
	8.9–26.6	7.6–25.5	6.0–25.6	6.9–20.3	7.6–22.5	10.3–26.7	19.3–42.7	11.7–32.9	16.4–43.9	18.0–36.0	12.8–35.4	12.8–38.0

*Mean ± Standard Error of Mean

**Inner limits of the percentiles P2.5–P97.5 with a probability of 95% confidence interval.

Extracted from: Samour J, Naldo J, Libanan N, Rahman H, Sakkir M (2011). Age-related hematology and plasma chemistry changes in captive Masai ostrich (*Struthio camelus massaicus*), *Comp Clin Pathol*, 20:659–667.

Reptiles

Haematology Reference Values in Free-Living Green Sea Turtles (*Chelonia mydas*)

Parameter	Adult Females	n	Adult Males	n
RBC (×10¹²/l)	0.40 ± 0.095* (0.28–0.64)**	25	0.34 ± 0.01 (0.24–047)	20
Hb (g/dl)	9.4 ± 0.3 (5.3–12.4)	25	9.6 ± 0.2 (7.7–11.6)	20
PCV (%)	34.5 ± 1.25 (22–52)	25	33.2 ± 1.3 (23.5–44)	20
MCV (fl)	894 ± 43.8 (601.6–1446.4)	25	974.5 ± 37.7 (670.2–1343.8)	20
MCH (pg)	242 ± 10.1 (156.3–342.9)	25	284.2 ± 8.1 (210.6–374.2)	20
MCHC (g/dl)	27.4 ± 0.7 (20–32.5)	25	29.7 ± 1.1 (22.8–45.5)	20
WBC (×10⁹/l)	1.88 ± 0.2 (0.2–4.3)	25	2.1 ± 0.2 (0.6–3.6)	20
Heterophils (×10⁹/l)	1.09 ± 0.12 (0.11–2.84)	25	1.38 ± 0.16 (0.46–2.92)	20
Eosinophils (×10⁹/l)	0.30 ± 0.06 (0.00–1.26)	25	0.30 ± 0.06 (0.00–0.91)	20
Basophils (×10⁹/l)	0.008 ± 0.003 (0.00–0.078)	25	0.007 ± 0.003 (0.00–0.041)	20
Lymphocytes (×10⁹/l)	0.41 ± 0.05 (0.02–0.87)	25	0.36 ± 0.04 (0.05–0.92)	20
Monocytes (×10⁹/l)	0.05 ± 0.01 (0.00–0.13)	25	0.06 ± 0.01 (0.00–0.16)	20
Azurophils (×10⁹/l)	0.014 ± 0.004 (0.00–0.060)	25	0.020 ± 0.007 (0.00–0.125)	20
Thrombocytes (×10⁹/l)	1.71 ± 0.43 (0.18–7.1)	17	1.67 ± 0.43 (0.1–8.2)	19
Fibrinogen (g/l)	2.61 ± 0.5 (0.8–6.5)	11	1.68 ± 0.26 (0.5–4.45)	15

*Mean ± Standard Error of Mean ** (Minimum–Maximum)
Extracted from: Samour JH, Howlett JC, Silvanose C, Hasbun CR, Al Ghais SM (1998). Normal haematology of free-living green sea turtles (*Chelonia mydas*) from the United Arab Emirates. *Comp Haem Int*, 8:102–107.

Haematology Reference Values in Free-Living Green Sea Turtles (*Chelonia mydas*) (Mean ± Standard Error of Mean) (Minimum–Maximum)

Parameter	Subadults	n	Juveniles	n
RBC (×10¹²/l)	0.38 ± 0.02* (0.25–0.59)**	14	0.42 ± 0.03 (0.25–0.68)	13
Hb (g/dl)	9.08 ± 0.40 (6.2–11.4)	14	8.8 ± 0.3 (7.1–10.7)	13
PCV (%)	33.5 ± 1.6 (24.5–41.5)	14	32.3 ± 1.5 (26.5–44)	13
MCV (fl)	905.5 ± 41.1 (627.1–1187.5)	14	827.6 ± 85.9 (561.2–1760)	13
MCH (pg)	248.5 ± 15.3 (159–332)	14	225.7 ± 21 (142.6–428)	13
MCHC (g/dl)	27.4 ± 1.0 (21–34.5)	14	274 ± 0.6 (24.3–30.3)	13
WBC (×10⁹/l)	2.80 ± 0.27 (0.95–4.5)	14	1.84 ± 0.24 (0.47–3.2)	13
Heterophils (×10⁹/l)	1.43 ± 0.27 (0.43–3.92)	13	1.29 ± 0.17 (0.27–2.18)	12
Eosinophils (×10⁹/l)	0.25 ± 0.06 (0.00–0.58)	13	0.11 ± 0.03 (0.00–0.34)	12
Basophils (×10⁹/l)	0.006 ± 0.003 (0.00–0.032)	13	0.008 ± 0.004 (0.00–0.04)	12
Lymphocytes (×10⁹/l)	0.33 ± 0.06 (0.10–0.93)	13	0.40 ± 0.08 (0.04–0.98)	12
Monocytes (×10⁹/l)	0.07 ± 0.02 (0.00–0.25)	13	0.12 ± 0.02 (0.00–0.30)	12
Azurophils (×10⁹/l)	0.013 ± 0.005 (0.00–0.042)	13	0.028 ± 0.011 (0.00–0.096)	12
Thrombocytes (×10⁹/l)	0.95 ± 0.19 (0.23–2.4)	12	0.71 ± 0.14 (0.1–1.2)	8
Fibrinogen (g/l)	2.7 ± 0.63 (1.0–5.63)	8	2.7 ± 0.8 (0.5–5.7)	6

*Mean ± Standard Error of Mean ** (Minimum–Maximum)
Extracted from: Samour JH, Howlett JC, Silvanose C, Hasbun CR, Al Ghais SM (1998). Normal haematology of free-living green sea turtles (*Chelonia mydas*) from the United Arab Emirates. *Comp Haem Int*, 8:102–107.

Haematology Reference Values in African Rock Pythons (*Python sebae*)

Parameter	Mean ± Standard Deviation (Rainy Season) n = 19	Range	Mean Standard Deviation (Dry Season) n = 14	Range
WBC (×10⁹/l)	17.18 ± 6.86	7.10–29.40	38.80 ± 19.11	15.20–69.00
Heterophils (×10⁹/l)	11.38 ± 5.67	4.26–21.76	26.39 ± 15.28	7.60–58.60
Lymphocytes (×10⁹/l)	5.39 ± 2.23	2.50–8.20	12.06 ± 9.74	4.15–33.00
Mono/Azuro (×10⁹/l)	0.20 ± 0.18	0.00–0.44	0.18 ± 0.32	0.00–0.00
Eosinophils (×10⁹/l)	0.18 ± 0.23	0.00–0.59	0.06 ± 0.15	0.00–2.16
Basophils (×10⁹/l)	0.016 ± 0.05	0.00–0.16	0.00 ± 0.00	0.00–1.08
RBC (×10¹²/l)	3.03 ± 0.57	2.16–4.17	1.79 ± 0.36	1.11–2.61
Hgb (g/dl)	6.85 ± 1.26	5.30–9.40	6.61 ± 1.39	3.30–8.70
PCV (%)	20.44 ± 3.09	16.00–26.00	19.86 ± 4.22	10.00–26.00
MCV (fl)	66.93 ± 5.60	57.00–74.10	111.84 ± 17.39	57.80–142.90
MCH (pg)	22.71 ± 1.29	21.10–24.50	37.27 ± 5.80	19.10–47.60
MCHC (g/dl)	33.91 ± 1.36	33.20–37.30	33.31 ± 0.13	33.00–33.50

Extracted from: Jegede HO, Omobowale TO, Okediran BS, Adegboye AA (2017). Hematological and plasma chemistry values for the African rock python (*Python sebae*). *Int J Vet Sci Med*, 5(2):181–186.

Haematology Reference Values in Green Iguanas (*Iguana iguana*)

Parameter	Males (n = 15)	Females (n = 15)	Juveniles (n = 6)
RBC (×10⁶μ/l)	1.3 ± 0.2* (1.0–1.7)**	1.4 ± 0.1 (1.2–1.8)	1.4 ± 0.1 (1.3–1.6)
Haemoglobin (g/dl)	8.6 ± 0.2 (6.7–10.2)	10.6 ± 1.2 (9.1–12.2)	9.6 ± 0.4 (9.2–10.1)
PCV (%)	34 ± 3.9 (29.2–38.5)	38 ± 3.7 (33–44)	38 ± 5.9 (30–47)
MCV (fl)	266 ± 27 (228–303)	270 ± 58 (235–331)	ND
MCHC (g/dl)	25.1 ± 2.3 (22.7–28)	27.9 ± 2.7 (24.9–31.0)	ND
Fibrinogen (mg/dl)	100 ± 46 (100–200)	100 ± 70 (100–300)	100 ± 82 (100–300)
WBC (×10³/μl)	15.1 ± 5.9 (11.1–24.6)	14.8 ± 6.0 (8.2–25.2)	16.3 ± 4.5 (8.0–22.0)
Heterophils (×10³/μl)	3.6 ± 2.3 (1.0–5.4)	3.2 ± 2.1 (0.6–6.4)	2.2 ± 1.2 (1.0–3.8)
Lymphocytes (×10³/μl)	9.7 ± 4.5 (5.0–16.5)	9.9 ± 4.66 (5.2–14.4)	12.9 ± 3.7 (6.2–17.2)
Monocytes (×10³/μl)	1.3 ± 0.9 (0.2–2.7)	1.2 ± 0.9 (0.4–2.3)	0.4 ± 0.09 (0.3–0.6)
Eosinophils (×10³/μl)	0.1 ± 0.2 (0.0–0.3)	0.1 ± 0.2 (0.0–0.4)	0.3 ± 0.09 (0.0–0.4)
Basophils (×10³/μl)	0.4 ± 0.3 (0.1–1.0)	0.5 ± 0.4 (0.2–1.2)	0.5 ± 0.2 (0.1–0.7)

*Mean ± Standard Deviation **(range)
Extracted from: Harr KE, Alleman AR, Dennis PM, Maxwell LK, Lock BA, Bennett A, Jacobson ER (2001). Morphological and cytochemical characteristics of blood cells and hematological and plasma biochemical reference ranges in green iguanas. *J Am Vet Med Assoc*, 218:915–921.

Haematology Reference Values in Boa Constrictors (*Boa constrictor*)

Parameters	Adult Females (n = 18)	Juvenile Females (n = 10)	Adult Males (n = 52)	Juvenile Males (n = 12)
Haematocrit (%)	21.7 ± 4.23* (15–26)**	23.3 ± 2.26 (20–27)	24.31 ± 3.87 (18–31)	23.43 ± 2.86 (20–26)
Haemoglobin (g/dl)	7.1 ± 2.21 (3.1–9.2)	7.43 ± 0.98 (6–8.8)	8.12 ± 1.41 (6–9.7)	7.74 ± 0.87 (6.7–8.8)
RBC (×10/^{12}/l)	0.64 ± 0.22 (0.2–1)	0.75 ± 0.6 (0.4–1)	0.72 ± 0.16 (0.3–1.1)	0.7 ± 0.11 (0.5–0.9)
MCV (fl)	341 ± 86.9 (212.7–456.1)	327.4 ± 111.9 (237–627)	346.3 ± 68.5 (245–472)	341.7 ± 79.9 (266–500)
MCH (pg)	116.8 ± 28 (81.8–175.3)	117.4 ± 1.5 (75–206.2)	346.3 ± 68.5 (245–472)	117.2 ± 1.48 (83–134)
MCHC (g/dl)	32.82 ± 4.8 (18–38.3)	32.8 ± 2.5 (30–37.1)	116.4 ± 2.34 (78–190.3)	32.79 ± 2.2 (30.4–35.4)
WBC (×10^9/l)	10.6 ± 6 (3.9–19.8)	8 ± 3.5 (2.8–13.2)	9 ± 4.54 (3.3–22.6)	9.67 ± 5.72 (5.2–17.1)
Heterophils (%)	25.1 ± 9 (30–41)	13.2 ± 7.2 (6–30)	21 ± 10.5 (4–48)	15 6. ± 63 (4–28)
Heterophils (×10^9/l)	2.6 ± 1.7 (0.3–6.6)	1 ± 0.5 (0.1–2.1)	1.97 ± 1.64 (0.2–8.8)	1.36 ± 0.92 (1.3–3.6)
Eosinophils (%)	1.5 ± 1.8 (0–6)	1.5 ± 1.7 (0–4)	1.25 ± 1.91 (0–9)	1.72 ± 3.22 (0–11)
Eosinophils (×10^9/l)	0.1 ± 0.2 (0–1.1)	0.13 ± 0.14 (0–0.33)	0.12 ± 0.26 (0–11)	0.16 ± 033 (0–1)
Lymphocytes (%)	36 ± 17.3 (6–67)	59.5 ± 13.8 (38–80)	44.5 ± 18.3 (12–86)	53.16 ± 16.1 (38–89)
Lymphocytes (×10^9/l)	3.6 ± 2.6 (0.7–9.5)	4.9 ± 2.7 (1–1.5)	3.73 ± 1.95 (0.6–9)	5.56 ± 5.25 (1–10)
Monocytes (%)	0.9 ± 1.6 (0–6)	1.4 ± 1.7 (0–5)	1.2 ± 1.25 (0–4)	1.5 ± 1.5 (0–4)
Monocytes (×10^9/l)	0.12 ± 0.2 (0.04–0.3)	0.1 ± 0.1 (0.07–0.6)	0.1 ± 0.16 (0–.0.7)	0.12 ± 0.11 (0.1–0.11)
Azurophils (%)	32.8 ± 15.5 (11–61)	20.2 ± 7.6 (6–32)	28.1 ± 12.3 (3–55)	23.5 ± 8.4 (6–40)
Azurophils (×10^9/l)	3.5 ± 2.7 (0.6–10)	1.4 ± 0.7 (0.7–3)	2.63 ± 1.88 (0–7)	2 ± 0.99 (1.3–4.1)
Basophils (%)	2.6 ± 3 (0–13)	4.2 ± 5.2 (0–18)	4.38 ± 6 (0–40)	4.25 ± 3.1 (0–8)
Basophils(×10^9/l)	0.26 ± 0.27 (0.07–0.7)	0.3 ± 0.5 (0–0.7)	0.38 ± 0.55 (0–26)	0.38 ± 0.41 (0.1–1)
Thrombocytes (×10^9/l)	15.3 ± 6.5 (12.2–24.7)	8.8 ± 6.4 (2.3–19.2)	13.7 ± 5.49 (5.6–24)	16.3 ± 1.74 (14–16)

*Average ± Standard Deviation **(range)

Extracted from: Sarmiento Fonseca JE, Jaramillo GH, Ramirez DM, Nieto Pico JE (2018). Establishment of hematological parameters in boas (*Boa constrictor*) in the Centro de Atención y Valoración de Fauna Silvestre del Valle de Aburra. *Int J Avian Wild Biol*, 3(2):146–150.

Haematology Reference Values in Free-Living Female Nile Monitors (*Varanus niloticus*)

Parameters	Adult Nile Monitor	Young Nile Monitor
RBC (×10^{12}/l)	1.13 ± 0.10* (0.57–1.70)**	1.90 ± 0.05 (1.61–2.20)
Hb (g/l)	88.9 ± 3.6 (69–125)	7.3 ± 3.1 (59–91)
PCV (l/l)	0.26 ± 0.006 (0.24–0.31)	0.26 ± 0.004 (0.24–0.28)
MCV (fl)	280.4 ± 29.95 (148.2–491.2)	137.4 ± 2.47 (126.8–149.1)
MCH (pg)	96.96 ± 12.35 (42.86–219.3)	38.7 ± 1.45 (30.4–46.15)
MCHC (g/l)	335.2 ± 9.7 (287.5–463)	282 ± 11.1 (227–350)
WBC (×10^9/l)	1.96 ± 0.05 (1.60–2.3)	1.9 ± 0.09 (1.50–2.4)
Lymphocytes (%)	67.33 ± 1.27 (58.00–74)	69.8 ± 0.86 (64.0–77.0)
Heterophils (%)	21.72 ± 1.57 (13.00–34.0)	16.8 ± 0.82 (11.00–22.0)
Monocytes (%)	9.28 ± 0.48 (6.00–13.0)	11.4 ± 0.62 (9.0–16.0)
Eosinophils (%)	1.00 ± 0.20 (0.00–3.00)	0.83 ± 0.21 (0.00–2.0)
Basophils (%)	0.67 ± 0.16 (0.00–2.00)	1.25 ± 0.28 (0.00–4.0)

*Mean ± Error of Mean **(Minimum–Maximum)
Extracted from: Moustafa MAM, Ismail MN-E, Mohammed AEA (2013). Hematologic and biochemical parameters of free-ranging female Nile monitors in Egypt. *J Wildl Dis*, 49(3):750–754.

Haematology Reference Values in Crossed Pit Vipers (*Bothrops alternatus*), Jararacussus (*B. jararacussu*), Brazilian Lance-heads (*B. moojeni*) and the Yarará Chica (*B. n. diporus*)

Parameter	*B. alternatus* (n = 50)	*B. jararacussu* (n = 50)	*B. moojeni* (n = 50)	*B. n. diporus* (n = 50)
RBC (×10^9/l)	660.6 ± 18.5*	642.3 ± 4.23	543.1 ± 9.2	667.3 ± 12.5
WBC (×10^9/l)	12.05 ± 1.85	10.64 ± 1.37	10.06 ± 1.25	13.52 ± 1.48
Haemoglobin (g/dl)	10.01 ± 2.42	12.13 ± 1.87	11.88 ± 1.65	12.16 ± 2.43
Haematocrit (%)	20.85 ± 4.75	21.89 ± 3.21	22.38 ± 2.32	22.78 ± 3.22
MCV (fl)	315.9 ± 36	340.88 ± 32	412.07 ± 38.7	341.37 ± 30.8
MCH (pg)	151.5 ± 21.4	188.85 ± 26.7	218.7 ± 33.8	182.22 ± 25
MCHC (%)	48.09 ± 6.1	55.38 ± 6.52	53.08 ± 7	53.38 ± 6.81
Lymphocytes (%)	51.40 ± 6.75	50.60 ± 3.27	51.05 ± 8.21	52 ± 6.85
Monocytes (%)	10.30 ± 1.87	11.20 ± 1.62	10.85 ± 2.05	11.90 ± 1.25
Azurophils (%)	9.80 ± 1.12	8.50 ± 2.10	8.60 ± 0.80	9.25 ± 2.01

(Continued)

Parameter	B. alternatus (n = 50)	B. jararacussu (n = 50)	B. moojeni (n = 50)	B. n. diporus (n = 50)
Heterophils (%)	12 ± 1.26	13.10 ± 1.34	12.80 ± 1.41	11.85 ± 14.3
Eosinophils (%)	15.20 ± 2.01	16.50 ± 0.95	15.70 ± 2.10	14.95 ± 3.01
Basophils (%)	1.3 ± 0.3	0.50 ± 0.2	0.9 ± 0.2	0.75 ± 0.1
Thrombocytes (×10⁹/l)	5.68 ± 0.12	5.20 ± 0.45	4.91 ± 0.28	6.14 ± 0.80

*Mean ± Standard Deviation
Extracted from: Troiano JC, Vidal JC, Gould EF, Heker J, Gould J, Vogt AU, Simoncini C, Amanti E, De Roodt A (2000). Hematological values of some *Bothrops* species (Ophidian-Crotalidae) in captivity. *J. Venom. Anim. Toxins*, 6(2):194–204.

Haematology Reference Values in Free-Living Nile Crocodiles (*Crocodylus niloticus*)

Parameter	Male (n = 27)	%	Females (n = 9)	%
PCV (%)	17.7 ± 1.9* (14–22)**		19.0 ± 2.1 (16–22)	
RBC (×10⁶/l)	0.57 ± 0.10 (4.7–8.2)		0.66 ± 0.15 (0.43–0.72)	
HB (g/dl)	6.90 ± 0.85 (4.7–8.2)		7.77 ± 1.23 (6.0–9.5)	
MCV (fl)	319 ± 57.9 (216.2–461.5)		298.4 ± 72.4 (200.0–465.1)	
MCH (pg)	124.3 ± 20.4 (83.8–182.1)		122.6 ± 39.0 (89.0–220.9)	
MCHC (g/dl)	39.1 ± 3.3 (29.0–45.0)		40.8 ± 3.4 (37.2–47.5)	
WBC (×10³/µl)	11.31 ± 5.02 (6.13–26.22)		10.95 ± 3.63 (3.75–15.53)	
Heterophils (×10³/µl)	2.10 ± 0.97 (0.45–3.66)	20.6 ± 9.1 (4.39)	1.96 ± 0.54 (1.33–2.92)	19.4 ± 7.0 (13–35)
Lymphocytes (×10³/µl)	7.31 ± 4.30 (3.3–17.83)	63.0 ± 11.3 (39–85)	6.59 ± 2.82 (1.65–10.02)	58.2 ± 11.7 (45–72)
Monocytes (×10³/µl)	0.10 ± 0.20 (0.0–0.79)	1.0 ± 2.1 (0.10)	0.07 ± 0.08 (0.0–0.19)	0.7 ± 0.9 (0–2)
Eosinophils (×10³/µl)	0.42 ± 0.46 (0.0–2.14)	3.8 ± 3.9 (0–17)	0.97 ± 0.66 (0.0–1.87)	9.1 ± 5.4 (2–16)
Basophils (×10³/µl)	0.64 ± 0.63 (0.0–2.90)	5.4 ± 4.0 (0–15)	0.86 ± 0.79 (0.0–2.49)	7.7 ± 6.1 (0–16)
Azurophils (×10³/µl)	0.63 ± 0.82 (0.0–3.93)	5.3 ± 5.0 (0–21)	0.47 ± 0.32 (0.0–0.99)	4.6 ± 2.9 (0–08)

*Mean ± Standard Deviation **(range)
Extracted from: Lovely CJ, Pittman JM, Leslie AJ (2007). Normal haematology and blood chemistry of wild Nile crocodiles (*Crocodylus niloticus*) in the Okavango Delta, Botswana. *J S Afr Vet Assoc*, 78(3):137–144.

Haematology Reference Values in Free-Living Nile Crocodiles (*Crocodylus niloticus*)

Parameter	Yearlings n = 22	%	Juvenile n = 11	%	Subadult n = 5	%
PCV (%)	17.3 ± 1.8 (14–21)		18.2 ± 1.5 (16–22)		20.2 ± 2.0 (17–22)	
RBC (×10⁶/l)	0.58 ± 0.11 (0.35–0.74)		0.58 ± 0.06 (0.47–0.66)		0.70 ± 0.21 (0.61–1.00)	
HB (g/dl)	6.73 ± 0.83 (4.7–8.2)		7.16 ± 0.67 (6.0–8.6)		8.62 ± 0.90 (7.1–9.5)	
MCV (fl)	311.0 ± 62.3 (216.2–461.5)		315.3 ± 41.0 (242.4–383.0)		311.0 ± 96.4 (200.0–465.1)	
MCH (pg)	120.1 ± 21.5 (83.8–182.1)		124.2 ± 16.2 (90.9–149.0)		134.2 ± 50.4 (89.0–220.9)	
MCHC (g/dl)	38.9 ± 3.6 (29.0–45.0)		39.4 ± 1.6 (37.2–42.2)		42.8 ± 3.2 (39.5–47.5)	
WBC (×10³/µl)	10.12 ± 3.63 (6.13–19.61)		13.53 ± 6.61 (3.75–26.22)		11.45 ± 2.83 (7.84–15.53)	
Heterophils (×10³/µl)	2.11 ± 0.85 (0.45–3.66)	22.0 ± 9.0 (4–39)	2.14 ± 0.67 (1.31–3.43)	19.4 ± 9.3 (8–35)	1.87 ± 0.47 (1.33–2.59)	16.8 ± 4.8 (13–25)
Lymphocytes (×10³/µl)	6.64 ± 3.08 (3.30–14.71)	64.1 ± 10.8 (49–85)	8.79 ± 5.25 (1.65–17.83)	61.9 ± 12.6 (39–83)	6.19 ± 2.11 (3.53–9.01)	53.0 ± 6.4 (45–58)
Monocytes (×10³/µl)	0.10 ± 0.17 (0.0–0.70)	1.2 ± 2.2 (0–10)	0.09 ± 0.24 (0.0–0.79)	0.4 ± 0.9 (0–3)	0.06 ± 0.08 (0.0–0.16)	0.6 ± 0.9 (0–2)
Eosinophils (×10³/µl)	0.27 ± 0.34 (0.0–1.41)	2.8 ± 3.1 (0–11)	0.68 ± 0.57 (0.20–2.14)	5.9 ± 4.9 (1–17)	1.35 ± 0.42 (0.72–1.87)	12.0 ± 3.7 (7–16)
Basophils (×10³/µl)	0.45 ± 0.36 (0.0–1.24)	0.45 ± 0.36 (0.0–1.24)	0.85 ± 0.86 (0.15–2.9)	0.85 ± 0.86 (0.15–2.9)	1.38 ± 0.68 (0.87–2.49)	1.38 ± 0.86 (0.87–2.49)
Azurophils (×10³/µl)	0.51 ± 0.52 (0.0–1.92)	0.51 ± 0.52 (0.0–1.92)	0.81 ± 1.18 (0.0–3.93)	0.81 ± 1.18 (0.0–3.93)	0.56 ± 0.20 (0.37–0.89)	0.56 ± 0.20 (0.37–0.89)

Haematology Reference Values in Free-Living and Captive Komodo Dragons (*Varanus komodoensis*)

Parameter	Free-Living	Captive Outdoor
Haematocrit	0.34* (0.25–0.40)** n = 31	0.33 (0.29–9.45) n = 19
Haemoglobin (g/l)	117 (95–159) n = 16	110 (97–125) n = 16
WBC (×10⁹/l)	2.3 (0.7–12.5) n = 2.3	6.0 (3.0–10.9) n = 21
Heterophils (×10⁹/l)	0.8 (0.1–4.6) n = 30	2.7 (0.7–5.0) n = 21
Lymphocytes (×10⁹/l)	1.2 (0.5–6.4) n = 30	3.2 (1.1–6.3) n = 21
Monocytes (×10⁹/l)	0.1 (0.0–1.5) n = 30	0.2 (0.0–1.1) n = 21
Azurophils (×10⁹/l)	0.0 (0.0–0.3) n = 30	0.1 (0.0–0.8) n = 21
Basophils (×10⁹/l)	0.0 (0.0–0.6)	0.0 (0.9–0.1) n = 21
Eosinophils (×10⁹/l)	0.0 (0.0–0.0) n = 30	0.0 (0.0–0.0) n = 21

*Median **(Minimum–Maximum)
Extracted from: Gillespie D, Frye FL, Stockham SL, Fredeking T (2000). Blood values in wild and captive Komodo dragon (*Varanus komodoensis*). *Zoo Biol*, 19:495–509.

Haematology Reference Values in Captive Panther Chameleons (*Furcifer pardalis*)

Parameters	Winter		Summer	
	Male	Female	Male	Female
PCV (%)	32.9 ± 9.1* (14.1–51.1)** n = 34	33.6 ± 11.0 (11.1–56.0) n = 32	26.3 ± 6.1 (13.8–38.7) n = 36	30.1 ± 7.9 (13.6–45.7) n = 37
RBC (×10^5/µl)	10.9 ± 3.0 (4.5–16.7) n = 34	11.3 ± 2.0 (7.2–15.2) n = 30	11.2 ± 1.9 (7.3–15.1) n = 36	0.6 ± 0.4 (6.4–13.7) n = 29
WBC (×10^3/µl)	7.3 ± 3.1 (1.0–13.6) n = 34	10.8 ± 5.1 (0.3–21.3) n = 30	7.3 ± 2.4 (2.3–12.2) n = 37	9.9 ± 5.1 (2.9–15.9) n = 35
Heterophils (%)	23.9 ± 6.3 (11.4–36.9) n = 37	26.3 ± 7.9 (10.7–46.9) n = 45	38.1 ± 11.7 (10.3–32.4) n = 37	24.2 ± 6.3 (11.3–38.6) n = 40
Azurophils (%)	8.8 ± 4.0 (0.3–16.7) n = 37	10.7 ± 3.9 (3.7–19.2) n = 43	10.6 ± 3.8 (2.8–18.1) n = 36	11.3 ± 4.4 (2.5–20.8) n = 40
Basophils (%)	0.0 ± 0.0 (0.0–0.0) n = 37	0.2 ± 0.2 (0.0–0.8) n = 45	0.1 ± 0.1 (0.0–0.0) n = 37	0.2 ± 0.2 (0.0–0.8) n = 40
Eosinophils (%)	0.0 ± 0.0 (0.0–0.0) n = 37	0.0 ± 0.0 (0.0–0.0) n = 45	0.0 ± 0.0 (0.0–0.0) n = 37	0.0 ± 0.0 (0.0–0.0)
Lymphocytes (%)	67.3 ± 7.8 (51.4–83.1) n = 37	63.2 ± 9.1 (40.8–80.4) n = 43	64.1 ± 7.7 (51.9–83.1) n = 37	64.1 ± 7.8 (50.7–79.7) n = 40
Blasts (%)	0.0 ± 0.0 (0.0–0.0) n = 37	0.01 ± 0.0 (0.0–0.2) n = 45	0.0 ± 0.0 (0.0–0.0) n = 37	0.01 ± 0.1 (0.0–0.2) n = 39

*Mean ± Standard Deviation **(range)
Extracted from: Laube A, Pendl H, Clauss M, Altherr B, Hatt J-M (2016). Plasma biochemistry and hematology reference values of captive panther chameleons (*Furcifer pardalis*) with special emphasis on seasonality and gender differences. *J Zoo Wildl Med*, 47(3):743–753.

Haematology Reference Values in Hermann's Tortoises (*Testudo hermanni*) and Spur-Thighed Tortoises (*Testudo graeca*)

Parameter	T graeca (n = 5)	T hermanni (n = 8)
RBC (×10^9/l)	758.40 ± 147*	649 ± 89.50
Haemoglobin (g/dl)	7.22 ± 1.66	6.13 ± 1.4
Haematocrit (%)	28.50 ± 4.43	27.87 ± 5.08
MCV (fl)	383.80 ± 36.59	435.22 ± 95.02
MCH (pg)	94.05 ± 7.99	95.40 ± 16.70
MCHC (g/dl)	24.61 ± 2.58	22.21 ± 3.01
Lymphocytes (%)	41.20 ± 3.56	44.37 ± 3.06
Monocytes (%)	23.60 ± 3.50	23.87 ± 2.23
Basophils (%)	11.20 ± 5.44	11.12 ± 2.29
Eosinophils (%)	19.60 ± 6.34	19.00 ± 2.00

*Mean ± Standard Deviation
Extracted from: Tosunoglu M, Tok CV, Gul C (2005). Hematological values in Hermann's tortoise (*Testudo hermanni*) and spur-thighed tortoise (*Testudo graeca*) from Thrace region (Turkey). *Int J Zool Res*, 1:11–14.

Amphibians

Haematology Reference Values in Bull Frogs (*Lithobates catesbeianus*) under Farming Conditions

Parameters	Males (n = 35)		Females (n = 35)	
	Active Period	Hibernation	Active Period	Hibernating
RBC ($\times 10^{12}$/l)	0.25 ± 0.09* (0.19–0.36)**	0.34 ± 0.14 (0.16–0.51)	0.24 ± 0.05 (0.18–0.33)	0.32 ± 0.07 (0.26–0.40)
WBC ($\times 10^{9}$/l)	9.93 ± 0.81 (6.25–12.40)	2.41 ± 0.79 (1.71–3.50)	9.03 ± 3.06 (6.87–13.10)	2.43 ± 1.03 (2.00–3.63)
Hgb (g/l)	73.0 ± 7.55 (65.0–81.0)	105.0 ± 13.1 (93.0–119.0)	60.7 ± 21.76 (38.0–88.0)	106.0 ± 14.7 (95.0–123.0)
PCV (%)	20.24 ± 2.00 (18.22–23.41)	29.78 ± 6.40 (22.90–38.13)	20.32 ± 5.81 (16.24–25.11)	29.91 ± 2.40 (26.02–35.56)
MCV (fl)	809.60 ± 88.91 (732.14–848.77)	875.88 ± 67.32 (832.31–957.07)	846.66 ± 41.33 (807.21–889.76)	934.68 ± 93.82 (874.20–1029.35)
MCH (pg)	290.00 ± 74.10 (226.02–335.12)	310.82 ± 36.94 (189.22–377.14)	253.92 ± 81.78 (193.30–309.28)	331.25 ± 45.50 (290.00–377.43)
MCHC (g/l)	360.67 ± 40.64 (318.7–407.8)	352.78 ± 44.94 (310.2–387.4)	300.72 ± 51.43 (253.7–320.2)	354.40 ± 37.68 (311.3–400.1)
Heterophils (%)	23.09 ± 10.19 (10.72–31.10)	29.33 ± 18.4 (17.50–55.21)	20.63 ± 6.72 (12.93–25.08)	29.21 ± 6.50 (24.26–36.59)
Lymphocytes (%)	41.86 ± 14.46 (27.50–56.41)	39.66 ± 9.10 (29.52–62.11)	52.87 ± 9.23 (44.61–62.50)	39.43 ± 5.20 (36.58–45.83)
Monocytes (%)	25.31 ± 8.18 (17.86–35.00)	25.07 ± 12.40 (17.31–40.50)	22.28 ± 4.39 (18.00–26.79)	25.33 ± 5.12 (20.83–30.77)
Eosinophils (%)	6.43 ± 3.90 (2.36–10.17)	4.61 ± 1.13 (1.19–5.43)	4.59 ± 0.69 (4.00–5.36)	3.88 ± 2.11 (2.44–6.41)
Basophils (%)	3.30 ± 1.52 (1.56–4.36)	1.17 ± 0.18 (0.59–0.89)	2.73 ± 1.45 (1.79–4.41)	2.23 ± 0.9 (0–3.17)

*Mean ± Standard Deviation **(Minimum–Maximum)

Extracted from: Peng F, Zhang R, Zhu X, Wamg H, Zhang S (2016). Hematology and serum chemistry of farmed bullfrog *Lithobates catesbeianus* during the active and hibernating periods. *J Vet Med Anim Health*, 8(11):176–182.

Haematology Reference Values in Wild-Caught Dubois' Tree Frogs (*Polypedates teraiensis*)

Parameters	Males	Females
RBC	0.59 ± 0.01* (0.58–0.61)**	062± 0.01 (0.61–0.65)
Hb (%)	5.95 ± 0.12 (5.8–6.1)	5.82 ± 0.29 (5.5–6.2)
PCV (%)	50.62 ± 1.10 (49.5–52.0)	51.55 ± 0.95 (50.5–52.5)
MCV (μ^3)	851.31 ± 27.12 (819.67–866.66)	822.41 ± 13.31 (803.07–833.33)
MCH (pg)	100.05 ± 3.66 (96.72–105.17)	96.71 ± 6.90 (90.76–105.76
MCHC (%)	11.74 ± 0.17 (11.71–11.95)	11.75 ± 0.73 (11.3–12.77)
WBC (mm^3)	12.12 ± 0.33 (11.8–12.5)	12.15 ± 0.69 (11.5–12.9)
Heterophils (%)	23.52 ± 1.30) (22.0–25.0)	26.66 ± 1.16 (25.0–27.6)
Lymphocytes (%)	55.52 ± 2.04 (53.1–58.0)	54.92 ± 0.47 (54.6–55.6)
Basophil (%)	3.5 ± 0.24 (10.0–12.5)	2.8 ± 1.1 (1.9–4.2)
Eosinophils (%)	11.02 ± 1.12 (10.0–12.5)	9.97 ± 0.17 (9.8–10.2)
Monocytes (%)	6.42 ± 0.51 (5.9–7.1)	5.7 ± 0.46 (5.0–6.0)

*Mean ± Standard Deviation **(Minimum–Maximum)
Extracted from: Mahapatra PK, Das M (2014). Hematology of wild-caught Dubois' tree frog *Polypedates teraiensis*. *Sci World J*, 2014(3):1–7.

Haematology Reference Values in California Tiger Salamanders (*Ambystoma californiense*)

Parameter	n	Mean ± Standard Deviation	Minimum–Maximum
PCV (%)	34	47.62 ± 7.49	26–64
RBC ($\times 10^6$/µl)	34	0.76 ± 2.11	0.08–9.50
Haemoglobin (g/dl)	32	11.26 ± 2.27	3.30–14.30
MCV (fl)	34	3,477.66 ± 1,772.34	60–7,200
MCH (pg)	32	872.14 ± 401.29	8.95–1,760
MCHC (g/dl)	32	39.03 ± 59.92	7.50–271.88
WBC ($\times 10^3$/µl)	34	0.55 ± 0.42	0.05–1.67
Heterophils (%)	34	25.21 ± 17.20	3–70
Heterophils ($\times 10^3$/µl)	34	0.11 ± 0.10	0.02–0.43
Lymphocytes (%)	34	48.65 ± 15.24	16–70
Lymphocytes ($\times 10^3$/µl)	34	0.26 ± 0.20	0.01–0.78
Monocytes (%)	34	1.34 ± 2.43	0–10
Monocytes ($\times 10^3$/µl)	34	0.00 ± 0.01	0–0.04
Eosinophils (%)	34	19.68 ± 14.77	4–63
Eosinophils ($\times 10^3$/µl)	34	0.14 ± 0.21	0–0.96
Basophils (%)	34	4.59 ± 4.77	0–15
Basophils ($\times 10^3$/µl)	34	0.02 ± 0.01	0–0.08

Extracted from: Brady S, Burgdorf-Moisuk A, Kass PH (2016). Hematology and plasma biochemistry intervals for captive-born California tiger salamanders (*Ambystoma californiense*). *J Zoo Wildl Med*, 47(3):731–735.

Fish

Haematology Reference Values in Atlantic Salmon (*Salmo salar*) under Farming Conditions (n = 30)

Parameter	23 Oct	20 Nov	11 Dec	15 Jan	19 Feb	26 Mar	20 May	Mean ± Standard Error of Mean
Haematocrit (%)	46*	49	44	47	48	45	47	47 ± 1
Haemoglobin (g/dl)	10.0	10.4	9.4	9.9	9.9	8.9	9.0	9.6 ± 0.2
RBC ($\times 10^{12}$/l)	1.00	1.10	0.99	0.00	0.98	0.85	0.85	0.97 ± 0.02
MCV ($\times 10^{15}$/l)	464	441	447	474	492	524	553	485 ± 7
MCH ($\times 10^6$/l)	100	94	95	101	101	104	106	100 ± 1
MCHC (g/dl)	21.7	21.6	21.3	21.4	20.6	20.	19.4	20.9 ± 0.3
WBC (%)	0.96	0.73	0.71	0.80	0.57	0.43	0.54	0.68 ± 0.05

*Mean ± Standard Error of Mean Pooled Samples
Extracted from: Sandnes K, Lie Ø, Waagbø R (1987). Normal ranges of some blood chemistry parameters in adult farmed Atlantic salmon, *Salmo salar*. *J Fish Biol*, 32:129–136.

Haematology Reference Values in Nile Tilapia (*Oreochromis niloticus*) under Farming Conditions* (n = 10 Each Group)

Parameter	EDTA			Heparin		
	CG	BG	EG	CG	BG	EG
PCV (%)	29.3 ± 2.3**	34.0 ± 8.5	32.8 ± 2.6	29.2 ± 2.6	33.3 ± 6.0	32.8 ± 2.3
RBC ($\times 10^6$/µl)	1.9 ± 0.3	1.9 ± 0.4	1.80 ± 0.47	2.22 ± 0.37	1.8 ± 0.5	1.9 ± 0.4
Hb (g/dl)	8.3 ± 1.0	9.0 ± 1.9	9.0 ± 1.1	8.5 ± 1.0	9.7 ± 1.8	8.8 ± 1.0
MCV (fl)	160.0 ± 27.2	186.7 ± 89.0	194.2 ± 58.9	137.1 ± 19.1	196.6 ± 58.6	178.2 ± 35.7
MCHC (%)	28.2 ± 2.3	27.0 ± 2.1	27.3 ± 2.7	28.5 ± 2.2	29.1 ± 2.1	26.7 ± 1.8

*Blood samples collected without anaesthesia (CG), anaesthesia with benzocaine (BG), anaesthesia with eugenol (EG) and stored in EDTA or Heparin tubes
**Mean ± Standard Deviation
Extracted from: Weinert AC, Volpato J, Costa Á, Antunes RR, De Oliveira AC, Scabelo Mattoso CR, Saito MR (2014). Hematology of Nile tilapia (*Oreochromis niloticus*) subjected to anesthesia and coagulant protocols. *Ciências Agrárias*, 36(6):4237–4250.

Haematology Reference Values in Farmed Rainbow Trout (*Oncorhynchus mykiss*), n = 20 (for Each Sex)

Parameter	12 +		15 +		18 +		21 +	
	Male	Female	Male	Female	Male	Female	Male	Female
MCV (fl)	314.3 ± 8.6*	317.9 ± 10.4	288.6 ± 9.65	295.7 ± 11.32	283.2 ± 12.9	300.8 ± 27	294. ± 19.14	293.8 ± 16.11
MCH (pg)	103.8 ± 4.5	101.3 ± 11.7	92.2 ± 8.1	94,8 ± 12	92.3 ± 8.5	98.5 ±13.7	95.9 ± 14.3	94.56 ± 8.3
MCHC (%)	33 ± 1.21	32 ± 1.4	32 ± 0.96	32.1 ± 1.3	32.6 ± 0.95	32.5 ± 1.44	32.6 ± 0.91	32.1 ± 1.63
RBC (×106/mm)	0.82 ± 0.1	0.79 ± 0.14	0.85 ± 0.2	0.84 ± 0.13	1.13 ± 0.19	0.97 ± 0.13	1.02 ± 0.17	0.99 ± 0.03
PCV (%)	22.51 ± 2.9	21.96 ± 4.4	24.1 ± 3.4	24.8 ± 2.1	28.9 ± 4.3	27.6 ± 3.1	29.6 ± 3.4	29.2 ± 2.22
Hb (g/dl)	7.3 ± 0.87	7.52 ± 1.24	7.74 ± 1.2	7.62 ± 0.67	10.1 ± 1.19	9.93 ± 1.2	9.7 ± 1.12	10.4 ± 1.01
WBC (mm3)	24.01 1.8	24.3 1.5	23.6 2.3	24.2 1.88	25.7 1.12	23 1.19	25.4 0.84	24.3 1.84
Lymphocytes (%)	68.2 ± 2.94	72.5 ± 3.4	71 ± 3.5	69.6 ± 3.14	70.3 ± 3.8	71 ± 3.33	70.25 ± 3.69	70.2 ± 3.43
Heterophils (%)	18.2 ± 1.8	19.1 ± 1.16	19.4 ± 2.14	20.5 ± 1.87	20.4 ± 2.2	19.2 ± 2.9	19 ± 2.45	20.3 ± 2.1
Eosinophils (%)	5.8 ± 1.2	5.4 ± 1.3	6 ± 1.58	6.3 ± 1.4	6 ± 1.16	6.2 ± 1.87	6.5 ± 1.12	5.5 ± 1.1
Monocyte (%)	4.2 ± 0.86	3.41 ± 0.77	3.6 ± 0.62	3.3 ± 0.46	3.75 ± 0.57	3.66 ±1.01	4.01 ± 0.94	4.00 ± 0.34

*Mean ± Standard Deviation

Extracted from: Charoo SQ, Chalkoo SR, Qureshi TA (2013). Trout hematology. *Int J Inn Sci Math*, 1(1):19–23.

Haematology Reference Values in Nile Tilapia (*Oreochromis niloticus*) under Farming Conditions* (n = 10 Each Group)

Parameter	EDTA			Heparin		
	CG	BG	EG	CG	BG	EG
WBC (µl)	49516.8 ± 6945.6**	51345.2 ± 10865.7	48065.9 ± 11593.5	42536.9 ± 10333.7	44329.6 ± 11293.6	39931/6 ± 11796.7
Neutrophils (µl)	19723.0 ± 4675.7	27691.90 ± 6611.30	19137.1 ± 7805.1	16818.9 ± 6057.6	24791.59 ± 8853.78	17126.6 ± 6860.7
Lymphocytes (µl)	26621.9 ± 3912.6	19711.1 ± 5501.2	24944.6 ± 6122.6	23436.3 ± 4669.8	16633.3 ± 4080.5	20088.1 ± 7456.4
Eosinophils (µl)	141.9 ± 229.8	0 ± 0	238.6 ± 261.3	41.2 ± 130.2	39.6 ± 125.1	0 ± 0
Basophils (µl)	0 ± 0	0 ± 0	0 ± 0	0 ± 0	35.7 ± 112.9	0 ± 0
Monocytes (µl)	3030.1 ± 2219.4	3942.2 ± 3619.4	3745.6 ± 1664.8	2281.7 ± 1601.3	2281.7 ± 2302	2714.9 ± 1783.7
Thrombocytes (µl)	34262.7 ± 6142.1	32888.7 ± 5914.3	33118.9 ± 8970.5	30006.4 ± 8087.5	31092/1 ± 9545.2	30006.6 ± 8333.8

*Blood samples collected without anaesthesia (CG), anaesthesia with benzocaine (BG), anaesthesia with eugenol (EG) and stored in EDTA or Heparin tubes
**Mean ± Standard Deviation
Extracted from: Weinert AC, Volpato J, Costa Á, Antunes RR, De Oliveira AC, Scabelo Mattoso CR, Saito MR (2014). Hematology of Nile tilapia (*Oreochromis niloticus*) subjected to anesthesia and coagulant protocols. *Ciências Agrárias*, 36(6):4237–4250.

BIBLIOGRAPHY

Allen JL, Jacobson ER, Harvey JW, Boyce W (1985). Hematologic and serum chemical values for young African elephants (*Loxodonta africana*) with variations for sex and age. *J Zoo Med*, 16(3):98–101.

Allender MC, Fry MM (2008). Amphibian hematology. *Vet Clin North Am Exot Anim Pract*, 11(3):463–480.

Alonso JC, Huecas V, Alonso JA, Abelenda M, Muñoz-Pulido R, Puerta ML (1991). Hematology and blood chemistry of adult white storks (*Ciconia ciconia*). *Comp Biochem Physiol*, 98(3–4):395–397.

Anderson ET, Minter LJ, Clarke III EO, Mroch III RM, Beasley JF, Harms CA (2011). The effects of feeding on hematological and plasma biochemical profiles in green (*Chelonia mydas*) and Kemp's Ridley (*Lepidochelys kempii*) sea turtles. *Vet Med Int*, 2011:1–7.

Andriansyah Candra D, Riyanto MA, Barry J, Radcliffe RW (2013). Hematology and serum biochemistry of Sumatran rhinoceroses (*Dicerorhinus sumatrensis*) in a rainforest sanctuary in Way Kambas National Park, Indonesia. *J Zoo Wildl Med*, 44(2):280–284.

Arikan H, Cicek K (2014). Haematology of amphibians and reptiles: A review. *North West J Zool*, 10(1):190–209.

Arnold J (2005). Hematology of the sandbar shark (*Carcharhinus plumbeus*): Standardization of complete blood count techniques for elasmobranchs. *Vet Clin Path*, 34(2):115–123.

Aubin DJSt, Forney KA, Chivers SJ, Scott MD, Danil K, Romano TA, Gulland FMD (2013). Hematological, serum, and plasma chemical constituents in pantropical spotted dolphins (*Stenella attenuata*) following chase, encirclement, and tagging. *Mar Mamm Sci*, 29(1):14–35.

Barbara JCA, Ferreira VL, Guida FJV, Prioste FES, Matushima ER, Raso TF (2017). Hematologic reference intervals for wild black vultures (*Coragyps atratus*). *Vet Clin Pathol*, 46(4):575–579.

Barnes TS, Goldizen AW, Coleman GT (2008). Hematology and serum biochemistry of the brush-tailed rock-wallaby (*Petrogale penicillata*). *J Wildl Dis*, 44(2):295–303.

Bielli M, Nardini G, Di Girolamo N, Savarino P (2015). Hematological values for adult eastern Hermann's tortoise (*Testudo hermanni boettgeri*) in semi-natural conditions. *J Vet Diagn Invest*, 27(1):68–73.

Black PA, McRuer DL, Horne LA (2011). Hematologic parameters in raptor species in a rehabilitation setting before release. *J Avian Med Surg*, 25(3):192–198.

Boily F, Beaudoin S, Meaures LN (2006). Hematology and serum chemistry of harp (*Phoca groenlandica*) and hooded seals (*Cystophora cristata*) during the breeding season in the Gulf of St. Lawrence, Canada. *J Wild Dis*, 42(1):115–132.

Bowerman WW, Stickle JE, Giesy JP (2000). Hematology and serum chemistries of nestling bald eagles (*Haliaeetus leucocephalus*) in the lower peninsula of MI, USA. *Chemosphere*, 41(10):1575–1579.

Burrell C, Zhang H, Li D, Wang C, Li C, Aitken-Palmer C (2017). Hematology, serum biochemistry and urinalysis in the adult giant panda (*Ailuropoda melanoleuca*). *J Zoo Wildl Med*, 48(4):1072–1076.

Burrell C, Li L, Marieke J, Andrea L, Elizabeth F, Copper A-P (2018). Hematology and serum biochemistry values of the red panda subspecies (*Ailurus fulgens styani*). *J Zoo Wildl Med*, 49(2):384–395.

Campbell TW (2004). Hematology of amphibians. In: Thrall MA, editor. *Veterinary Hematology and Clinical Chemistry*. Lippincott, Williams & Wilkins, Philadelphia, PA, 291–297.

Campbell TW (2006). Clinical pathology of reptiles. In: Mader DR, editor. *Reptile Medicine and Surgery*, 2nd edition. Saunders, St Louis, MO, 453–470.

Campbell TW (2015). *Exotic Animal Hematology and Cytology*, 4th edition. Wiley-Blackwell, Hoboken, NJ.

Campbell TW, Grant K (2010). *Clinical Cases in Avian and Exotic Animal Hematology and Cytology*. Wiley-Blackwell, Hoboken, NJ.

Celdrán J, Polo FJ, Peinado VI, Viscor G, Palomeque J (1994). Haematology of captive herons, egrets, spoonbill, ibis and gallinule. *Comp Biochem Physiol A*, 107(2):337–341.

Clark P, Boardman W, Raidal S (2009). *Atlas of Clinical Avian Hematology*. Wiley-Blackwell, Hoboken, NJ.

Clauss TM, Dove ADM, Arnold JE (2008). Hematologic disorders of fish. *Vet Clin Exot Anim Pract*, 11(3):445–462.

Claver JA, Quaglia AIE (2009). Comparative morphology, development, and function of blood cells in nonmammalian vertebrates. *J Exot Pet Med*, 18(2):87–97.

Cornell LH (1983). Hematology and clinical chemistry values in the killer whale, *Orcinus orca* L. *J Wildl Dis*, 19(3):259–264.

Currier MJ, Russell KRJ (1982). Hematology and blood chemistry of the mountain lion (*Felis concolor*). *J Wildl Dis*, 18(1):99–104.

Dal'Bo GAD, Sampaio FG, Losekann ME, De Queiroz JF, Barreto Luiz AJ, Gyárfas Wolf VH, Gonçalves VT, Carra ML (2015). Hematological and morphometric blood value of four cultured species of economically important tropical food fish. *Neotrop Ichthyol*, 13(2):439–446.

D'Aloia MA, Samour JH, Howlett JC, Bailey TA, Naldo J (1994). Haemopathological responses to chronic inflammation in the houbara bustard (*Chlamydotis undulata macqueenii*). *Comp Haem Inter*, 4(4):203–206.

D'Aloia MA, Samour JH, Howlett JC, Bailey TA, Naldo J (1995). Normal haematology and age-related findings in rufous-crested bustards (*Eupodotis ruficrista*). *Comp Haem Inter*, 5:10–12.

D'Aloia MA, Samour JH, Bailey TA, Naldo J, Howlett JC (1996). Normal haematology of the white bellied (Eupodotis *senegalensis*), little black (*Eupodotis afra*) and Heuglin's (*Neotis heuglinii*) bustards. *Comp Haem Inter*, 6(1):46–49.

Dove ADM, Arnold J, Clauss TM (2010). Blood cells and serum chemistry in the world's largest fish: The whale shark *Rhincodon typus*. *Aquat Biol*, 9(2):177–183.

Dujowich M, Mazet JK, Zuba JR (2005). Hematologic and biochemical reference ranges for captive California condors (*Gymnogyps californianus*). *J Zoo Wild Med*, 36(4):590–597.

Eatwell K, Hedley J, Barron R (2014). Reptile haematology and biochemistry. *Pract*, 36(1):34–42.

Ellman MM (1997). Hematology and plasma chemistry of the inland bearded dragon, *Pogona vitticeps*. *Bull ARAV*, 7(4):10–12.

Fagbenro OA, Adeparusi EO, Jimoh WA (2013). Haematological profile of blood of African catfish (*Clarias gariepinus*, Burchell, 1822) fed sunflower and sesame meal-based diets. *J Fish Aquat Sc*, 8:80–86.

Fazio F (2019). Fish hematology analysis as an important tool of aquaculture: A review. *Aquaculture*, 500:237–242.

Fudge AM (2000). Avian complete blood count. In: Fudge AM, editor. *Laboratory Medicine: Avian and Exotic Pets*. WB Saunders, Philadelphia, PA, 9–18.

Fudge AM (2000). Disorders of the avian leucocytes. In: Fudge AM, editor. *Laboratory Medicine: Avian and Exotic Pets*. WB Saunders, Philadelphia, PA, 19–27.

Fukuzawa AH, Vellutini BC, Lorenzini DM, Silva Jr PI, Mortara RA, José da Silva MC, Daffre S (2008). The role of hemocytes in the immunity of the spider *Acanthoscurria Gomesiana*. *Devel Comp Imm*, 32:716–725.

Gallo L, Vanstreels RET, Cook RA, Karesh WB, Uhart M (2019). Hematology, plasma chemistry, and trace elements reference values of free-ranging adult Magellanic penguins (*Spheniscus magellanicus*). *Polar Biol*, 42(4):733–742.

García del Campo AL, Huecas V, Fernández A, Puerta ML (1991). Hematology and blood chemistry of macaws, *Ara rubrogenys*. *Comp Biochem Physiol*, 100(4):943–944.

Gentz EJ (2007). Medicine and surgery of amphibians. *ILAR J*, 48(3):255–259.

Grant KR (2015). Fish hematology and associated disorders. *Clin Lab Med*, 35(3):681–701.

Greig DJ, Gulland FMD, Rios CA, Ailsa Hall AJ (2010). Hematology and serum chemistry in stranded and wild-caught harbor seals in central California: Reference intervals, predictors of survival, and parameters affecting blood variable. *J Wild Dis*, 46(4):1172–1184.

Græsli AR, Fahlman Å, Evans AL, Bertelsen MF, Arnemo JM, Nielsen SS (2014). Haematological and biochemical reference intervals for free-ranging brown bears (*Ursus arctos*) in Sweden. *BMC Vet Res*, 10:183.

Haile Y, Chanie M (2014). Comparative aspects of the clinical hematology of birds: A review. *Br Poult Sci*, 3(3):88–95.

Hall AJ (1998). Blood chemistry and hematology of gray seal (*Halichoerus grypus*) pups from birth to postweaning. *J Zoo Wildl Med*, 29(4):401–407.

Han JI, Jang HJ, Na KJ (2016). Hematologic and serum biochemical reference intervals of the Oriental white stork (*Ciconia boyciana*) and the application of an automatic hematologic analyzer. *J Vet Sci*, 7(3):399–405.

Hart MG, Samour J, Spratt DM, Savage B, Hawkey CM (1991). An analysis of haematological findings on a feral population of Aldabra giant tortoises (*Geochelone gigantea*). *Comp Haem Inter*, 1(3):145–149.

Hellgren EC, Rogers LL, Seal US (1993). Serum chemistry and hematology of black bears: Physiological indices of habitat quality or seasonal patterns? *J Mammal*, 74(2):304–315.

Howlett JC, Samour JH, D'Aloia M-A, Bailey TA, Naldo J (1995). Normal haematology of captive adult kori bustards (*Ardeotis kori*). *Comp Haem Inter*, 5(2):102 –105.

Howlett JC, Samour JH, Bailey TA, Naldo JL (1998). Age-related haematology changes in captive-reared kori bustards (*Ardeotis kori*). *Comp Haem Inter*, 8(1):26–30.

Howlett JC, Bailey TA, Samour JH, Naldo JL, D'Aloia MA (2002). Age-related hematologic changes in captive-reared houbara, white-bellied and rufous crested bustards. *Wild Dis*, 38(4):804–816.

Hawkey CM (1975). *Comparative Mammalian Haematology: Cellular Components and Blood Coagulation of Captive Wild Animals*, 1st edition. Butterworth-Heinemann, Oxford, UK.

Hawkey CM, Hart MG, Knight JA, Samour JH, Jones DM (1982). Haematological findings in healthy and sick African grey parrot (*Psittacus erithacus*). *Vet Rec*, 111(25–26):580–582.

Hawkey CM, Ashton DG, Hart MG, Cindery RN, Jones DM (1983). Normal and clinical haematology in the yak (*Bos grunniens*). *Res Vet Sci*, 34(1):31–36.

Hawkey CM, Samour JH, Ashton DG, Hart MG, Cindery RN, Ffinch JM, Jones DM (1983). Normal and clinical haematology of captive cranes (Gruiformes). *Avian Pathol*, 12(1):73–84.

Hawkey CM, Hart MG, Samour JH (1984). Age-related haematological changes and haemopathological responses in Chilean flamingo (*Phoenicopterus chilensis*). *Avian Pathol*, 13(2):223–229.

Hawkey CM, Hart MG, Samour JH, Knight JA, Hutton RE (1984). Haematological findings in healthy and sick captive rosy flamingo (*Phoenicopterus ruber ruber*). *Avian Pathol*, 13(2):163–172.

Hawkey CM, Hart MG, Samour HJ (1985). Normal and clinical haematology of greater and lesser flamingos (*Phoenicopterus roseus* and *Phoeniconaias minor*). *Avian Pathol*, 14(4):537–541.

Hawkey CM, Samour JH, Henderson GM, Hart MG (1985). Haematological findings in captive Gentoo penguins (*Pygoscelis papua*) with bumblefoot. *Avian Pathol*, 14(2):251–256.

Hawkey CM, Samour J (1988). The value of clinical haematology in exotic birds. In: Jacobson ER, Kollias GV, editors. *Exotic Animals, Contemporary Issues in Small Animal Practice*. Churchill Livingstone, London, UK, 109–141.

Hawkey CM, Horsley DT, Keymer IF (1989). Haematology of wild penguins (Sphenisciformes) in the Falkland Islands. *Avian Pathol*, 18(3):495–502.

Heatley JJ, Johnson M (2009). Clinical technique: Amphibian hematology: A practitioner's guide. *J Exot Pet Med*, 18(1):14–19.

Hernández M, Margalida A (2010). Hematology and blood chemistry reference values and age-related changes in wild bearded vultures (*Gypaetus barbatus*). *J Wildl Dis*, 46(2):390–400.

Howell S, Hoffman K, Bartel L, Schwandt M, Morris J, Fritz J (2003). Normal hematologic and serum clinical chemistry values for captive chimpanzees (*Pan troglodytes*). *Comp Med*, 53(4):413–423.

Hong M, Hwang D, Cho S (2008). Hemocyte morphology and cellular immune response in termite (*Reticulitermes speratus*). *J Insect Sci*, 18(2):46.

Ibañez AE, Najle R, Larsen K, Pari M, Figueroa A, Montalti D (2015). Haematological values of three Antarctic penguins: Gentoo (*Pygoscelis papua*), Adélie (*P adeliae*) and chinstrap (*P antarcticus*). *Polar Res*, 34(1):25718.

Ihrig M, Tassinary LG, Bernacky B, Keeling ME (2001). Hematologic and serum biochemical reference intervals for the chimpanzee (*Pan troglodytes*) categorized by age and sex. *Comp Med*, 51(1):30–37.

Jones MP (2015). Avian hematology. *Vet Clin North Am Exot Anim Pract*, 18(1):51–61.

Keller KA, Sanchez-Migallon Guzman D, Paul-Murphy J, Byrne BA, Owens SD, Kass PH, Scott Weber III EP (2012). Hematological and biochemical plasma values of free-ranging Western pond turtles (*Emys marmorata*) with comparison to a captive population. *J Herpetol Med Surg*, 22(3–4):99–106.

Kock A, Mihok SO, Wambua J, Mwanzia J, Saigawa K (1999). Effects of translocation on hematologic parameters of free-ranging black rhinoceros (*Diceros bicornis michaeli*) in Kenya. *J Zoo Wildl Med*, 30(3):389–396.

Kolesnikovas CKM, Niemeyer C, Teixerira RHF, Nunes ALV, Rameh de Albuquerque LC, Sant'Anna SS, Catao-Dias JL (2012). Hematological and plasma biochemical values of hyacinth macaws (*Anodorhynchus hyacinthinus*). *J Avian Med Surg*, 26(3):125–129.

Kuhn-Nentwig L, Kopp S, Nentwig W, Haenni B, Streitberger K, Schürch S, Johann Schaller J (2014). Functional differentiation of spider hemocytes by light and transmission electron microscopy, and MALDI-MS-imaging. *Dev Comp Immunol*, 43(1):59–67.

Larsson AHMA, Do Espírito Santo PL, Mirandola RMS, Fedullo JDL, Ito FH, Itikawa PH, Pessoa RB (2015). Hematologic parameters of captive lions (*Panthera leo*) and Siberian tigers (*Panthera tigris altaica*). *Acta Sci Vet*, 43:1311.

Le Souëf AT, Holyoake CS, Vitali SD, Warren KS (2013). Hematologic and plasma biochemical reference values for three species of black cockatoos (*Calyptorhynchus* species). *J Avian Med Surg*, 27(1):14–22.

Leonard R, Ruben Z (1986). Hematology reference values for peripheral blood of laboratory rats. *Lab Anim Sci*, 36(3):277–281.

Lewbart GA editor. (2011). *Invertebrate Medicine*, 2nd edition. Wiley-Blackwell, Hoboken, NJ.

Low M, Eason D, Elliott G, McInnes K, Paul-Murphy J (2006). Hematologic and biochemical reference ranges for the kakapo (*Strigops habroptilus*): Generation and interpretation in a field-based wildlife recovery program. *J Avian Med Surg*, 20(2):80–88.

Mader DR (2000). Normal hematology of reptiles. In: Feldman BF, Zinkl JG, Jain NC, editors. *Veterinary Hematology*. Lippincott Williams & Wilkins, Philadelphia, PA, 1126–1132.

Marinescu B, Isvoranu G, Anghelache L, Cionca F (2014). Hematology references for three laboratory mice strains. *Roum Arch Microbiol Immunol*, 73(1–2):30–43.

Marshall KL (2008). Rabbit hematology. *Vet Clin North Am Exot Anim Pract*, 11(3):551–567.

Martinez Silvestre AM, Dominguez MAR, Mateo JA, Pastor J, Marco I, Lavin S, Cuenca R (2004). Comparative haematology and blood chemistry of endangered lizards (*Gallotia* species) in the Canary Islands. *Vet Rec*, 155(9):266–269.

McClure HM, Keeling ME, Guilloud NB (1972). Hematologic and blood chemistry data for the gorilla (*Gorilla gorilla*). *Int J Primatol*, 18(3):300–316.

McClure HM, Keeling ME, Guilloud NB (1972). Hematologic and blood chemistry data for the orangutan (*Pongo pygmaeus*). *Int J Primatol*, 18(3):284–299.

McCracken H, Hyatt AD, Slocombe RF (1994). Two cases of anemia in reptiles treated with blood transfusions: (1) hemolytic anemia in a diamond python caused by erythrocytic virus; (2) nutritional anemia in a bearded dragon. *Proc Assoc Rept Amph Vet*, 47–51.

Miller M, Buss P, Wanty R, Parsons S, van Helden P, Olea-Popelka F (2015). Baseline hematologic results for free-ranging white rhinoceros (*Ceratotherium simum*) in Kruger National Park, South Africa. *J Wildl Dis*, 51(4):916–922.

Mitchell EB, Johns J (2008). Avian hematology and related disorders. *Vet Clin North Am Exot Anim Pract*, 11(3):501–522.

Moreau B, Vié JC, Cotellon P, De Thoisy I, Motard A, Raccurt CP (2003). Hematological and serum biochemistry values in two species of free-ranging porcupines (*Coendou prehensilis, Coendou melanurus*) in French Guiana. *J Zoo Wildl Med*, 34(2):159–162.

Naldo N, Libanan N, Samour J (2009). Health assessment of a spiny tailed lizard (*Uromastyx* spp) population in Abu Dhabi, United Arab Emirates. *J Zoo Wild Med*, 40(3):445–452.

Nardini G, Leopardi S, Bielli M (2013). Clinical hematology in reptilian species. *Vet Clin Exot Anim Pract*, 16(1):1–30.

Nazifi S, Vesal N (2003). Hematological values of healthy rose-ringed parakeets (*Psittacula krameri*). *J Appl Anim Res*, 24(2):165–168.

Nazifi S, Nabinejad A, Sepehrimanesh M, Poorbaghi SL, Farshneshani F, Rahsepa M (2008). Haematology and serum biochemistry of golden eagle (*Aquila chrysaetos*) in Iran. *Comp Clin Pathol*, 17(3):197–201.

Nentwig LK, Nentwig W (2013). The immune system of spiders. In: Nentwig W, editor. *Spider Ecophysiology*. Springer-Verlag, Berlin Heidelberg, Germany, 81–91.

Nirmalan G, Nair SG, Simo KJ (1967). Hematology of the Indian elephant (*Elephas maximus*). *Can J Physiol Pharmacol*, 45(6):985–991.

Nuno S, Serra P, Fernandes ML, Pacheco C, Franco C, Rosa G (2006). Hematology and blood parasites of juvenile black storks *Ciconia nigra* in Portugal. *Biota*, 7:83–88.

Ochiai Y, Baba A, Hiramatsu M, Toyota N, Watanabe T, Yamashita K, Yokota H, Iwano H (2018). Blood biochemistry and hematological changes in rats after administration of a mixture of three anesthetic agents. *J Vet Med Sci*, 80(2):387–394.

Padrtova R, Lloyd CG (2009). Hematologic values in healthy gyr × peregrine falcons (*Falco rusticolus × Falco peregrinus*). *J Avian Med Surg*, 23(2):108–113.

Parrino V, Capello T, Costa G, Cannava C, Sanfilippo M, Fazio F, Fasulo S (2018). Comparative study of haematology of two teleost fish (*Mugil cephalus* and *Carassius auratus*) from different environments and feeding habits. *Eur Zool J*, 85(1):193–199.

Pendl H, Samour J (2016). Hematology analyses. In: Samour J, editor. *Avian Medicine*, 3rd edition. Elsevier, St. Louis, MO, 77–100.

Perpiñán D (2017). Chelonian haematology 1. Collection and handling of samples. *Pract*, 39(5):194–202.

Perpiñán D (2017). Chelonian haematology 2. Identification of blood cells. *Pract*, 39(6):274–283.

Polo FJ, Celdran JF, Peinado VI, Viscor G, Palomeque J (1992). Hematological values for four species of birds of prey. *Condor*, 94(4):1007–1013.

Polo FJ, Peinado VI, Viscor G, Palomeque J (1998). Hematologic and plasma chemistry values in captive psittacine birds. *Avian Dis*, 42(3):523–535.

Priddel D, Wheeler R (2006). Hematology and blood chemistry of a Bryde's whale, *Balaenoptera edeni*, entrapped in the Manning river, New South Whale, Australia. *Mar Mamm Sci*, 14(1):72–81.

Puerta M, Pulido MR, Huescas V, Abelenda M (1989). Hematology and blood chemistry of chicks of white and black storks (*Ciconia ciconia* and *Ciconia nigra*). *Comp Biochem Phys*, 94(2):201–204.

Rafaj RB, Tončić J, Vickovi I, Šoštarić B (2011). Haematological and biochemical values of farmed red deer (*Cervus elaphus*). *Vet Archiv*, 81(4):513–523.

Rayhel L, Aitken-Palmer C, Joyner P, Cray C, Lizárraga CA, Ackerman B, Crowe C (2015). Hematology and biochemistry in captive white-naped cranes (*Grus vipio*). *J Zoo Wildl Med*, 46(4):747–754.

Rosskopf WR (2000). Disorder of reptilian leucocytes and erythrocytes. In: Fudge AD, editor. *Laboratory Medicine Avian and Exotic Pets*. WB Saunders, Philadelphia, PA, 19–27.

Samour JH, Jones DM, Pugsley SL, Fitzgerald AK (1983). Blood sampling techniques in penguins (Sphenisciformes). *Vet Rec*, 113(15):340.

Samour JH, Risley D, March T, Savage B, Nieva O, Jones DM (1984). Blood sampling techniques in reptiles. *Vet Rec*, 114(19):472–476.

Samour JH, Howlett JC, Hart MG, Bailey TA, Naldo J, D'Aloia MA (1994). Normal haematology of the houbara bustard (*Chlamydotis undulata macqueenii*). *Comp Haem Int*, 4(4):198–202.

Samour JH, D'Aloia MA, Howlett JC (1996). Normal haematology of captive saker falcons (*Falco cherrug*). *Comp Haem Int*, 6(1):50–52.

Samour JH, Howlett JC, Silvanose C, Hasbun CR, Al-Ghais SM (1998). Normal haematology of free-living green sea turtles (*Chelonia mydas*) from the United Arab Emirates. *Com Haem Int*, 8(2):102–107.

Samour JH, Naldo JL, John SK (2005). Normal haematological values in gyr falcons (*Falco rusticolus*). *Vet Rec*, 157(26):844–847.

Samour J (2005). Diagnostic value of hematology. In: Harrison GJ, Lightfoot T, editors. *Clinical Avian Medicine*. Spix Publishing, Lake Worth, FL, 587–609.

Samour J (2008). Haematology of bustards. In: Bailey TA, editor. *Diseases and Medical Management of Houbara Bustards and Other Otididae*. National Avian Research Center, Environment Agency, Abu Dhabi, United Arab Emirates, 87–96.

Samour J, Naldo J, Rahman H, Sakkir M (2010). Hematologic and plasma biochemical reference values in Indian peafowl (*Pavo cristatus*). *J Avian Med Surg*, 24(2):99–106.

Samour JH, Naldo JL, Libanan NL, Habeeb Rahman PK, Sakkir M (2011). Age-related hematology and plasma chemistry changes in captive Masai ostriches (*Struthio camelus massaicus*). *Comp Clin Path*, 19(4):659–667.

Seguel M, Muñoz F, Keenan A, Perez-Venegas DJ, DeRango E, Paves H, Gottdenker N, Müller A (2016). Hematology, serum chemistry, and early hematologic changes in free-ranging South American fur seals (*Arctocephalus australis*) at Guafo Island, Chilean Patagonia. *J Wildl Dis*, 52(3):663–668.

Schlenke TA, Morales J, Govind S, Clark AG (2007). Contrasting infection strategies in generalist and specialist wasp parasitoids of *Drosophila melanogaster*. *PLOS Pathog*, 3(10):1486–1501.

Schmidt EM, Lange RR, Ribas JM, Daciuk BM, Montiani-Ferreira F, Paulillo AC (2009). Hematology of the red-capped parrot (*Pionopsitta pileata*) and Vinaceous Amazon parrot (*Amazona vinacea*) in captivity. *J Zoo Wildl Med*, 40(1):15–17.

Scope A, Schwendenwein I, Enders F, Gabler C, Seidl E, Filip T, Soklaridis U (2000). Hematologic and clinical chemistry reference values in red lories (*Eos* spp). *Avian Dis*, 44(4):885–890.

Siperstein LJ (2008). Ferret hematology and related disorders. *Vet Clin North Am Exot Anim Pract*, 11(3):535–550.

Small MF, Baccus JT, Mink JN, Roberson JA (2005). Hematologic responses in captive white-winged doves (*Zenaida asiatica*), induced by various radiotransmitter attachments. *J Wildl Dis*, 41(2):387–394.

Smith SA, Zimmerman K, Moore DM (2015). Hematology of the domestic ferret (*Mustela putorius furo*). *Clin Lab Med*, 35(3):609–616.

Soares T, Napoleão TH, Ferreira FRB, Paiva PMG (2015). Hemolymph and hemocytes in tarantula spiders: Physiological roles and potential sources of bioactive molecules. In: Jenkins OP, editor. *Advances in Animal Science and Zoology*, Vol. 8. Nova Science Publishers Inc, Hauppauge, NY, 113–129.

Stacy BA, Whitaker N (2000). Hematology and blood biochemistry of captive mugger crocodiles (*Crocodylus palustris*). *J Zoo Wild Med*, 31(3):339–347.

Stacy NI, Alleman AR, Sayler KA (2011). Diagnostic hematology of reptiles. *Clin Lab Med*, 31(1):87–108.

Stacy NI, Fredholm DV, Rodriguez C, Castro L, Harvey JW (2017). Whip-like heterophil projections in consecutive blood films from an injured gopher tortoise (*Gopherus polyphemus*) with systemic inflammation. *Vet Quart*, 37(1):162–165.

Stang-Voss C (1974). On the ultrastructure of invertebrate hemocytes: An interpretation of their role in comparative hematology. In: Hanna MG, Cooper EL, editors. *Contemp Top Immunobiol*. Springer, Boston, MA, 65–76.

Sykes JM 4th, Klaphake E (2015). Reptile hematology. *Clin Lab Med*, 35(3):661–680.

Szabo Z, Klein A, Jakab C (2014). Hematologic and plasma biochemistry reference intervals of healthy adult barn owls (*Tyto alba*). *Avian Dis*, 58(2):228–231.

Teixeirai MA, Chaguri LCAG, Carissim AS, De Souza NL, Mori CMC, Gomes VMW, Neto AP, Nonoyama K, Merusse JLB (2000). Hematological and biochemical profiles of rats (*Rattus norvegicus*) kept under microenvironmental ventilation system. *Braz J Vet Res Anim Sci*, 37(5):341–347.

Telford SR, Jacobson ER (1993). Lizard erythrocytic virus in East African chameleons. *J Wildl Dis*, 29(1):57–63.

Travis EK, Vargas FH, Merkel J, Gottdenker Miller NRE, Parker PG (2006). Hematology, serum chemistry, and serology of Galapagos penguins (*Spheniscus mendiculus*) in the Galapagos Islands, Ecuador. *J Wild Dis*, 42(3):625–632.

Troiano JC, Vidal JC, Gould JE, Gould E (1997). Haematological reference intervals of the South American rattlesnake (*Crotalus durissus terrificus*, Laurentis 1768) in captivity. *Comp Haem Int*, 7(2):109–112.

Troiano JC, Altahus R, Scaglione MC, Scaglione L (1998). Osmotic fragility and size of erythrocytes in *Caiman latirostris* and caiman *Crocodylus jacare* (Crocodylia – Alligatoridae) under captive conditions. *Com Haem Int*, 8:50–52.

Troiano JC, Silva MC (1998). Hematological reference values from Argentine terrestrial turtle (*Chelonidis chilensis chilensis*). *Analecta Vet*, 18(1–2):47–51.

Troiano JC, Vidal JC, Gould E, Malinskas G, Gould E, Scaglione LM, Scaglione M, Heker JJ, Simoncini C, Dinapoli HA (1999). Haematological and blood chemical values from *Bothrops ammodytoides* (Ophidia – crotalidae) in captivity. *Com Haem Inter*, 9(1):31–35.

Troiano JC, Vidal JC, Gould EF, Heker K, Gould J, Vogt AU, Simoncini C, Amantinie CE, De Roodt A (2000). Hematological values of some *Bothrops* species (Ophidia - Crotalidae) in captivity. *J Venom Anim Toxins*, 6. 10.1590/S0104-79302000000200005.

Troiano JC, Vidal JC, Uriarte E, Gould E, Gould J, Heker J, Simoncini C, Tapia G (2000). Osmotic fragility and erythrocyte size in *Iguana* (Sauria – Iguanidae) in captivity. *Comp Haem Inter*, 10(1):14–18.

Troiano JC (2005). Reptilian hematology. *Reptilia*, 51:79–83.

Troiano JC, Gould E, Gould I (2006). Hemolytic action of *Naja naja atra* cardiotoxin on erythrocytes from different animals. *J Venom Anim Toxins incl Trop Dis*, 12(1):44–58.

Troiano JC, Gould E, Gould I (2008). Hematological reference intervals in Argentine lizard *Tupinambis merianae* (Sauri – Teiidae). *Comp Clin Pathol*, 17(2):93–97.

Troiano JC, Gould E (2011). Interspecific differences in the haematological references intervals in *Tupinambis* genus from Argentina (Sauria-Teiidae). *Comp Clin Patho*, 20(4):309–331.

Van der Knaap WPW, Adema CM, Sminia T (1993). Invertebrate blood cells: Morphological and functional aspects of the haemocytes in the pond snail *Lymnaea stagnalis*. *Comp Haem Int*, 3(1):20–26.

Videan EN, Fritz J, Murphy J (2008). Effects of aging on hematology and serum clinical chemistry in chimpanzees (*Pan troglodytes*). *Am J Primatol*, 70(4):311–414.

Wallace RS, Teare JA, Diebold E, Michaels M, Willis MJ (1995). Hematology and plasma chemistry values in free-ranging Humboldt penguins (*Spheniscus humboldti*) in Chile. *Zoo Biol*, 14(4):311–316.

Wilhelm FD, Eble GJ, Kassner G, Caprario FX, Dafré AL, Ohira M (1992). Comparative hematology in marine fish. *Comp Biochem Physiol Comp Physiol*, 102(2):311–321.

Williams DL (1999). Sample taking in invertebrate veterinary medicine. *Vet Clin North Am Exot Anim Pract*, 2(3):777–801.

Wilson GR, Hoskins L (1975). Haematology and blood chemistry of the red kangaroo *Megaleia rufa* in captivity. *Aust Vet J*, 51(3):146–149.

Woolford L, Wong A, Sneath HL, Long T, Boyd SP, Lanyon JM (2015). Hematology of dugongs (*Dugong dugon*) in Southern Queensland. *Vet Clin Pathol*, 44(4):530–541.

Zimmerman K, Moore DM, Smith SA (2015). Hematological assessment in pet guinea pigs (*Cavia porcellus*): Blood sample collection and blood cell identification. *Vet Clin North Am Exot Anim Pract*, 18(1):33–40.

INDEX

Abnormal blood cell, 2

African eagle owls (*Bubo africanus*), haematology reference values in, 226

African elephant (*Loxodonta africana*), 25 lymphocytes and platelets from, 152

African grey parrot (*Psittacus erithacus*) *Haemoproteus psittaci* from, 196 red blood cell agglutination in, 69

African penguin (*Spheniscus demersus*) *Babesia peircei* from, 207 heterophil from, 119 monocytosis in, 164

African rock python (*Python sebae*) haematology reference values in, 235 *Hepatozoon sebai* from, 200

Agglutination, 65–66

Albino axolotl (*Ambystoma mexicanum*), blood sample collection from, 18

Aldabra giant tortoise (*Aldabrachelys gigantea*) anisocytosis in, 75 heterophil and eosinophil from, 99 monocytosis in, 165 red blood cells from, 43, 55

Alligators, blood sample collection from, 5

Alpaca, clinically normal captive (*Vicugna pacos*), neutrophils in a blood smear from, 88

Alpaca, conscious (*Vicugna pacos*), blood sample collection from, 12

Alpini, Prospero, 2

Amazon parrots (*Amazona* spp.), haematology reference values in, 229

American alligator (*Alligator mississippiensis*), eosinophil from, 95

American bull frog (*Lithobates catesbeianus*), blood sample collection from, 18

American/green iguana (*Iguana iguana*), 31

American rosy flamingos (*Phoenicopterus ruber*), haematology reference values in, 223

American white pelicans (*Pelecanus erythrorhynchos*), haematology reference values in, 223

Amphibians blood sample collection from, 5 haematology reference values in, 240–242

Anaesthetised Southern koala (*Phascolarctos cinereus victor*), blood sample collection from, 13

Andean condor (*Vultur gryphus*), blood sample collection from, 14

Angulate tortoise (*Chersina angulata*), *Hepatozoon fitzsimonsi* from, 202

Anisocytosis, 49, 65, 66

Arabian camel, healthy (*Camelus dromedarius*), lymphocytes from, 137

Arabian oryx, severely ill, one-day-old (*Oryx leucoryx*), red blood cell deformation in, 72

Arboreal prehensile porcupines (*Coendou prehensilis* and *Coendou melanurus*), 4

Argentine black and white tegu (*Salvator merianae*), lymphocytes from, 143

Artefacts, 28

Asian elephant, conscious (*Elephas maximus*), blood sample collection from, 10, 12

Asian elephants (*Elephas maximus indicus*), *Trypanosoma evansi* in, 185

Asian lion, anaesthetised (*Panthera leo persica*), blood sample collection from, 11

Asian short-clawed otter (*Aonyx cinerea*), blood sample collection from, 8

Atlantic bottle nosed dolphins (*Tursiops truncatus*), haematology reference values in, 220

Atlantic guitarfish (*Rhinobatos lentiginosus*), blood smear from, 102

Atlantic salmon (*Salmo salar*), 5 haematology reference values in, 243

Auricular veins, 4

Australian fur seal, anaesthetised (*Arctocephalus pusillus doriferus*), blood sample collection from, 14

Australian water dragon (*Itellagama lesueurii*) heterophils from, 121 lymphocyte from, 143

Australian white ibises (*Threskiornis molucca*), haematology reference values in, 224

Axolotl (*Ambystoma mexicanum*) lymphocyte from, 124, 144 monocyte from, 160 red blood cells of, 43

Azurophil, 32, 33, 165–170

Bactrian camel (*Camelus bactrianus*) basophil from, 89 hypochromia and anisocytosis in, 59 neutrophils from, 86 red blood cells from, 44

Bald eagle (*Haliaeetus leucocephalus*), basophil and frequent misshapen erythrocytes from, 132

Barbary macaque (*Macaca sylvanus*), neutrophils from, 110

Barn owls (*Tito alba*), haematology reference values in, 226

Basophilic stippling, 26

Baw baw frog (*Philoria frosti*)
 monocyte from, 78, 159
 red blood cells, white blood cells and thrombocytes from, 43

Bearded dragon (*Pogona vitticeps*), 154
 heterophil from, 154
 lymphocytes from, 154

Biconcave discoid red blood cells, 28

Bighorn sheep (*Ovis canadensis*), neutrophilia from, 113

Birds
 blood sample collection from, 4–5
 haematology reference values in, 222–233

Black and white ruffed lemur (*Varecia variegata*)
 marked crenation in an EDTA sample from, 54
 monocytes from, 160

Blackbuck (*Antilope cervicapra*), 26

Black curassow (*Crax alector*), 5

Black-necked crowned crane (*Balearica pavonina*), 5

Black-pointed tegu (*Tupinambis teguizin*)
 azurophil from, 166
 heterophil from, 97

Black rhinoceros (*Diceros bicornis*)
 blood sample collection from, 10, 11
 haematology reference values in, 218

Black spider monkey (*Ateles paniscus*), microspherocytes from, 68

Blesbok (*Damaliscus pygargus phillipsi*), eosinophil with blue-grey cytoplasm from, 88

Blood cells, 39
 abnormalities in lymphocytes associated with disease, 145–156
 active erythropoiesis in small mammals, birds and reptiles, 51–53
 granulocytes
  common artefacts affecting, 102–105
  pathological responses involving, 106–132
  species variation in normal granulocytes, 79–102
 invertebrates, cellular appearance in haemolymph of, 179
 lymphocytes morphology, normal variation in, 132–145
 Mediterranean spur-thighed tortoise, erythropoiesis in, 63–64
 normal azurophil morphology, 165–170

normal monocyte morphology, 156–165
 platelet and thrombocytes, variation in, 171–175
  associated with disease, 175–178
 red blood cell morphology, abnormal variations in, 56
 red blood cells, common artefacts affecting, 53–55
 red blood cell shape, normal variation in, 44–51
 red blood cell size, normal species variation in, 39–44

Blood film, 32–33
 red blood cells, 28–29

Bloodletting, 2

Blood parasites, 183–211

Blood sample collection
 birds, 4–5
 fish, 5
 invertebrates, 6
 large mammals, 4
 reptiles and amphibians, 5
 small mammals, 4

Blood sampling, 6–20

Blotched genet (*Genetta maculata*), red blood cell from, 46

Blue and gold macaw (*Ara ararauna*)
 haematology reference values for, 227–228
 heterophilia from, 116

Blue catfish (*Ictalurus furcatus*), lymphocytes from, 156

Blue cranes (*Anthropoides paradisea*), haematology reference values in, 225–226

Blue fronted amazon (*Amazona aestiva*), heterophil from, 120

Blue tit (*Cyanistes caeruleus*), *Trypanosoma everetti* from, 187

Blue-tongued skink (*Tiliqua scincoides*)
 heterophils from, 121
 red blood cells from, 75

Blue waxbill (*Uraeginthus angolensis*)
 Plasmodium circumflexum schizont from, 193
 Trypanosoma bouffardi from, 187
 Trypanosoma everetti from, 187

B lymphocytes, 33

Boa constrictor (*Boa constrictor*)
 arenavirus infection in, 155
 basophil from, 95, 100
 blood smear from, 100
 haematology reference values in, 236
 poikilocytes and erythroplastid from, 65
 polychromatic erythroblasts from, 65

Boobook owls (*Ninox boobook*), haematology reference values in, 226

Boomslang snake (*Dispholidus typus*)

macrocytic and microcytic polychromatic erythroblast from, 74
 red blood cells from, 74

Bornean orangutan (*Pongo pygmaeus*), neutrophil from, 81

Bottlenose dolphin (*Tursiops truncatus*), 40
 blood sample collection from, 14
 lymphocyte from, 136
 neutrophils from, 87
 red blood cells from, 40

Bovine eperythrozoonosis, 209

Brachial plexus of the pectoral flipper, 4

Brazilian lanceheads (*B. moojeni*), haematology reference values in, 237–238

Broad-shelled river turtle (*Chelodina expansa*), monocyte from, 159

Brown capuchin monkey (*Sapajus apella*)
 lymphocyte from, 134
 red blood cells from, 50

Brown kiwi (*Apteryx mantelli*), *Hepatozoon kiwii* from, 200

Brown lemur (*Eulemur fulvus*), neutrophil and eosinophil from, 83

Brown pelicans (*Pelecanus occidentalis*), haematology reference values in, 223

Brown python (*Liasis fuscus*)
 azurophil, 167
 erythropoiesis in, 53
 heterophil from, 98, 167
 thrombocytes from, 167, 174

Brown rat (*Rattus norvegicus*), 4
 neutrophil from, 84

Brush tailed rock wallaby (*Petrogale penicillata*), 4

Brushtail possum, common (*Trichosurus vulpecula vulpecula*), lymphocyte from, 138

Brush turkey (*Alectura lathami*), 5

Budgerigar (*Melopsittacus undulatus*)
 binucleated polychromatic erythroblast from, 52
 erythroplastid from, 51
 lymphocyte from, 139
 mature red blood cells, basophilic, polychromatic and orthochromatic erythroblasts from, 52
 polychromatic red blood cells from, 139
 red blood cells from, 41
 thrombocyte, 139

Buffon's macaw (*Ara ambiguus*)
 heterophils from, 116
 toxic granules from, 116

Buffy fish owl (*Ketupa ketupu*), lymphocyte and thrombocyte from, 139

Bull frogs (*Lithobates catesbeianus*), haematology reference values in, 241

Burgundy goliath birdeater
(*Theraphosa stirmi*)
blood sample collection from, 20
granulocyte from, 179
theraphosid haemolymph from, 179
Buzzards (*Buteo buteo*)
blood smear from, 102
haematology reference values in,
224–225
heterophil from, 92
red blood red cells from, 42

Cabot's ring, 51
Caiman (*Caimaninae*), *Hepatozoon
caimani* from, 202
California sea lions (*Zalophus
californianus*)
haematology reference values in, 220
neutrophil from, 112
California tiger salamanders (*Ambystoma
californiense*), haematology
reference values in, 242
Camelidae, 25
Canadian beaver (*Castor canadensis*), red
blood cells from, 40
Canary, healthy (*Serinus canaria*),
heterophil and eosinophil
from, 92
Cape fur seal (*Arctocephalus pusillus*),
blood sample
collection from, 11
Cape turtle dove (*Streptopelia capicola*),
Haemoproteus columbae
macrogametocyte from, 195
Captive alpaca, clinically normal (*Vicugna
pacos*), lymphocytes and platelets
from, 137
Captive bottlenose dolphin (*Tursiops
truncatus*), monocyte
from, 158
Captive killer whales (*Orcinus orca*),
haematology reference
values in, 219
Captive panther chameleons (*Furcifer
pardalis*), haematology reference
values in, 240
Captive Tasmanian devil, clinically normal
(*Sarcophilus harrissi*), monocyte
from, 158
Capuchin monkey, newly captured (*Sapajus
apella*), microfilarian in blood
from, 183
Capybara (*Hydrochoerus hydrochaeris*), 26
intracellular haemoglobin
crystallisation in, 48
micro sickles from, 47
monocyte from, 157
pseudoeosinophilic neutrophil
from, 83
showing a Foà-Kurloff body, 135, 136

Caracaras (*Caracara cheriway*),
haematology reference
values in, 224–225
Carpet python (*Morelia spilota*)
azurophilia in, 167
heterophils from, 122
Casiragua (*Proechimys guairae*),
thrombocytosis from, 175
Cats (*Felis catus*), *Trypanosoma cruzi*
in, 185
Cattle (*Bos taurus*)
Babesia bigemina from, 205
Babesia divergens from, 205
Babesia major from, 205
Ehrlichia phagocytophilia from, 209
Theileria parva from, 207
Cattle egret, immature (*Bubulcus ibis*),
basophilic erythroblasts in the
blood of, 52
Caudal vein, 5
Caval venipuncture, 4
Cephalic vein, 4
Cheetah (*Acinonyx jubatus*)
agranular neutrophils and eosinophil
from, 80
blood sample collection from, 9
eosinophilia in, 128
Hepatozoon sp. from, 199
monocytes from, 162
neutrophils from, 108, 109
Chilean flamingos (*Phoenicopterus
chilensis*)
haematology reference values in, 223
heterophil from, 90
Chilean rose tarantula, anaesthetised
(*Grammostola rosea*), blood
sample collection from, 20
Chimpanzee (*Pan troglodytes*)
haematology reference values in, 218
neutrophil from, 81, 105, 110
Chinchilla (*Chinchilla lanigera*)
blood sample collection from, 8
haematology reference values in, 216
Chinese alligator (*Alligator sinensis*)
eosinophil from, 99
thrombocyte and lymphocyte from, 175
white blood cell in, 105
Chinese water dragon (*Physignathus
cocincinus*)
azurophil from, 169–170
heterophils from, 122, 170
Cirl bunting (*Emberiza cirlus*), *Isospora
normanlevinei* from, 211
Clinically normal axolotl (*Ambystoma
mexicanum*), eosinophil from, 101
Clinically normal ferret (*Mustela putorius
furo*), lymphocyte from, 136
Collared peccary (*Pecari tajacu*)
blood sample collection from, 12
neutrophils from, 112

Common bottlenose dolphin (*Tursiops
truncatus*), 4
Common box turtle (*Terrapene carolina*),
eosinophil from, 101
Common fruit fly/vinegar fly (*Drosophila
melanogaster*), 25
Common hill mynah (*Gracula religiosa*),
lymphocyte and thrombocytes
from, 139
Common moorhen (*Gallinula chloropus*),
lymphocytes from, 138
Common pheasant (*Phasianus colchicus*),
heterophils from, 120
Common squirrel monkey (*Saimiri
sciureus*), neutrophilia from, 111
Corn snake (*Pantherophis guttatus*)
blood sample collection from, 17
Hepatozoon guttata from, 201
Cotton-headed tamarin (*Saguinus oedipus*)
monocyte and neutrophil from, 156
target cells and stomatocytes in, 70
Cow (*Bos taurus*)
anisocytosis and erythroblasts in, 61
monocytosis in, 113, 162
neutrophils from, 85, 112, 113
Crab-eating macaque (*Macaca fascicularis*)
lymphocyte from, 145
monocyte from, 161
Crab-eating macaque, healthy (*Macaca
fascicularis*), neutrophil and
lymphocyte from, 82
Crab-eating mongoose (*Herpestes urva*),
neutrophils in, 110
Cranial vena cava, 4
Crenated red blood cells, 53
Crocodiles, blood sample collection from, 5
Crossed pit vipers (*Bothrops alternatus*),
haematology reference values in,
237–238
Crowned crane (*Balearica regulorum*), 28
Babesia balearicae from, 206
haematology reference values in,
225–226
Haemoproteus antigonis from, 195
Haemoproteus balearicae from, 196
Crowned hornbill (*Tockus
alboterminatus*), 5
Cutaneous ulnar/basilic vein, 4

Dark plain-backed pipit (*Anthus
leucophrys*), *Plasmodium rouxi*
schizont from, 193
Demoiselle cranes (*Grus virgo*),
haematology reference values in,
225–226
Diamond python (*Morelia spilota*)
haemolytic anaemia in, 77
higher magnification showing two
mitotic RBCs, 77
Discocytic red blood cells, 45

Döhle bodies, 32
Domestic cat (*Felis catus*)
  autoagglutination and marked
    polychromasia in, 66
  chronic lymphocytic leukaemia from,
    149, 150
  chronic myeloid in, 126
  chronic myeloid leukaemia in, 126, 127
  eosinophil from, 79, 128, 129
  erythrocyte refractile (ER) bodies in an
    unstained blood film from, 75
  hypochromic anaemia in, 56
  lymphoblastic overflow in the
    blood of, 148
  lymphoblasts from, 150
  lymphocytes from, 146
  lymphocytosis from, 149
  mast cells from, 130
  neutrophils from, 79, 108
  profound reticulocytosis in, 66
  red blood cells from, 50, 76
  unclassified lymphoid leukaemia
    from, 149
Domestic cow (*Bos taurus*), red blood
    cells from, 39
Domestic dog (*Canis lupus familiaris*)
  Babesia canis acute infection in, 204
  Babesia canis chronic infection in, 203
  Babesia gibsoni from, 203
  blood cells from, 127
  blood film showing the presence of
    elliptocytes from, 57
  bone marrow preparation from, 77
  chronic myeloid leukaemia from, 127
  Dirofilaria immitis (heart worm)
    microfilarian from, 184
  eosinophil from, 107, 129
  haemoglobin crystals in the blood of, 48
  Hepatozoon canis from, 199
  iron deficiency anaemia in, 56
  Leishmania donovani in, 189
  lymphoblasts from, 149
  lymphocytes from, 147, 149
  marked anisocytosis, neutrophilia and
    thrombocytosis in, 56
  mast cells from, 131
  metamyelocyte from, 128
  monoblasts from, 163
  monocyte from, 106
  myeloblasts from, 125–127
  neutrophils from, 80, 106, 107, 125–128,
    131, 147
  normoblast from, 128
  plasma cells from, 156
  platelets from, 171, 176–177
  poikilocytes and schistocytes from, 71
  promyelocytes and neutrophil from, 126
  red blood cells from, 39
  regenerative hypochromic
    anaemia in, 56

rouleaux formation in
  with myeloma, 76
  with neoplasia, 76
target cells from, 69
thrombocytosis from, 176
Trypanosoma cruzi in, 185
Trypanosoma evansi in, 185
white blood cells in, 104
Domestic duck (*Anas platyrhynchos
    domesticus*)
  basophilic erythroblasts from, 60
  heterophil from, 91
  hypochromic anaemia in, 59
  red blood cells from, 59
Domestic fowl (*Gallus gallus domesticus*)
  Aegyptianella pullorum from, 208
  heterophilia and monocytosis in, 118
  heterophilia in, 118–119
  heterophils from, 90
  monocytes from, 164
  monocytosis and heterophilia in, 164
  plasma, thrombocytes-rich, 172
  Plasmodium gallinaceum from, 193
Domestic goat (*Capra aegagrus hircus*)
  marked anisocytosis in, 61
  red blood cells from, 39
Domestic goose (*Anser anser*), heterophil
    from, 91
Domestic pig (*Sus scrofa domesticus*), red
    blood cells from, 54
Domestic sheep (*Ovis aries*), lymphoid
    leukaemia in, 151
Domestic turkey (*Meleagris gallopavo*)
  heterophils and lymphocyte from, 91
  thrombocytes from, 172
Donkey, healthy domestic (*Equus asinus*),
    red blood cells from, 41
Dorsal aorta vein, 5
Dorsal cervical/supravertebral sinus, 5
Dorsal tail vein, 4
Dugongs (*Dugong dugon*), 4
Dumeril's boa (*Acrantophis dumerili*),
    blood sample collection from, 18
Dutch Golden Age, 2

Eastern fiddler ray (*Trygonorrhina
    guaneria*)
  blood smear from, 78
  lymphocyte from, 145
  thrombocyte from, 178
Eastern fox snake (*Pantherophis gloydi*)
  azurophils from, 166, 167
  heterophilia in, 122, 166
  polychromatic erythroblast from, 166
Eastern grey kangaroo (*Macropus giganteus*),
    neutrophil in a blood film from, 88
EDTA, *see* Ethylenediaminetetraacetic acid
Egyptian vultures (*Neophron percnopterus*),
    haematology reference values in,
    224–225

Eider duck, three-year-old male (*Somateria
    mollissima*), lymphocyte from, 153
Eland (*Taurotragus oryx*), *Anaplasma
    marginale* from, 208
Eland, healthy (*Taurotragus oryx*),
    neutrophil from, 85
Elasmobranchii fish, 25
Elephant, blood sample collection from, 12
Embryonic haemocytes, 25
Emperor goose (*Anser canagicus*),
    lymphocytes from, 140
Emu (*Dromaius novaehollandiae*)
  heterophils from, 103
  red blood cells from, 42
Eosinophils, 31, 32
Eosin Y, 31
ER bodies, *see* Erythrocyte refractile bodies
Erythroblasts, 23, 83
Erythrocyte refractile (ER) bodies, 27, 66
Erythropoiesis, 23
  in Mediterranean spur-thighed tortoise,
    63–64
  in small mammals, birds and reptiles,
    51–53
Ethylenediaminetetraacetic acid (EDTA), 2,
    5, 28, 29
Eurasian beaver (*Castor fiber*), 4
Eurasian crane (*Grus grus*), 210
  Eimeria reichenowi from, 210
Eurasian oystercatcher (*Haematopus
    ostralegus*), heterophils and
    eosinophil from, 94
European beaver (*Castor fiber*), blood
    sample collection from, 8
European bison (*Bison bonasus*),
    Trypanosoma theileri in, 186
European eagle owls (*Bubo bubo*),
    haematology reference
    values in, 226
European hamster (*Cricetus cricetus*), red
    blood cells from, 61
European hedgehog (*Erinaceus
    europaeus*), 4
  lymphocyte from, 136, 152

Faber, Giovanni, 2
Fallow deer, eight-week-old (*Dama dama*),
    red blood cells in, 73
Farmed rainbow trout (*Oncorhynchus
    mykiss*), haematology reference
    values in, 244
Ferret (*Mustela putorius furo*), 4
  blood collection from, 7–8
  haematology reference values in, 215
  neutrophil in a blood smear of, 87
Fish, haematology reference values in,
    243–245
Florida soft shell turtle (*Apalone ferox*),
    heterophils and monocytes
    from, 124

Foà-Kurloff body, 135
Free-living and captive komodo dragons (*Varanus komodoensis*), haematology reference values in, 239
Free-living female nile monitors (*Varanus niloticus*), haematology reference values in, 237
Free-living Nile crocodiles (*Crocodylus niloticus*), haematology reference values in, 238, 239
Free-ranging lions (*Panthera leo*), haematology reference values in, 217
Friesian cow (*Bos taurus*), red blood cells from, 76
Fruit fly (*Drosophila melanogaster*), 6

Gaboon viper (*Bitis gabonica*)
    disrupted azurophil from, 168
    lymphocytoid azurophil from, 168
    monocytoid azurophil from, 168
Galah cockatoos (*Eolophus roseicapilla*), haematology reference values in, 227
Galilei, Galileo, 2
Gaur (*Bos gaurus*)
    basophil from, 89
    monocytes from, 162
Gaur calf, 10-week-old (*Bos gaurus*), red blood cells in, 72
General anaesthesia, 4
Gerbils, haematology reference values in, 216
Giant anteater (*Myrmecophaga tridactyla*), echinocytes and polychromatic erythroblast from, 60
Giant green tree frog (*Litoria infrafrenata*), multiple abnormal lymphocytes from, 155
Giant panda (*Ailuropoda melanoleuca*), *Hepatozoon* sp. from, 199
Giant rat (*Cricetomys gambianus*), *Grahamella* sp. from, 209
Gila monster (*Heloderma suspectum*)
    basophilic erythroblasts from, 62
    binucleated macrocytes and mature red cells from, 63
    cell in mitosis from, 63
    heterophil from, 104
    polychromatic megaloblast from, 63
Gluteal caudal vein, 4
Goat, domestic (*Capra aegagrus hircus*)
    platelets from, 171
    red blood cells from, 71
Golden eagle (*Aquila chrysaetos*)
    abnormal erythropoiesis in, 62
    haematology reference values in, 224–225

microcytic erythroblasts and erythroblast in mitosis from, 62
    thrombocytes from, 177
Golden/Syrian hamster (*Mesocricetus auratus*), 4
Goldfish (*Carassius auratus*), 33
    neutrophils from, 125
Gopher frog (*Lithobates capito*), eosinophil and neutrophil from, 101
Gopher tortoise (*Gopherus polyphemus*)
    erythrocyte and heterophils from, 132
    lymphocyte from, 143
Goshawk (*Accipiter gentilis*), *Leucocytozoon toddi* macrogametocyte from, 191
Granulocytes, 31–32
    common artefacts affecting, 102–105
    pathological responses involving, 106–132
    species variation in, 79–102
Granulocytopoiesis, 23
Granulopoiesis, 32
Greater flamingos (*Phoenicopterus roseus*), haematology reference values in, 223
Greater Indian hill mynah bird (*Gracula religiosa intermedia*), red blood cell morphology in, 49
Greater plated lizard, healthy (*Gerrhosaurus major*), red blood cells from, 53
Greater sulphur-crested cockatoos (*Cacatua g. galerita*), haematology reference values in, 227
Great pied hornbill (*Buceros bicornis*), eosinophil from, 94
Great pond snail (*Lymnaea stagnalis*), 6
Greek or spur-thighed tortoise (*Testudo graeca*), metamyelocyte from, 120
Greek tortoise (*Testudo graeca*)
    lymphocyte from, 142
    thrombocyte and lymphocyte from, 174
Green anaconda (*Eunectes murinus*), *Hepatozoon serpentium* from, 201
Green iguana (*Iguana iguana*)
    azurophils from, 170
    basophil from, 97
    blood sample collection from, 16
    cells with bilobed nuclei in, 97
    eosinophil from, 94, 96
    haematology reference values in, 235
    heterophils from, 96, 121
    monocyte from, 159
    monocytoid azurophil from, 166
    red blood cells from, 42
    ruptured heterophil from, 96
Green sea turtle (*Chelonia mydas*)
    blood sample collection from, 20
    haematology reference values in, 234

Green turtle (*Chelonia mydas*), blood smear from, 99
Green-winged macaws (*Ara chloropterus*), haematology reference values for, 227–228
Grey crowned crane (*Balearica regulorum*), red blood cells from, 55
Grey heron, healthy (*Ardea cinerea*), red blood cell precursors from, 52
Greylag goose (*Anser anser*), heterophils, eosinophil, lymphocyte from, 102
Grey mouse lemur (*Microcebus murinus*)
    acute lymphoblastic leukaemia in, 150
    lymphoblasts from, 150, 151
    lymphocytes from, 151
Grey-necked crowned crane (*Balearica regulorum*), 5
Grey parrots (*Psittacus erithacus*)
    blood sample collection from, 15
    haematology reference values in, 229
    thrombocytes from, 177
Grey squirrel (*Sciurus carolinensis*)
    Hepatozoon sciuri from, 198
    neutrophil from, 84
Guaira spiny rat (*Proechimys guairae*)
    neutrophils from, 114
    thrombocytosis from, 175
Guinea pig (*Cavia porcellus*), 4
    basophil from, 90
    blood collection from, 7
    Foà-Kurloff body from, 135
    haematology reference values in, 216
    monocytes from, 163
    platelets from, 171
    pseudoeosinophilic neutrophil from, 84
    red blood cells from, 53
Gyr falcons (*Falco rusticolus*), haematology reference values in, 230–231

Haematology, 1–2
Haematology reference values
    in amphibians, 241–242
    in birds, 222–233
    in fish, 243–245
    in mammals, 215–221
    in reptiles, 234–240
Haematopoiesis, 23
Haemocytes, 25
Haemoglobin crystallisation, 26
Haemolymph, 6, 25
    cellular appearance of invertebrates, 179
Haemoproteus from yellow-footed tortoise (*Chelonoidis denticulatus*), 201
Haemostatic blood cells, 35
Hamsters, haematology reference values in, 216
Hawksbill sea turtle (*Eretmochelys imbricata*)
    heterophils from, 123

polychromatic erythroblast from, 74
red blood cells from, 73
Healthy black rat (*Rattus rattus*),
    lymphocyte from, 135
Healthy ferret (*Mustela putorius furo*),
    neutrophil from, 80
Healthy gaur (*Bos gaurus*), monocyte and
    lymphocyte from, 157
Healthy human (*Homo sapiens*), neutrophils
    from, 81
Healthy Indian elephant (*Elephas maximus
    indicus*), bilobed mononuclear
    cells from, 157
Healthy ostrich (*Struthio camelus*),
    lymphocyte from, 139
Healthy red-eared terrapin (*Trachemys
    scripta elegans*), basophils
    from, 100
Healthy three-striped night monkey
    (*Aotus trivirgatus*), neutrophils
    from, 82
Heinz bodies, 26–27, 50, 75
Hermann's tortoise (*Testudo hermanni*)
    haematology reference values in, 204
    thrombocytes from, 173
Heterophils, 31, 32
Hippopotamus, female common
    (*Hippopotamus amphibious*),
    neutrophil from, 86
History of haematology, 2
    microscope, invention of, 2–3
Hog badger (*Arctonyx collaris*),
    echinocytes, target cells and
    hypochromia in, 70
Home's hinge-back tortoise, decalcified
    (*Kinixys homeana*), blood sample
    collection from, 18
Honeyeater (*Lychenostomus versicolor*),
    mononuclear cell
    from, 158
Hooded crane (*Grus monacha*)
    basophil from, 93
    eosinophil from, 119
    heterophils from, 119, 141
    lymphocytes from, 119
    thrombocyte from, 119
Hooke, Robert, 2
Horn shark (*Heterodontus francisci*),
    heterophil and lymphocytes
    from, 144
Horse (*Equus caballus*)
    Babesia caballi from, 204
    eosinophil from, 88
    with lymphosarcoma, 148
    malignant mononuclear cell from, 164
    neutrophils from, 113
    red blood cells and neutrophil from, 41
    Theileria equi from, 204
    Trypanosoma brucei in, 185
    Trypanosoma equiperdum in, 185

Houbara bustards (*Chlamydotis
    macqueenii*), haematology
    values in, 232
House mouse (*Mus musculus*), 4
House sparrow (*Passer domesticus*)
    Haemoproteus passeris from, 197
    Lankesterella sp. from, 203
Howell-Jolly bodies, 26, 50
Human (*Homo sapiens*) plasma, platelet-
    rich, 172
Hyacinth macaws (*Anodorhynchus
    hyacinthinus*), haematology
    reference values for, 227–228
Hypervitaminosis A, 68

Indian brown mongoose (*Herpestes fuscus*),
    echinocytes from, 70
Indian elephant (*Elephas maximus indicus*)
    haematology reference values in, 217
    red blood cells from, 40
    trilobed monocyte from, 157
Indian peacock soft-shelled turtle
    (*Nilssonia hurum*), heterophil and
    eosinophil from, 99
Indian peafowl (*Pavo cristatus*),
    haematology reference
    values in, 230
Indian python (*Python molurus*)
    azurophil from, 168
    heterophil from, 98
    lymphocyte from, 153
    Pirhemocyton sp. from, 210
    polychromatic erythroblast from, 53
    red blood cells from, 42
    thrombocytes and lymphocytes
    from, 174
Interdigital veins, 4
Invertebrates
    blood sample collection from, 6
    cellular appearance in haemolymph
    of, 179
Iron deficiency anaemia, 56, 57

Jackdaw (*Corvus monedula*), 5
Jaguar (*Panthera onca*), neutrophils
    from, 79
Jamaican coney (*Geocapromys brownii*)
    monocyte from, 163
    neutrophilia in, 115
Janssen, Hans, 2
Janssen, Zacharias, 2
Jararacussu (*B. jararacussu*), haematology
    reference values in, 237–238
Javan fish owl (*Ketupa ketupu*), basophil
    from, 93
Jugular vein, 4
Jugular venipuncture, 4, 5
Juvenile gyrfalcon (*Falco rusticolus*), blood
    sample collection
    from, 15

Juvenile olive baboon (*Papio anubis*)
    lymphocyte from, 132–134
    monocyte from, 132

Keas (*Nestor notabilis*), haematology
    reference values in, 229
Kestrel (*Falco tinnunculus*), *Babesia shortti*
    from, 206
Killer whale (*Orcinus orca*), 4
Kingfisher (*Alcedinidae*), *Haemoproteus
    enucleator* from, 197
Koi carp (*Cyprinus carpio*), blood sample
    collection from, 19
Koi/nishikigoi (*Cyprinus carpio*),
    neutrophils from, 124
Kookaburras (*Dacelo* spp.), 28
Kori bustard (*Ardeotis kori*)
    haematology reference values in, 232
    heterophil from, 91

Lappet-faced vulture (*Torgos tracheliotos*),
    eosinophilia in, 130
Large mammals, blood sample collection
    from, 4
Large tree shrew (*Tupaia tana*), neutrophils
    from, 115
Larval haemocytes, 25
Lateral coccygeal vein, 4
Laysan duck or Laysan teal (*Anas
    laysanensis*), 141
Leopard, normal (*Panthera pardus*),
    crenation and anisocytosis
    from, 54
Leopard tortoise (*Stigmochelys pardalis*),
    Haemoproteus testudinalis
    from, 202
Lesser flamingos (*Phoeniconaias minor*),
    haematology reference
    values in, 223
Lesser mouse-deer or lesser Malay
    chevrotain (*Tragulus
    kanchil*), 25
Lion (*Panthera leo*)
    blood sample collection from, 8
    neutrophils from, 107
Lion-tailed macaque (*Macaca silenus*)
    iron deficiency anaemia resulting from
    chronic blood loss in, 57
    red blood cells in, 57
Lithium heparin, 5
Little corellas (*Cacatua sanguinea*),
    haematology reference
    values in, 227
Little eye, 2
Little owl (*Athene noctua*), *Leucocytozoon
    danilewskyi* macrogametocyte
    in, 189
Lizard (*Gerrhosaurus major*)
    blood sample collection from, 5
    microfilarian from, 184

Llama, healthy (*Lama glama*), red blood cell from, 51
Loggerhead sea turtle (*Caretta caretta*), basophil from, 100
Long-haired spider monkey (*Ateles belzebuth*), monocytes from, 161
Lumpsucker or lumpfish (*Cyclopterus lumpus*), monocyte from, 170
Lymphocytes, 33
    abnormalities in, 145
    normal variation in lymphocytes morphology, 132–145
Lymphopenia, 33
Lymphopoiesis, 33

Macaroni penguin (*Eudyptes chrysolophus*), *Leucocytozoon tawaki* macrogametocyte from, 191
Maguari stork (*Ciconia maguari*), haematology reference values in, 224
Malayan tapir, conscious (*Tapirus indicus*), blood sample collection from, 10
Mallard duck (*Anas platyrhynchos*)
    lymphoblasts from, 153
    lymphocytosis in, 154
Mammals, haematology reference values in, 215–221
Manchurian cranes (*Grus japonensis*), haematology reference values in, 225–226
Mara (*Dolichotis patagonum*), 26
    red blood cells in, 48
Marabou storks (*Leptoptilos crumenifer*)
    distorted *Haemoproteus crumeniferus* from, 196
    haematology reference values in, 224
Markhor (*Capra falconeri*), 26
    platelet from, 177
    red blood cells from, 48
Marmoset, healthy common (*Callithrix jacchus*)
    Heinz bodies and few reticulocytes in the blood of, 50
    neutrophils and lymphocyte from, 82
Masked palm civet (*Paguma larvata*), lymphocyte from, 135, 146
Matchstick cells, 47
Matchstick red blood cells, 26
May-Grünwald-Giemsa stain, 2
May-Grünwald-Giemsa technique, 47
MCHC, *see* Mean cell haemoglobin concentration
MCV, *see* Mean cell volume
Mean cell haemoglobin concentration (MCHC), 28
Mean cell volume (MCV), 39
Medial metatarsal/caudal tibial vein, 4
Medial saphenous vein, 4

Mediterranean spur-thighed tortoise (*Testudo graeca*)
    basophilic and polychromatic erythroblasts from, 65
    erythropoiesis in, 63, 64
Mexican red-knee tarantula (*Brachypelma smithi*), 6
Mice, haematology reference values in, 216
Microscope, invention of, 2–3
Microspherocytes, 65, 68
Mikado pheasant (*Syrmaticus mikado*), *Leucocytozoon* sp. macrogametocyte from, 190
Military macaws (*Ara militaris*), haematology reference values for, 227–228
Mississippi alligator (*Alligator mississippiensis*), heterophil from, 99
Mongolian gerbil (*Meriones unguiculatus*), 4
Monk parakeet/Quaker parrot (*Myiopsitta monachus*), lymphocytes and monocyte from, 140
Monocyte morphology, normal, 156–165
Monocytes, 33
Montpellier snake (*Malpolon monspessulanus*)
    azurophil from, 169
    heterophils from, 123
Moorhen (*Gallinula chloropus*), thrombocytes from, 173
Moose (*Alces alces*)
    neutrophil from, 84
    platelet anisocytosis in, 171
    red blood cell from, 60
Motorbike frog (*Litoria moorei*)
    eosinophil from, 101, 124
    lymphocyte from, 124
    neutrophil from, 124
    thrombocyte from, 124, 175
Mouse (*Mus musculus*), neutrophil from, 87
Mouse lemur, healthy adult (*Microcebus murinus*), polychromasia and polychromatic erythroblast in, 51
Mozambique spitting cobra (*Naja mossambica*), 105
    Haemoproteus mesnili from, 201
    Hepatozoon najae from, 201
Myeloblast (granuloblast), 32

Nanoject II Auto-Nanoliter Injector, 6
Necrotic granulocytes, 97
Nene or Hawaiian goose (*Branta sandvicensis*), lymphocyte from, 140
Neutrophil, 32
Neutrophil/heterophil granulocytes, 32
New Zealand white rabbits (*Oryctolagus cuniculus*), haematology reference values in, 215

Night herons (*Nycticorax nycticorax*), haematology reference values in, 224
Nile (River) hippopotamus (*Hippopotamus amphibius*), blood sample collection from, 9
Nile crocodile (*Crocodylus niloticus*)
    blood sample collection from, 16
    heterophils and thrombocyte from, 103, 104
    Oswaldofilaria versterae microfilarian from, 184
Nile monitor lizard (*Varanus niloticus*), heterophil from, 98
Nile tilapia (*Oreochromis niloticus*), haematology reference values in, 243, 245
Nilgai (*Boselaphus tragocamelus*), neutrophils from, 113
Normal blood cell, 1–2
Normal kori bustard (*Ardeotis kori*), thrombocytes from, 172
North American (*Meleagris gallopavo*), red blood cells, 45

Occhiolino, 2
Okapi, captive (*Okapia johnstoni*), blood sample collection from, 9
Olive baboon (*Papio anubis*), platelets from, 172
Orangutan, healthy (*Pongo pygmaeus*), lymphocyte from, 134
Ostrich (*Struthio camelus*), 5, 28
    heterophil myelocyte from, 117
    heterophils from, 117
    monocyte from, 158
    thrombocytes from, 173

Palm civet (*Paguma larvata*)
    monocyte from, 161
    red blood cells from, 44
Pelger-Huët-like phenomenon, 87
Père David deer (*Elaphurus davidianus*), 46
Peregrine falcons (*Falco peregrinus*), haematology reference values in, 230–231
Peripheral blood, 23
Philippine crocodile (*Crocodylus mindorensis*)
    blood sample collection from, 17
    eosinophil from, 95
    lymphocytes from, 143
Pigeon, healthy (*Columba livia*), heterophils and thrombocytes from, 91
Pig-tailed macaques (*Macaca nemestrina*), haematology reference values in, 221
Pilot whale (*Globicephala macrorhynchus*), 41

Pink/humpback salmon (*Oncorhynchus gorbuscha*), red blood cells from, 44

Plains vizcacha (*Lagostomus maximus*)
myeloblast from vizcacha in, 115
neutrophilia from, 150, 175
thrombocytosis from, 175

Plains zebra, etorphine anaesthetised (*Equus quagga*), blood sample collection from, 11

Platelets, 35
variations in platelet and thrombocytes, 171–175
associated with disease, 175–178

Platypus (*Ornithorhynchus anaticus*), lymphocyte from, 137

Polar bear (*Ursus maritimus*), neutrophils from, 80

Polychromasia, 51, 66

Polychromatic erythroblast, 51

Polychromatic red blood cells, 23

Pool frog (*Pelophylax lessonae*), *Trypanosoma rotatorium* from, 189

Porkfish (*Anisotremus virginicus*), neutrophil and monocyte from, 125

Port Jackson shark (*Heterodontus portusjacksoni*)
clinically normal (*Heterodontus portusjacksoni*), thrombocyte from, 175
heterophil and monocyte from, 160
lymphocyte from, 145

Prairie marmot (*Cynomys ludovicianus*), target cells and hypochromia in, 70

Proerythroblast, 23

Propylene glycol, 27

Przewalski's horses (*Equus przewalskii*)
accentuated rouleaux formation in the blood of, 50
haematology reference values in, 220

Pseudoeosinophilic neutrophil, 83, 84

Pseudoeosinophils, 83

Puffback shrike (*Dryoscopus cubla*), *Leucocytozoon balmorali* macrogametocyte from, 190

Quaker parrot (*Myiopsitta monachus*), heterophil from, 153

Rabbit (*Oryctolagus cuniculus*), 4
blood sample collection from, 6, 7
comparison with true eosinophil from, 83
neutrophils from, 83
Trypanosoma nabiasi from, 188

Racing pigeon (*Columba livia*), blood sample collection from, 15

Rainbow boa (*Epicrates cenchria*)
Arenavirus infection in, 155
Trypanosoma constrictor from, 188

Rainbow trout (*Oncorhynchus mykiss*), blood sample collection from, 20

Rats (*Rattus* spp.)
blood sample collection from, 7
haematology reference values in, 216
lymphocytic leukaemia in, 152
Trypanosoma lewisi in, 186

Raven (*Corvus corax*), 5

Red-bellied tamarin (*Saguinus labiatus*)
anisocytosis, polychromasia and polychromatic erythroblast in, 59
monocytes from, 160
reticulocytes in the blood of, 51

Red-billed blue magpie (*Urocissa erythrorhyncha*), *Chandlerella sinensis* microfilarian from, 184

Red blood cell morphology, abnormal variations in, 56
erythropoiesis in a Mediterranean spur-thighed tortoise, 63–64

Red blood cells, 25
artefacts, 28
blood film, 28–29
common artefacts affecting, 53–55
inclusion bodies, 26–28
maturation, 23
normal species variation in red blood cell size, 39–44
normal variation in red blood cell shape, 44–53
relationship between red blood cell size and number, 41
reversible changes in shape of, 26
rouleaux formation, 28

Redbreasted goose, clinically normal (*Branta ruficollis*), monocyte from, 158

Red deer, healthy adult (*Cervus elaphus*), red blood cells from, 45

Red-eared slider (*Trachemys scripta elegans*), blood sample collection from, 17, 19

Red-eared terrapin (*Trachemys scripta elegans*)
lymphocyte from, 142
red blood cells from, 45

Red-fronted macaws (*Ara rubrogenys*), haematology reference values in, 228–229

Red-fronted or Jardine's parrot (*Poicephalus gulielmi*), eosinophil from, 132

Red kangaroo (*Megaleia rufa*), 4

Red-necked wallaby (*Macropus rufogriseus*)
basophilic erythroblast and marked agglutination in the wallaby of, 68
lymphocytes from, 134

red blood cells from, 49
regenerative haemolytic anaemia in, 68

Red-rumped swallow (*Cecropis daurica*), *Babesia rustica* from, 207

Red salmon, red blood cells of, 43

Red slender loris (*Loris tardigradus*)
lymphocytes from, 145, 146
monocyte from, 160
neutrophil from, 110

Reeve's muntjac (*Muntiacus reevesi*), oxygenated blood from, 46

Regenerative hypochromic anaemia, 56

Reindeer, normal (*Rangifer tarandus*), red blood cells and neutrophils from, 45

Reptiles
blood sample collection from, 5
haematology reference values in, 215–240

Reptilian red blood cells, 45

Reticulated python (*Malayopython reticulatus*)
azurophils from, 165
lymphocyte from, 142
thrombocytes from, 178

Reticulocytes, 24

Rhesus macaque, healthy (*Macaca mulatta*), neutrophils from, 82

Rhinoceros iguana, healthy (*Cyclura cornuta*), lymphocytes, thrombocyte and monocyte from, 143

Rickettsia-like organisms, 210

Right jugular vein, 4

Ring-necked parakeet (*Psittacula krameri*), *Haemoproteus handai* from, 195

Ring-tailed coati (*Nasua nasua*)
anisocytosis, poikilocytosis, hypochromasia and stomatocytosis in, 71
binucleated erythroblast from the coati in, 67
eosinophil from, 130
hyperplasia and eosinophil from, 130
lymphocyte with, 146
marked anisocytosis, 67
marked reticulocytosis in the coati in, 67
neutrophils from, 109
regenerative haemolytic anaemia in, 66

Roan antelope (*Hippotragus equinus*)
extreme red cell deformation in, 72
neutrophil from, 84
severe regenerative anaemic in, 71

Roe deer (*Capreolus capreolus*), haematology reference values in, 220–221

Romanowsky stains, 24, 26, 31–32

Roseate spoonbill, normal (*Platalea leucorodia*), 55

Rosy flamingo (*Phoenicopterus roseus*)
heterophilia in, 117

lymphocyte from, 139
thrombocytes from, 172
Rothschild's giraffe, conscious (*Giraffa camelopardalis rothschildi*), blood sample collection from, 13
Rouleaux formation, 28
Royal python (*Python regius*), red blood cells from, 73
Russian tortoise (*Agrionemys horsfieldii*), blood sample collection from, 17

Sacred ibis (*Threskiornis aethiopicus*), four basophilic erythroblasts from, 62
Saker falcon (*Falco cherrug*)
   Babesia shortti from, 207
   haematology reference values in, 230–231
Sand lizard (*Lacerta agilis*), *Karyolysus latus* macrogametocyte from, 203
Sand tiger shark (*Carcharias taurus*), blood sample collection from, 19
Sarus crane (*Grus antigone*)
   basophil from, 103
   haematology reference values in, 225–226
   heterophils from, 102, 118
   lymphocyte from, 141
   monocytes, heterophils and lymphocyte from, 164
   regenerative anaemia in, 61
Saw shelled turtle (*Myuchelys latisternum*), lymphoid leukaemia in, 155
Scarlet-chested sunbird (*Chalcomitra senegalensis*), *Haemoproteus sequeirae* in, 195
Scarlet ibises (*Eudocimus ruber*), haematology reference values in, 224
Scarlet macaws (*Ara macao*), haematology reference values in, 228–229
Schneider's toad/rococo toad (*Rhinella schneideri*), azurophil or monocyte from, 170
Scimitar horned oryx (*Oryx dammah*), blood film from, 72, 73
Seal (*Phoca vitulina*), 4
Secretary birds (*Sagittarius serpentarius*), haematology reference values in, 224–225
Senegal chameleon (*Chamaeleo senegalensis*), blood sample collection from, 17
Severe macaws (*Ara severus*), haematology reference values in, 228–229
Seychelles house snake (*Boaedon geometricus*), *Hepatozoon musotae* from, 200
Sheep (*Ovis aries*), 25
   Babesia motasi from, 206
   Babesia ovis from, 206

eosinophilia and basophilia in, 128
Eperythrozoon ovis from, 209
Theileria lestoquardi from, 208
Shingleback skink, healthy (*Tiliqua rugosa*), 98
Short-beaked echidna (*Tachyglossus aculeatus*), lymphocyte from, 137, 138
Short-tailed chinchilla (*Chinchilla chinchilla*), 4
Siberian sturgeon (*Acipenser baerii*), blood smear from, 79
Sickled red blood cells, 45, 46
Silvery marmoset (*Mico argentatus*)
   erythroblasts and leucocytosis in, 57
   polychromatic and orthochromic erythroblast from the marmoset in, 58
Small diffuse basophilic granules, 26
Small mammals, blood sample collection from, 4
Small to medium marsupial species, blood sample collection from, 13
Snakes, blood sample collection from, 5
Snowy owl (*Bubo scandiacus*), *Plasmodium* sp. from, 194
Sockeye, clinically normal (*Oncorhynchus nerka*), red blood cells of, 43
Sockeye salmon, healthy wild (*Oncorhynchus nerka*), blood smear from, 79
Sodium citrate tubes, blood samples storage in, 5
South American rattlesnake (*Crotalus durissus*), heterophil and eosinophil from, 95
South American sea lion (*Otaria flavescens*)
   blood film from, 87
   blood sample collection from, 14
Southern koala (*Phascolarctos cinereus victor*), lymphocyte from, 138
Sparrow hawk (*Accipiter nisus*), *Haemoproteus nisi* from, 194
Spectacled owls (*Pulsatrix perspicillata*), haematology reference values in, 226
Spiny-tailed lizard (*Uromastyx aegyptia*)
   blood sample collection from, 14
   Haemocystidium apigmentada from, 202
   Hepatozoon sp. from, 202
Spotted eagle ray (*Aetobatus narinari*), blood film from, 102
Spotted marsh frog (*Limnodynastes tasmaniensis*)
   megathrombocyte from, 178
   monocyte from, 78, 178
Spur-thighed tortoise (*Testudo graeca*)
   cell containing a large nuclear remnant, 64
   haematology reference values in, 240

normal erythroblast and an erythroblast in mitosis, 64
polychromatic erythroblast in, 63
red blood cells from, 64
Square red blood cells, 26
Squirrel monkey (*Saimiri sciureus*), *Plasmodium brasilianum* macrogametocyte from, 111
Stone curlews (*Burhinus oedicnemus*), haematology values in, 231
Submandibular vein is, 4
Sudan plated lizard (*Gerrhosaurus major*), heterophil from, 97
Sugar glider (*Petaurus breviceps*), 4
Sulphur-crested cockatoo (*Cacatua galerita*), 105
Sulphur-crested cockatoo (*Cacatua sulphurea*)
   autoagglutination in an, 69
   basophils from, 93
   heterophil from, 93
   red blood cells showing nuclear bleeding, 54
Swamp deer (*Rucervus duvaucelii*), red blood cells from, 46

Taipan snake (*Oxyuranus microlepidotus*)
   azurophil and heterophil from, 165
   intracellular haemoglobin crystallisation in, 49
   lymphocyte and thrombocytes from, 141, 142
   thrombocytes from, 174
Tammar wallabies (*Macropus eugenii*), haematology reference values in, 221
Tawny owl (*Strix aluco*)
   distorted *Haemoproteus* sp. from, 197
   haematology reference values in, 226
   Haemoproteus sp. from, 194, 197
   Leucocytozoon danilewskyi from, 189
   Plasmodium circumflexum microgametocyte from, 192
   Trypanosoma avium from, 188
Teleost fish, 23
Terrapins, blood sample collection from, 5
Terrier dog (*Canis lupus familiaris*), eosinophil with cytoplasmic vacuoles from, 88
Thiosulfate compounds, 27
Three-striped night monkey (*Aotus trivirgatus*)
   abnormal erythropoiesis in, 58
   lymphocytes and neutrophil from, 134
   mitotic erythroblast from, 58
   neutrophilia in, 111
   neutrophil myelocyte from, 111
   red blood cells from, 58
Three-toe salamander (*Amphiuma tridactylum*), 25

Thrombocytes, 35
Tiger (*Panthera tigris*)
   anisocytosis in, 49
   blood sample collection from, 9
T lymphocytes, 33
Tortoises, blood sample collection from, 5
Triangular red blood cells, 26
Turkey vultures (*Cathartes aura*), haematology
   reference values in, 224–225
Turtles, blood sample collection from, 5
Two-toed sloth (*Choloepus didactylus*),
   basophil from, 89

Van Leeuwenhoek, Antoine Philips, 2
Vena cava, 4
Vena cava caudalis, 5
Vena cutanea ulnaris superficialis, 4
Vena jugularis dextra, 4
Ventral tail vein, 4, 5
Vervet monkey (*Chlorocebus pygerythrus*),
   Hepatocystis kochi from, 198

Wandering albatross (*Diomedea exulans*),
   Hepatozoon albatrossi from, 199
Wattled starling (*Creatophora cinerea*),
   Plasmodium sp. trophozoites from
   post-mortem examination of, 194
West African cave bat (*Miniopterus inflatus*),
   Polychromophilus melanipherus
   macrogametocytes from, 192
West African crowned crane (*Balearica
   pavonina*), heterophils and
   lymphocyte from, 120

Western long-beaked echidna, healthy
   (*Zaglossus bruijnii*), platelets
   from, 172
Western lowland gorilla, anaesthetised
   (*Gorilla gorilla gorilla*), placement
   of an intravenous catheter in, 13
White blood cells, 31
   azurophils, 33
   blood film, 32–33
   granulocytes, 31–32
   lymphocytes, 33
   monocytes, 33
White-eared bulbul (*Pycnonotus leucotis*),
   Trypanosoma pycnonoti from, 187
White-headed saki monkey (*Pithecia
   pithecia*), neutrophil from, 81
White rhinoceros (*Ceratotherium simum*),
   26, 27
   haemoglobin crystals from, 47
   haematology reference values in, 218
   matchstick cells from, 47
   neutrophils from, 87
White stork (*Ciconia ciconia*)
   haematology reference values in, 224
   heterophil and eosinophil from, 92
Whooper swan (*Cygnus cygnus*),
   hypochromia, anisocytosis,
   microcytosis and poikilocytosis in, 60
Wild-caught Dubois' tree frogs (*Polypedates
   teraiensis*), haematology reference
   values in, 242
Wolf (*Canis lupus*)
   neutrophil from, 106

autoimmune haemolytic anaemia in, 66
Wood pigeon, healthy (*Columba
   palumbus*), red blood cells
   from, 55
Wright-Giemsa stain, 2

Yak (*Bos grunniens*), neutrophil from, 86
Yarará chica (*B. n. diporus*),
   haematology reference
   values in, 237–238
Yellow-bellied slider (*Trachemys scripta
   scripta*), basophils from, 94
Yellow-footed tortoise (*Chelonoidis
   denticulatus*), distorted
   Hepatozoon and *Haemoproteus*
   parasites from, 201
Yellow fronted Amazon parrot (*Amazona
   ochrocephala*), heterophils
   from, 116
Yellow-naped macaws (*Primolius
   auricollis*), haematology reference
   values in, 228–229
Yellow-necked spurfowl (*Pternistis
   leucoscepus*), *Leucocytozoon
   naevei* macrogametocyte
   from, 191

Zambian fruit bat (*Epomophorus
   gambianus*), *Hepatocystis
   epomophori* from, 198
Zambian house snake (*Boaedon
   fuliginosus*), *Hepatozoon boodoni*
   in, 200